# Lecture Notes in Mechanical Engineering

**Lecture Notes in Mechanical Engineering (LNME)** publishes the latest developments in Mechanical Engineering—quickly, informally and with high quality. Original research reported in proceedings and post-proceedings represents the core of LNME. Volumes published in LNME embrace all aspects, subfields and new challenges of mechanical engineering. Topics in the series include:

- Engineering Design
- Machinery and Machine Elements
- Mechanical Structures and Stress Analysis
- Automotive Engineering
- Engine Technology
- Aerospace Technology and Astronautics
- Nanotechnology and Microengineering
- Control, Robotics, Mechatronics
- MEMS
- Theoretical and Applied Mechanics
- Dynamical Systems, Control
- Fluid Mechanics
- Engineering Thermodynamics, Heat and Mass Transfer
- Manufacturing
- Precision Engineering, Instrumentation, Measurement
- Materials Engineering
- Tribology and Surface Technology

To submit a proposal or request further information, please contact the Springer Editor of your location:

**China:** Ms. Ella Zhang at ella.zhang@springer.com
**India:** Priya Vyas at priya.vyas@springer.com
**Rest of Asia, Australia, New Zealand:** Swati Meherishi at swati.meherishi@springer.com
**All other countries:** Dr. Leontina Di Cecco at Leontina.dicecco@springer.com

To submit a proposal for a monograph, please check our Springer Tracts in Mechanical Engineering at https://link.springer.com/bookseries/11693 or contact Leontina.dicecco@springer.com

**Indexed by SCOPUS. All books published in the series are submitted for consideration in Web of Science.**

More information about this series at https://link.springer.com/bookseries/11236

Justyna Trojanowska · Agnieszka Kujawińska ·
Jose Machado · Ivan Pavlenko
Editors

# Advances in Manufacturing III

## Volume 2 - Production Engineering: Research and Technology Innovations, Industry 4.0

Springer

*Editors*
Justyna Trojanowska
Faculty of Mechanical Engineering
Poznan University of Technology
Poznan, Poland

Agnieszka Kujawińska
Faculty of Mechanical Engineering
Poznan University of Technology
Poznan, Poland

Jose Machado
Department of Mechanical Engineering
University of Minho
Guimarães, Portugal

Ivan Pavlenko
Sumy State University
Sumy, Ukraine

ISSN 2195-4356 ISSN 2195-4364 (electronic)
Lecture Notes in Mechanical Engineering
ISBN 978-3-030-99309-2 ISBN 978-3-030-99310-8 (eBook)
https://doi.org/10.1007/978-3-030-99310-8



This Springer imprint is published by the registered company Springer Nature Switzerland AG
The registered company address is: Gewerbestrasse 11, 6330 Cham, Switzerland

# Preface

This volume of Lecture Notes in Mechanical Engineering gathers selected papers presented at the 7th International Scientific-Technical Conference MANUFACTURING 2022, held in Poznan, Poland, on May 16–19, 2022. The conference was organized by the Faculty of Mechanical Engineering, Poznan University of Technology, Poland.

The aim of the conference was to present the latest achievements in the broad field of mechanical engineering and to provide an occasion for discussion and exchange of views and opinions. The conference covered topics in:

- mechanical engineering
- production engineering
- quality engineering
- measurement and control systems
- biomedical engineering

The organizers received 165 contributions from 23 countries around the world. After a thorough peer review process, the committee accepted 91 papers for conference proceedings prepared by 264 authors from 23 countries (acceptance rate around 55%). Extended versions of selected best papers will be published in the following journals: *Management and Production Engineering Review, Bulletin of the Polish Academy of Sciences: Technical Sciences, Materials, Applied Sciences.*

The book **Advances in Manufacturing III** is organized into five volumes that correspond to the main conference disciplines mentioned above.

**Advances in Manufacturing III - Volume 2 - Production Engineering: Research and Technology Innovations, Industry 4.0** gathers research and practical solutions aiming at increase the efficiency of production processes. Chapters are devoted to the analysis of the latest trends in using immersive technologies and to the applications of Industry 4.0 solutions in manufacturing. Written by scientists and practitioners around the world, this book is intended to promote the exchange of views and experiences and contribute to the dissemination and effective application. It includes 25 chapters, prepared by 71 authors from 8 countries.

We would like to thank the members of the International Program Committee for their hard work during the review process.

We acknowledge all people that contributed to the staging of MANUFACTURING 2022: authors, committees and sponsors. Their involvement and hard work were crucial to the success of the MANUFACTURING 2022 conference.

May 2022

Justyna Trojanowska
Agnieszka Kujawińska
Jose Mendes Machado
Ivan Pavlenko

# Organization

## Steering Committee

### General Chair

| | |
|---|---|
| Adam Hamrol | Poznan University of Technology, Poland |

### Chairs

| | |
|---|---|
| Olaf Ciszak | Poznan University of Technology, Poland |
| Magdalena Diering | Poznan University of Technology, Poland |
| Justyna Trojanowska | Poznan University of Technology, Poland |

## Scientific Committee

| | |
|---|---|
| Ahmad Majdi Abdul-Rani | University of Technology PETRONAS, Malaysia |
| Dario Antonelli | Politecnico di Torino, Italy |
| Katarzyna Antosz | Rzeszow University of Technology, Poland |
| Jorge Bacca-Acosta | Fundación Universitaria Konrad Lorenz, Colombia |
| Zbigniew Banaszak | Koszalin University of Technology, Polska |
| Andre D. L. Batako | Liverpool John Moores University, UK |
| Stefan Berczyński | West Pomeranian University of Technology in Szczecin, Poland |
| Alain Bernard | Ecole Centrale de Nantes, France |
| Kristina Berladir | Sumy State University, Ukraine |
| Christopher Brown | Worcester Polytechnic Institute, USA |
| Anna Burduk | Wrocław University of Science and Technology, Poland |
| Marcin Butlewski | Poznan University of Technology, Poland |
| Lenka Čepová | VSB Technical University of Ostrava, Czech Republic |

| | |
|---|---|
| Nadežda Čuboňová | University of Žilina, Slovakia |
| Somnath Chattopadhyaya | Indian School of Mines, India |
| Olaf Ciszak | Poznań University of Technology, Polska |
| Jacek Diakun | Poznań University of Technology, Poland |
| Magdalena Diering | Poznan University of Technology, Poland |
| Ewa Dostatni | Poznan University of Technology, Poland |
| Jan Duda | Cracow University of Technology, Poland |
| Davor Dujak | J.J. Strossmayer University of Osijek, Croatia |
| Milan Edl | University of West Bohemia, Czech Republic |
| Brigita Gajšek | University of Maribor, Slovenia |
| Mosè Gallo | University of Naples Federico II, Italy |
| Adam Gałuszka | Silesian University of Technology, Poland |
| Józef Gawlik | Cracow University of Technology, Poland |
| Adam Gąska | Cracow University of Technology, Poland |
| Adam Górny | Poznan University of Technology, Polska |
| Filip Górski | Poznan University of Technology, Poland |
| Adam Hamrol | Poznan University of Technology, Poland |
| Mukund Harugade | Padmabhooshan Vasantraodada Patil Institute of Technology, India |
| Ivan Hudec | Slovak University of Technology in Bratislava, Slovakia |
| Jozef Husár | Technical University of Košice, Slovakia |
| Aminul Islam | Technical University of Denmark, Denmark |
| Vitalii Ivanov | Sumy State University, Ukraine |
| Andrzej Jardzioch | West Pomeranian University of Technology in Szczecin, Poland |
| Sabahudin Jašarević | University of Zenica, Bosnia and Herzegovina |
| Małgorzata Jasiulewicz-Kaczmarek | Politechnika Poznańska, Poland |
| Wojciech Kacalak | Technical University of Koszalin, Poland |
| Sławomir Kłos | University of Zielona Góra, Polska |
| Lucia Knapčíková | Technical University of Košice, Slovakia |
| Damian Krenczyk | Silesian University of Technology, Poland |
| Jolanta Krystek | Silesian University of Technology, Poland |
| Józef Kuczmaszewski | Lublin University of Technology, Poland |
| Agnieszka Kujawińska | Poznan University of Technology, Poland |
| Ivan Kuric | University of Žilina, Slovak Republic, Slovak Republic |
| Piotr Łebkowski | AGH, University of Science and Technology, Poland |
| Stanisław Legutko | Poznan University of Technology, Polska |
| Sławomir Luściński | Kielce University of Technology, Poland |
| José Machado | University of Minho, Portugal |
| Vijaya Kumar Manupati | National Institute of Technology Warangal, India |

| | |
|---|---|
| Uthayakumar Marimuthu | Kalasalingam Academy of Research and Education, India |
| Jorge Martín Gutiérrez | Universidad de La Laguna, Spain |
| Józef Matuszek | University of Bielsko-Biala, Poland |
| Arkadiusz Mężyk | Silesian University of Technology, Poland |
| Dariusz Mikołajewski | Kazimierz Wielki University, Poland |
| Andrzej Milecki | Poznań University of Technology, Poland |
| Piotr Moncarz | Stanford University, USA |
| Manurl Francisco Morales-Contreras | Universidad Pontificia Comillas, Spain |
| Keddam Mourad | USTHB, Algeria |
| Leticia Neira | UANL, México |
| Przemysław Niewiadomski | University of Zielona Góra, Polska |
| Erika Ottaviano | University of Cassino and Southern Lazio, Cassino |
| Eren Özceylan | Gaziantep University, Turkey |
| Razvan Ioan Pacurar | Technical University of Cluj-Napoca, Romania |
| Justyna Patalas-Maliszewska | University of Zielona Góra, Poland |
| Ivan Pavlenko | Sumy State University, Ukraine |
| Lucjan Pawłowski | Lublin University of Technology; Polish Academy of Science, Poland |
| Dragan Peraković | University of Zagreb, Croatia |
| Alejandro Pereira | Universidade de Vigo, Spain |
| Marko Periša | University of Zagreb, Croatia |
| Pierluigi Rea | University of Cagliari, Italy |
| Ján Piteľ | Technical University of Košice, Slovakia |
| Izabela Rojek | Kazimierz Wielki University, Poland |
| Alessandro Ruggiero | University of Salerno, Italy |
| Dominik Rybarczyk | Poznań University of Technology, Polska |
| Krzysztof Santarek | Warsaw University of Technology, Poland |
| Jarosław Sęp | Rzeszow University of technology, Poland |
| Bożena Skołud | Silesian University of Technology, Polska |
| Jerzy Andrzej Sładek | Cracow University of Technology, Poland |
| Dorota Stadnicka | Rzeszów University of Technology, Poland |
| Beata Starzyńska | Poznan University of Technology, Poland |
| Manuel Francisco Suárez Barraza | Universidad de las Américas Puebla, Mexico |
| Marcin Suszyński | Poznan University of Technology, Polska |
| Marek Szostak | Poznan University of Technology, Poland |
| Erfan Babaee Tirkolaee | Istinye University, Turkey |
| Justyna Trojanowska | Poznan University of Technology, Poland |
| Leonilde Varela | University of Minho, Portugal |
| Katarzyna Węgrzyn-Wolska | Engineering School of Digital Technologies, France |
| Gerhard-Wilhelm Weber | Poznan University of Technology, Poland |

| | |
|---|---|
| Edmund Weiss | Calisia University, Poland |
| Dorota Więcek | University of Bielsko-Biala, Poland |
| Michał Wieczorowski | Poznan University of Technology, Poland |
| Hanna Włodarkiewicz-Klimek | Poznan University of Technology, Poland |
| Szymon Wojciechowski | Poznan University of Technology, Poland |
| Ralf Woll | BTU Cottbus-Senftenberg, Germany |
| Jozef Zajac | Košice University of Technology, Slovakia |
| Nermina Zaimovic-Uzunovic | University of Zenica, Bosnia and Herzegovina |

## Program Committee

Karla Alvarado-Ramírez
José Angel
Dario Antonelli
Katarzyna Antosz
Erfan Babaee Tirkolaee
Jorge Bacca-Acosta
Diana Baila
Prashanth Bandari
Lucía Barcos-Redín
Andre Batako
Petr Beneš
Kristina Berladir
Marcin Białek
Mikołaj Bilski
Matej Borovinšek
Anna Borucka
Cristina Borzan
Sara Bragança
Daniel Brissaud
Christopher Brown
Paweł Buń
Anna Burduk
Bartłomiej Burlaga
Jacek Buśkiewicz
Marcin Butlewski
Massimiliana Carello
Fernando Castillo
Robert Cep
Miroslav Cisar
Olaf Ciszak
Felipe Contreras
Tomáš Coranič
Eric Costa
Margareta Coteata
Dorota Czarnecka-Komorowska
Reggie Davidrajuh
Mario del Valle
Boris Delibašić
Jakub Demčák
Jacek Diakun
Magdalena Diering
Grzegorz Domek
Ewa Dostatni
Radosław Drelich
Jan Duda
Luboslav Dulina
Eduard Franas
Paweł Fritzkowski
Brigita Gajšek
Mosè Gallo
Bartosz Gapiński
Katarzyna Gawdzińska
Adam Gąska
Dominik Gojdan
Arkadiusz Gola
Fernando González-Aleu
Adam Górny
Filip Górski
Cezary Grabowik
Marta Grabowska
Jakub Grabski
Damian Grajewski
Patrik Grznár
Aleksander Gwiazda
Michal Hajžman
Adam Hamrol

Mukund Harugade
Amalija Horvatić-Novak
Stella Hrehová
Jozef Husár
Vitalii Ivanow
Carmen Jaca
Rajat Jain
Michał Jakubowicz
Małgorzata Jankowska
Andrzej Jardzioch
Małgorzata Jasiulewicz-Kaczmarek
Anna Karwasz
Jakub Kaščak
Sławomir Kłos
Lucia Knapčíková
Adam Koliński
Elif Kongar
Boris Kostov
Kateryna Kovbasiuk
Arkadiusz Kowalski
Tomasz Kowaluk
Damian Krenczyk
Grzegorz Królczyk
Józef Kuczmaszewski
Agnieszka Kujawińska
Panagiotis Kyratsis
Georgios Lampropoulos
Peter Lazorík
Stanislaw Legutko
Andrzej Loska
Czesław Łukianowicz
José Machado
Marek Macko
Ján Majerník
Vitalii Maksymenko
Damjan Maletič
Matjaž Maletič
Vijaya Kumar Manupati
Sven Maricic
Jorge Martín Gutiérrez
Thomas Mathia
Maciej Matuszewski
Dariusz Mazurkiewicz
Jakub Michalski
Dariusz Mikołajewski
Janusz Mleczko
Ladislav Morovic
Keddam Mourad
Leticia Neira
Zdeněk Neusser
Magdalena Niemczewska-Wójcik
Przemysław Niewiadomski
Czesław Niżankowski
Nejc Novak
Filip Osiński
Erika Ottaviano
Eren Ozceylan
Ancuta Pacurar
Kyratsis Panagiotis
Jason Papathanasiou
Waldemar Paszkowski
Justyna Patalas-Maliszewska
Ivan Pavlenko
Paweł Pawlus
Jarosław Pempera
Dragan Peraković
Alejandro Pereira
Dariusz Plinta
Florin Popister
Evangelos Psomas
Paulina Rewers
Francisco Gabriel Rodriguez-González
Michał Rogalewicz
Izabela Rojek
Mirosław Rucki
Biserka Runje
Dominik Rybarczyk
Michał Rychlik
Milan Saga
Alžbeta Sapietova
Filip Sarbinowski
Holger Schlegel
Armin Schleinitz
Dariusz Sędziak
Robert Sika
Zbyněk Šika
Varinder Singh
Bożena Skołud
António Lucas Soares
Ewa Stachowska

Dorota Stadnicka
Sergiu Dan Stan
Roman Starosta
Beata Starzyńska
Krzysztof Stępień
Tomasz Stręk
Manuel Suárez-Barraza
Marcin Suszynski
Grażyna Sypniewska-Kamińska
Marek Szostak
Jiří Tengler
Samala Thirupathi
Francisco Torres
Justyna Trojanowska
Piotr Trojanowski
Krzysztof Tyszczuk
M. Uthayakumar
Leonilde Varela
Nikola Vitkovic
Adrián Vodilka
Sachin Waigaonkar
Tomasz Walczak
Łukasz Warguła
Gerhard-Wilhelm Weber
Katarzyna Węgrzyn-Wolska
Radosław Wichniarek
Michał Wieczorowski
Dariusz Więcek
Hanna Włodarkiewicz-Klimek
Szymon Wojciechowski
Adam Woźniak
Ryszard Wyczółkowski
Nermina Zaimovic-Uzunovic
Ivan Zajačko
Magdalena Żukowska
Krzysztof Żywicki

## Special Sessions

### Manufacturing and Management Approaches and Tools for Companies in Era of Digitalization

#### Special Session Organizing Committee

| | |
|---|---|
| Leonilde Varela | University of Minho, Portugal |
| Justyna Trojanowska | Poznan University of Technology, Poland |
| Vijaya Kumar Manupati | Mechanical Engineering Department, India |
| Paulina Rewers | Poznan University of Technology, Poland |

### Operational Innovation and Excellence in Organizations of 21st Century. Lean Thinking, Kaizen Philosophy, and Toyota Production System

| | |
|---|---|
| Manuel F. Suárez-Barraza | Universidad de las Américas Puebla, Mexico |
| Manuel F. Morales-Contreras | Universidad Pontificia Comillas, Spain |
| Agnieszka Kujawińska | Poznan University of Technology, Poland |
| Marta Grabowska | Poznan University of Technology, Poland |

### Cyber-Physical Production Systems: From Design to Applications

| | |
|---|---|
| Katarzyna Antosz | Rzeszow University of Technology, Poland |
| José Machado | University of Minho, Portugal |
| Erika Ottaviano | University of Cassino and Southern Lazio, Italy |
| Pierluigi Rea | University of Cagliari, Italy |

## Augmented, Virtual and Mixed Reality Systems

| | |
|---|---|
| Jozef Husár | Technical University of Košice, Slovak Republic |
| Jakub Kaščak | Technical University of Košice, Slovak Republic |
| Paweł Buń | Poznan University of Technology, Poland |

## Intelligent Methods Supporting Manufacturing Systems Efficiency

| | |
|---|---|
| Anna Burduk | Wroclaw University of Science and Technology, Poland |
| Andre Batako | Liverpool John Moores University, UK |
| Dorota więcek | University of Bielsko-Biala, Poland |
| Ivan Kuric | University of Žilina, Slovak Republic |

## Sustainability on Production in the Aspect of Industry 4.0

| | |
|---|---|
| Ewa Dostatni | Poznan University of Technology, Poland |
| Izabela Rojek | Kazimierz Wielki University in Bydgoszcz, Poland |
| Jacek Diakun | Poznan University of Technology, Poland |
| Dariusz Mikołajewski | Kazimierz Wielki University in Bydgoszcz, Poland |

## Design and Rapid Manufacturing of Customized Medical Products

| | |
|---|---|
| Filip Górski | Poznan University of Technology, Poland |
| Magdalena Żukowska | Poznan University of Technology, Poland |
| Razvan Pacurar | Technical University of Cluj-Napoca, Romania |

## Metrology and Statistical Analysis of Measurement and Control Systems

| | |
|---|---|
| Alejandro Pereira Domínguez | University of Vigo, Spain |
| Lenka Čepová | VSB - Technical University of Ostrava, Czech Republic |
| Magdalena Diering | Poznan University of Technology, Poland |
| Bartosz Gapiński | Poznan University of Technology, Poland |

## Analytical and Computational Methods in (Bio-) Mechanical and Material Engineering

| | |
|---|---|
| Tomasz Stręk | Poznan University of Technology, Poland |
| Roman Starosta | Poznan University of Technology, Poland |
| Nejc Novak | University of Maribor, Slovenia |
| Pavel Polach | University of West Bohemia, Czech Republic |

# Contents

# Production Engineering

# Modeling Production Systems as Modular Systems: A Petri Nets Based Approach

Reggie Davidrajuh[1](✉), Damian Krenczyk[2], and Bozena Skolud[2]

[1] Faculty of Science and Technology, University of Stavanger, Kristine Bonnevies vei 22, 4021 Stavanger, Norway
reggie.davidrajuh@uis.no

[2] Faculty of Mechanical Engineering, Silesian University of Technology, Konarskiego 18A, 44-100 Gliwice, Poland

**Abstract.** Modern Production systems are large, and the mathematical models of these become huge. Hence, the models become useless as they are challenging to analyze and perform model verification. Moreover, some of these production systems are inherently discrete (e.g., discrete production systems), while others can be discretized (e.g., Robotics). Petri nets are useful for modeling discrete production systems. But the resulting Petri nets suffer from huge size and infinite state space. To eliminate huge size, modular Petri nets are proposed. This paper looks into a modular Petri net and uses it to model a production system (focusing on automatically guided vehicles). The scope of this paper is to show the benefits of modularization, e.g., smaller modular components that can be independently developed and analyzed. Also, the modular Petri net is implemented in software GPenSIM. Hence, real-life and industrial production systems (such as car manufacturing, smart microgrids) can be modeled and analyzed.

**Keywords:** Production systems · Modular models · Petri nets · Modular petri nets · GPenSIM

## 1 Introduction

Modern Production systems are oversized. Therefore, we need to develop mathematical models to evaluate and find deficiencies (e.g., bottlenecks). However, mathematical models of these tend to be huge, so that analyzing these for performance improvement becomes impossible. Moreover, some of these production systems are inherently discrete, e.g., discrete production systems such as car manufacturing. At the same time, some other productions systems are continuous systems that can be discretized (e.g., Robotics).

Petri nets are useful for modeling discrete production systems due to their graphical self-documentation property. Also, Petri nets possess a simple mathematical background, and there exist many analytical tools [1, 2]. However, the Petri net models suffer from huge size and infinite state space, even for simple production systems [3]. To eliminate huge size, modular Petri nets are proposed.


J. Trojanowska et al. (Eds.): MANUFACTURING 2022, LNME, pp. 3–12, 2022.
https://doi.org/10.1007/978-3-030-99310-8_1

This paper looks into modular Petri nets; literature review (Sect. 2) provides a short survey on modular Petri nets. And then, in Sect. 3, the focus is given to the newest modular Petri nets, proposed by the first author of this paper [4]. Section 4 uses the modular Petri nets to model a production system that uses automatically guided vehicles. Section 5 (conclusion) summarizes the benefits of modularization, e.g., smaller modular components that are robust, independently developed, and analyzed.

## 2 Literature Review

Firstly, a formal definition of Petri nets is presented. And then, a concise literature study on Modular Petri Nets is given.

### 2.1 Petri Nets

Petri net formalism consists of many classes of Petri nets. The simplest and original one is the P/T (Place-Transition) Petri nets. Figure 1 presents the definition for P/T Petri nets.

**The P/T Petri Net** is defined as a four-tuple:

$$PTN = (P, T, A, M_0),$$

where:

$P$ is a finite set of places, $P = \{p_1, p_2, \ldots, p_{n_p}\}$.
$T$ is a finite set of transitions, $T = \{t_1, t_2, \ldots, t_{n_t}\}$.
$P \cap T = \emptyset$.
$A$ is the set of arcs (from places to transitions and from transitions to places).
$A \subseteq (P \times T) \cup (T \times P)$.
The default arc weight $W$ of $a_{ij}$
($a_{ij} \in A$, an arc going from $p_i$ to $t_j$ or from $t_i$ to $p_j$)
is one, unless noted otherwise.
$M$ is the row vector of markings (tokens) on the set of places.
$M = \left[M(p_1), M(p_2), \ldots, M(p_{n_p})\right] \in N^{n_p}$, |
$M_0$ is the initial marking.
Due to the markings, a $PTN = (P, T, A, M)$,
is also called **a marked P/T Petri Net**.

**Fig. 1.** P/T Petri nets [4].

### 2.2 Modular Petri Nets

Ref. [5, 6, 7] are the earlier works on modular Petri nets (MPN). These works provide a methodology for the compression of Petri net modules. In [5, 6], modules have to be event graphs with input and output ports as transitions. In [7], the module possesses a clear-cut interface. Ref. [8, 9] uses object-oriented programming for realizing Object-oriented Petri Nets.

Ref. [10] presents uses fusion places and fusion transitions for modularization. This study also proves the preservation of the properties of the model in the modules. Ref. [11] reveals generic modules that are reusable. This study focuses on manufacturing systems. Ref. [12] is on facing uncertainties in model building and uses reconfigurable modules as a solution.

Ref. [13] presents "Exhost-PIPE" as a tool for modular Petri nets. Ref. [14] focuses on molecular nets. Refs. [15, 16] are application papers, focusing on modularization of "Spanish National Health System." Ref. [17] is also an application, on control of traffic intersections. Whereas, [18] uses transitions as interfaces between the modules for communication between the modules.

Ref. [19] is about elimination of redundancy in virtual enterprises. And [20] presents Smart Factory Networks (SFN) modeling.

## 3 Modular Petri Nets

Refs. [3, 4] propose the newest modular Petri Nets. The modular Petri nets proposed in these two works are implemented in the General-purpose Petri Net Simulator (GPenSIM) [21, 22]. Thereby, models of real-life systems can be developed, simulated, and investigated with GPenSIM as modular Petri Net models.

This section presents a summary of the newest Modular Petri Nets.

### 3.1 MPN = Petri Modules + Inter-modular Connectors

MPN consists of at least one Petri module and zero or more Inter-Modular connectors (IMC). Figure 2 presents the formal definition for MPN.

**A Modular Petri Net** is defined as a two-tuple:

$$MPN = (M, C)$$

where:

$M = \sum_{i=1}^{m} \Phi_i$ (one or more Petri Modules)
$C = \sum_{j=0}^{n} \Psi_j$ (zero or more Inter-Modular Connectors)

**Fig. 2.** Modular Petri nets [4].

Figure 3 and 4 define Petri module and Inter-Modular connector (IMC), respectively.

**Formal Definition of Petri Module**. A Petri Module is defined as a six-tuple:

$$\Phi = (P_{L\Phi}, T_{IP\Phi}, T_{L\Phi}, T_{OP\Phi}, A_{\Phi}, M_{\Phi 0}),$$

where:

$T_{IP\Phi} \subseteq T$: $T_{IP\Phi}$ is known as the input ports of the module.
$T_{L\Phi} \subseteq T$: $T_{L\Phi}$ is known as the local transitions of the module.
$T_{OP\Phi} \subseteq T$: $T_{OP\Phi}$ is known as the output ports of the module.
$T_{IP\Phi}, T_{L\Phi},$, and $T_{OP\Phi}$ are all mutually exclusive:
$T_{IP\Phi} \cap T_{L\Phi} = T_{L\Phi} \cap T_{OP\Phi} = T_{OP\Phi} \cap T_{IP\Phi} = \emptyset$.
$T_{\Phi} = T_{IP\Phi} \cup T_{L\Phi} \cup T_{OP\Phi}$ (the transitions of the module).
$P_{L\Phi} \subseteq P$ is known as the set of local places of the module. Since a module has only local places, $P_{\Phi} \equiv P_{L\Phi}$.
$\forall p \in P_{L\Phi}$,
$\bullet p \in (T_{\Phi} \cup \emptyset)$. Input transitions of local places are either the transitions of the module or none (none means a local place can be a source, without any input transition).
$p\bullet \in (T_{\Phi} \cup \emptyset)$. Output transitions of local places are either the transitions of the module or none (none means a local place can be a sink, without any output transition).
This means, local places cannot have direct connections with external transitions.
$\forall t \in T_{L\Phi}$,
$\bullet t \in (P_{L\Phi} \cup \emptyset)$. Input places of local transitions are either the local places or none (none here means that a local transition can be a cold start (a source), without any input places).
$t\bullet \in (P_{L\Phi} \cup \emptyset)$. Output places of local transitions either the local places or none (none means the local transition is a sink, without any output places).
$\forall t \in T_{IP\Phi}$
$\bullet t \in (P_{L\Phi} \cup P_{IM} \cup \emptyset)$. (input places of input ports can be local places or places in inter-modular connectors or can be even an empty set)
$t\bullet \in (P_{L\Phi} \cup \emptyset)$. (output places of input ports can only be local places, or empty set)
$\forall t \in T_{OP\Phi}$
$\bullet t \in (P_{L\Phi} \cup \emptyset)$. (input places of output ports can be local places or an empty set)
$t\bullet \in (P_{L\Phi} \cup P_{IM} \cup \emptyset)$. (output places of output ports can be local places or places in inter-modular connectors or empty set).
$A_{\Phi} \subseteq (P_L \times T_{\Phi}) \cup (T_{\Phi} \times P_L)$: where $a_{ij} \in A_{\Phi}$ is known as the internal arcs of the module.
$M_{\Phi 0} = [\, M(p_L)]$ is the initial markings in the local places.

**Fig. 3.** Petri module [4].

**Formal Definition of Inter-modular Connector**. An Inter-modular Connector (IMC) is defined as a four tuple:

$$\Psi = (P_\Psi, T_\Psi, A_\Psi, M_{\Psi 0})$$

where:

$P_\Psi \subseteq P$: $P_\Psi$ is the set of places in the IMC (known as the IM-places).

$\forall p \in P_\Psi$,

$\bullet p \in (T_{OP} \cup T_\Psi \cup \emptyset)$ (input transitions of IM places are either the output ports of modules, IM transitions of this IMC, or none).

$p\bullet \in (T_{IP} \cup T_\Psi \cup \emptyset)$ (output transitions of IM places are either the input ports of modules, IM transitions of this IMC, or none). This means, IM places cannot have direct connections with local transitions, or IM transitions of other IMCs.

$\forall p \in P_\Psi, \forall i\ p \notin P_{\Phi_i}$ (an IM-place cannot be a local place of any Petri module).

$T_\Psi \subseteq T$: $T_\Psi$ is the transitions of the IMC (known as the IM-transitions).

$\forall t \in T_\Phi$,

$\bullet t \in (P_\Psi \cup \emptyset)$. (input places of IM-transitions are either the IM-places of this IMC, or none (cold start)).

$t\bullet \in (P_\Psi \cup \emptyset)$. (output places of IM-transitions either the IM-places of this IMC, or none (sink)).

$\forall t \in T_\Psi, \forall i\ t \notin T_{\Phi_i}$ ( an IM-transition cannot be a transition of any Petri module).

$A_\Psi \subseteq \left(P_\Psi \times (T_\Psi \cup T_{IP})\right) \cup \left((T_\Psi \cup T_{OP}) \times P_\Psi\right)$: where $a_{ij} \in A_\Psi$ is known as the IMC arcs.

$M_{\Psi 0} = [M(p_\Psi)]$ is the initial markings in the IM-places.

**Fig. 4.** Inter-modular connector [4].

## 4 Application Example: A Production Systems with AGVs

Figure 5 shows a manufacturing system. This manufacturing system uses Automated Guided Vehicles (AGV) as the means of transportation between workstations. The system possesses six workstations (named S1-S6). The lanes between the workstations are marked as T12 to T61.

We want to model the system to find the optimal number of AGVs needed. The problem is a variation of the circular train problem given in [23], and also presented in [24]. In these works, neither the modular model nor the solution is given.

The production system has the following characteristics:

- The vehicles move in both clockwise and counterclockwise directions.
- Between the workstations, there is only one lane meaning one vehicle only can travel on the lane between two workstations. However, any number of vehicles can stay in a workstation.
- A vehicle can only change its direction (clockwise to counterclockwise and vice versa) at the workstation $S1$.

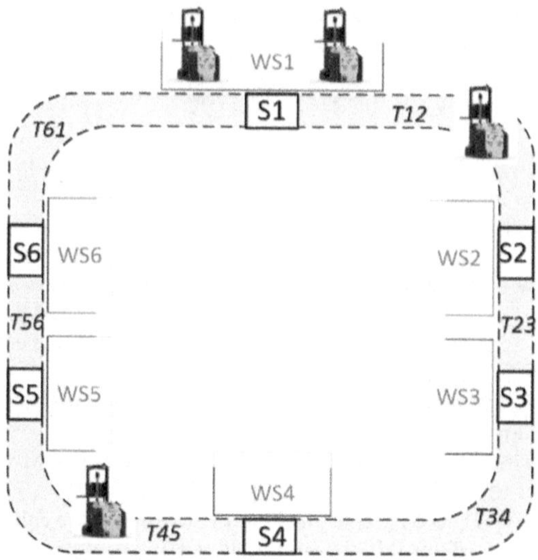


**Fig. 5.** Production system involving AGVs.

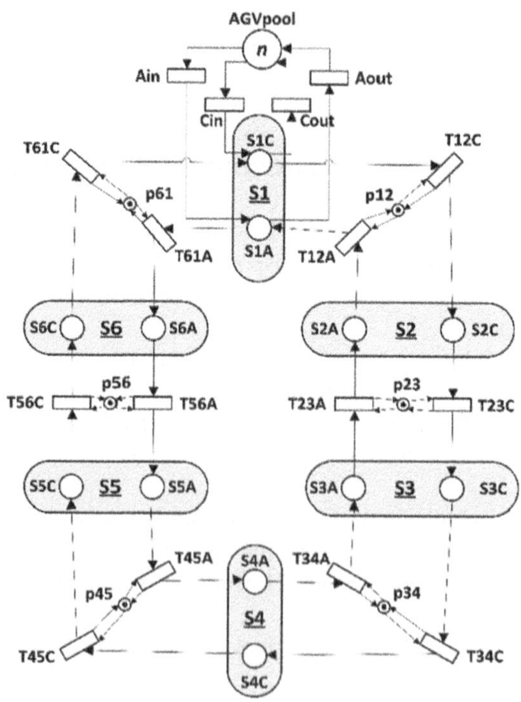


**Fig. 6.** *Monolithic* Petri net model of a production system involving AGVs.

### 4.1 Petri Net Model (Monolithic Model)

Figure 6 shows the monolithic model (one whole non-modular model) of the production system.

In this model:

- *Vpool* possesses initial tokens (number of vehicles). For example, *Cin* induces vehicles in a clockwise direction. On the other hand, *Ain* introduces vehicles in an anti-clockwise direction at workstation *S1*.
- Places *SiC* and *SiA* represent a workstation *Si*. *SiC* for the vehicles in clockwise direction, and *SiA* for anti-clockwise direction.
- *TmnC* represents the travel of a vehicle from *Sm* to *Sn* in clockwise direction. Similarly, and *TmnA* represents anti-clockwise travel.
- *pmn* is a locking mechanism for the lane *Lmn*, so that either *TmnA* or *TmnC* fires (transports an AGV), making use of the track *Lmn*.
- Finally, *Cout* is for removing an AGV from service in the clockwise movement. At the same time, *Aout* removes an AGV from service in anti-clockwise movement. An AGV can be removed from service only at *S1C* or *S1A*.

The P/T Petri net model is shown in Fig. 2. This Petri net can be simplified by the application of colored tokens:

- *Cin* and *Ain* can be combined and represented by one transition *Tin*. *Tin* introduces AGVs both directions. Likewise, *Cout* and *Aout* can be combined into one transition *Tout*.
- *SiA* and *SiC* can be combined into one place *Si*. Thereby, there is no need for two logical places to represent a physical station.

Thus, by the application of colored Petri nets, some places and transitions can be removed from the model, making the model compact.

The monolithic Petri net model (see Fig. 6) has apparent problems:

- Extensible: the model is not extensible; if we need to add more stations, the model must be redesigned.
- Comprehensibility: Due to many elements, it is not easy to understand the model.
- Robustness: the model assumes that once a vehicle starts from one workstation, it will end up in the next workstation. There is no way we can include delays in the lane or stoppage in the lane.

### 4.2 Modular Petri Net Model

Figure 7 shows a Petri module that represents the movement of vehicles between two workstations (workstations *S0* and *S1*). In this module, the internals include delays and stoppage during the movement. Furthermore, any other details can be independently added to this module without affecting other modules and the rest of the whole model.

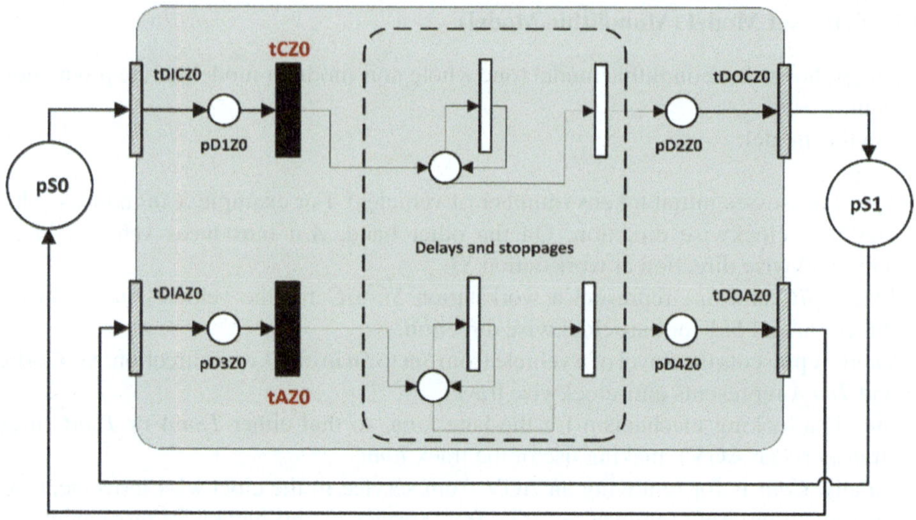


**Fig. 7.** Petri module representing movement between two workstations *S0* and *S1*.

Hence, this model is extensible (as more details can be added independently), also robust (can cope with any changes).

Figure 8 shows the overall model. This modular model is obtained simply by adding the modules together. The IMCs are the places that represent the workstations. The overall model is also easy to understand (comprehensible). Hence, the modular model satisfies all three characteristics (extensible, comprehensible, and robust) of modular model building.

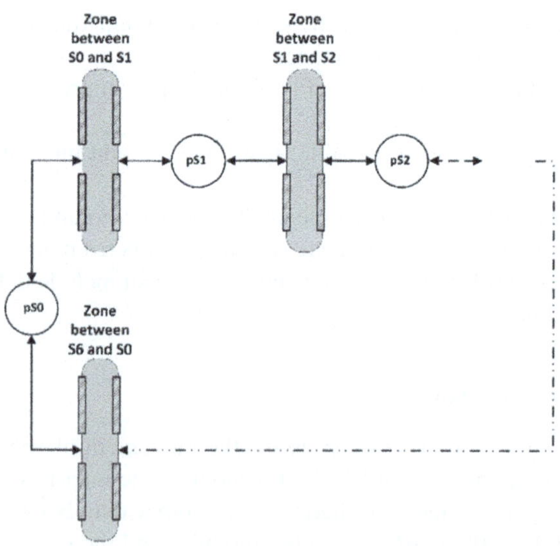


**Fig. 8.** The modular Petri model of the production system (internals of the modules are not shown).

## 5 Conclusion

This paper presents the newest modular Petri net, devised by the first author of this paper [25, 3]. Also, this paper takes an application example of how modular models can be developed based on the newest modular Petri nets. Also, the modular Petri net is implemented in software General-purpose Petri net Simulator (GPenSIM) so that real-life and industrial production systems (such as car manufacturing and smart microgrids) can be modeled and analyzed. The modular Petri net model possesses all the necessary characteristics of modular models, such as extensible, comprehensible, and robust.

Due to space and time limitations, this paper does not show the other properties of modular model development, such as independent development of the modules & testing and the analysis of the modules independently. This is considered further work.

## References


1. Peterson, J.L.: Petri Net Theory and the Modeling of Systems. Prentice Hall PTR (1981)
2. Skolud, B., Krenczyk, D., Davidrajuh, R.: Solving repetitive production planning problems. an approach based on activity-oriented petri nets. In: Graña, M., López-Guede, J.M., Etxaniz, O., Herrero, Á., Quintián, H., Corchado, E. (eds.) SOCO/CISIS/ICEUTE -2016. AISC, vol. 527, pp. 397–407. Springer, Cham (2017). https://doi.org/10.1007/978-3-319-47364-2_38
3. Davidrajuh, R.: Design of a new modular petri nets. In: Petri Nets for Modeling of Large Discrete Systems. AA, pp. 95–105. Springer, Singapore (2021). https://doi.org/10.1007/978-981-16-5203-5_7
4. Davidrajuh, R.: Petri Nets for Modeling of Large Discrete Systems. Springer, Singapore (2021). https://doi.org/10.1007/978-981-16-5203-5
5. Savi, V.M., Xie, X.: Liveness and boundedness analysis for Petri nets with event graph modules. In: Jensen, K. (ed.) ICATPN 1992. LNCS, vol. 616, pp. 328–347. Springer, Heidelberg (1992). https://doi.org/10.1007/3-540-55676-1_19
6. Claver, J., Harhalakis, G., Proth, J., Savi, V., Xie, X.: A step-wise specification of a manufacturing system using Petri nets. In: Conference Proceedings 1991 IEEE International Conference on Systems, Man, and Cybernetics, pp. 373–378. IEEE (1991)
7. De Jong, G.G., Lin, B.: A communicating Petri net model for the design of concurrent asynchronous modules. In: 31st Design Automation Conference, pp. 49–55. IEEE (1994)
8. Wang, L.C.: Object-oriented Petri nets for modelling and analysis of automated manufacturing systems. Comput. Integr. Manuf. Syst. **9**(2), 111–125 (1996)
9. Wang, L.C., Wu, S.Y.: Modeling with colored timed object-oriented Petri nets for automated manufacturing systems. Comput. Ind. Eng. **34**(2), 463–480 (1998)
10. Christensen, S., Petrucci, L.: Modular analysis of Petri nets. Comput. J. **43**(3), 224–242 (2000)
11. Tsinarakis, G.J., Tsourveloudis, N., Valavanis, K.P.: Modular Petri net based modeling, analysis, synthesis and performance evaluation of random topology dedicated production systems. J. Intell. Manuf. **16**(1), 67–92 (2005)
12. Lee, H., Banerjee, A.: A modular Petri net based architecture to model manufacturing systems exhibiting resource and timing uncertainties. In: 2009 IEEE International Conference on Automation Science and Engineering, pp. 525–530. IEEE (2009)
13. Bonnet-Torrès, O., Domenech, P., Lesire, C., Tessier, C.: Exhost-pipe: pipe extended for two classes of monitoring Petri nets. In: International Conference on Application and Theory of Petri Nets, pp. 391–400. Springer, Heidelberg (2006)

14. Blätke, M.A., Meyer, S., Marwan, W.: Pain signaling-a case study of the modular Petri net modeling concept with prospect to a protein-oriented modeling platform. In: Proceedings of the 2nd International Workshop on Biological Processes & Petri Nets (BioPPN 2011), Newcastle upon Tyne, United Kingdom, pp. 1–19 (2011)
15. Mahulea, C., García-Soriano, J.M., Colom, J.M.: Modular Petri net modeling of the Spanish health system. In: Proceedings of 2012 IEEE 17th International Conference on Emerging Technologies & Factory Automation (ETFA 2012), pp. 1–8. IEEE (2012)
16. Mahulea, C., Mahulea, L., Soriano, J.M.G., Colom, J.M.: Modular Petri net modeling of healthcare systems. Flex. Serv. Manuf. J. **30**(1–2), 329–357 (2018)
17. dos Santos Soares, M., Vrancken, J.: A modular Petri net to modeling and scenario analysis of a network of road traffic signals. Control. Eng. Pract. **20**(11), 1183–1194 (2012)
18. Slota, A., Zajkac, J., Uthayakumar, M.: Synthesis of Petri net based model of a discrete event manufacturing system for nonlinear process plan. Manag. Prod. Eng. Rev. **7**(2), 62–72 (2016)
19. Davidrajuh, R.: Distributed workflow based approach for eliminating redundancy in virtual enterprising. J. Supercomput. **63**(1), 107–125 (2013)
20. Berger, S., Bogenreuther, M., Häckel, B., Niesel, O.: Modeling availability risks of IT threats in smart factory networks–a modular Petri net approach. In: Twenty Seventh European Conference on Information Systems (ECIS2019), Stockholm Uppsala, Sweden, pp. 1–17. Association for Information Systems, AIS Electronic Library (AISeL) (2019)
21. Davidrajuh, R.: Modeling Discrete-Event Systems with GPenSIM. Springer, Cham (2018)
22. GPenSIM: General-purpose Petri net simulator. Technical report (2019). http://www.davidrajuh.net/gpensim. Accessed 20 July 2020
23. Hrúz, B., Zhou, M.: Modeling and Control of Discrete-Event Dynamic Systems: with Petri nets and Other Tools. Springer, London (2007). https://doi.org/10.1007/978-1-84628-877-7
24. Davidrajuh, R.: Models of real-life systems. In: Petri Nets for Modeling of Large Discrete Systems. AA, pp. 29–50. Springer, Singapore (2021). https://doi.org/10.1007/978-981-16-5203-5_3
25. Davidrajuh, R.: Developing a toolbox for modeling and simulation of elevators. Int. J. Simul. Syst. Sci. Technol. **20**(S1), 1.1–1.6 (2019)

# Solving Scheduling Problems in Case of Multi-objective Production Using Heuristic Optimization

Kamil Musiał[1], Artem Balashov[1], Anna Burduk[1](✉), Andre Batako[2], and Andrii Safonyk[3]

[1] Faculty of Mechanical Engineering, Wroclaw University of Science and Technology, 27 Wybrzeże Wyspiańskiego Street, 50-370 Wrocław, Poland
anna.burduk@pwr.edu.pl

[2] Faculty of Engineering and Technology, Liverpool John Moores University, 70 Mount Pleasant, Liverpool L3 5UX, UK

[3] Department of Automation, Electrical Engineering and Computer-Integrated Technologies, National University of Water and Environmental Engineering, Rivne 33000, Ukraine

**Abstract.** The paper raises the issue of production scheduling for various types of employees in a large manufacturing company. Till now, the decision-making process has been based on a human factor and the foreman's know-how, what was error prone. The presented work aimed at developing a new employee scheduling system which might be considered as a special case of the job shop problem from the set of employee scheduling problems. That would make it possible to minimize the costs of employees' work and the cost of the overall production process. Solving the problem of optimization is offered by Tabu Search and Genetic algorithms. The modification process of algorithms and the verification of algorithm performance are reported in the paper.

**Keywords:** Production scheduling · Decision-making processes · Genetic algorithm · Tabu search · Meta-heuristics · Intelligent optimization methods of production systems

## 1 Introduction

Production process scheduling is a very complex task, which complexity results from the necessity of taking into account many different factors simultaneously. Thus, in order to remain competitive, companies are forced to improve and find new methods of the process organization. Traditional methods are usually based on the knowledge and experience of process engineers, with the possible use of basic computer tools such as simple spreadsheets. These techniques are typical particularly for small and medium-sized enterprises in which employees' know-how is the basis for the proper functioning of processes.

In more advanced and often larger enterprises, the ERP (Enterprise Resources Planning) software is used to support resource planning and management. This solution is


J. Trojanowska et al. (Eds.): MANUFACTURING 2022, LNME, pp. 13–24, 2022.
https://doi.org/10.1007/978-3-030-99310-8_2

often fully sufficient for the needs of a given enterprise. However, it requires the purchase of appropriate software and training in its use, as well as often technical support after the purchase what can be very expensive. Moreover, this kind of systems appears to be more and more versatile, what is often an advantage. However, in case of production processes which require individual solutions, it can be an obstacle. Thus, with the science and technology development, more and more methods are based on intelligent solutions, i.e. heuristic algorithms [1–6].

Among the discrete optimization algorithms, two types of methods can be distinguished: exact methods, which allow to find the optimal solution, and approximate methods that do not guarantee to provide the optimal solution. With the current size of production issues, approximate methods are much more often used [7–10].

It is impossible to single out the best method for a given type of problem. The following are widely used in production scheduling problems: greedy algorithm, genetic algorithm, Tabu Search, Scatter Search and Simulated Annealing. In the literature a given problem is often solved with the help of two or more algorithms with different results [2, 8, 10, 11].

Finding the most beneficial solution to this type of problem can be possible thanks to the use of artificial intelligence methods [4, 11, 12]. They allow to find the solutions close to optimal in a relatively short time, what makes them more and more popular. The application of intelligent algorithms makes it possible to reduce time, labour input and human errors in processes. Therefore, they can be used to support decision-making processes [9, 10].

The paper examines the case where various parts are produced by two types of employees with different skills and permissions. The aim of the study was to build a program that would allow to obtain a lower cost of a production process by planning a lower number of employees. The research was carried-out on the example of a large manufacturing company.

The work consists of five section. Section 1 describes the need to write this work. Section 2 thoroughly discusses the analysed production problem and the proposed approach to the solution. Section 3 presents the algorithms used to propose the solution to the problem. Section 4 presents the results of the operation of algorithms and Sect. 5 summarizes the work and presents conclusions.

## 2 Case Study

In the paper we consider a decision-making problem, inspired by a real company, which is a global manufacturer of tools and components to produce electric machines. Up to now, all decision-making processes have been based only on a human factor. Over time an observation was made that such an approach often generates financial losses. Then, a possibility was noticed to reduce costs by an implementation of new decision-making methods.

In the studied production process, each product has its own sequence of processes that should be carried out in a specific order. However, observing a big manufactory with a medium to large serial production we might consider the processes schedule independently by assuming a parallel part production, instead of sequential one (see Fig. 1).

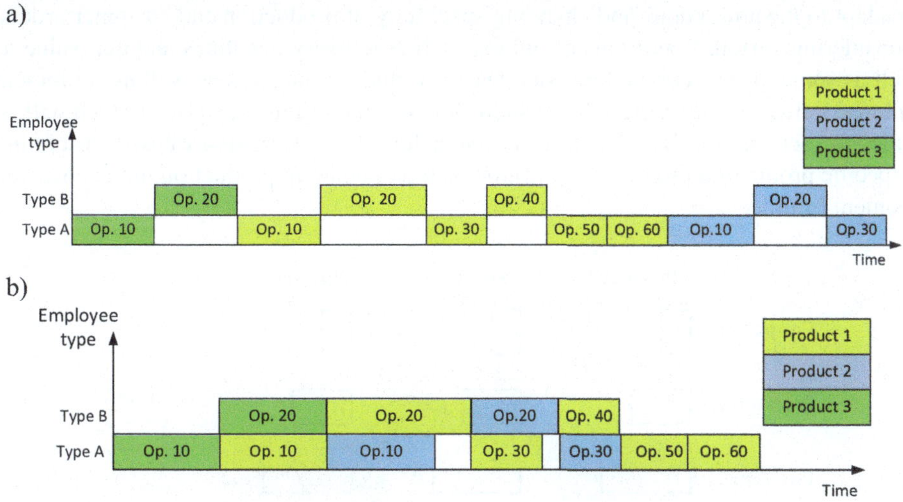


**Fig. 1.** Parallel part manufacturing assumption.

## 2.1 Decision Making Problem

One of the areas subjected to observation was the main production process on one production hall. The company produces 80 products in 28 production groups what means that several products are assigned to each group and cannot be produced parallelly. Each Friday company analyses customers' orders and prepares a production plan for the whole nearest week as well as it calculates the number of employees needed to fulfil this plan (consisting of up to 80 products in various volumes). For the described production process two types of employees are necessary. Both have different skills, permissions and are allowed to use different workstations.

As a simplified example of different production plans, we can observe another situation (see Fig. 1), where we have 6 processes, three of which require employee type 1, and other three requiring an employee type 2. Assuming, in this case, that the factory has enough machines and workstations for each and every employee, we can consider two scheduling plans – the one of minimizing overall production time (upper graph) by considering a new employee for each process, and the one of "minimizing production costs" (lower graph) by assigning only two employees of different types to work through these processes sequentially. "Minimizing production costs" is in quotation marks here because the unit costs of processes remain constant, although auxiliary expenses (such as insurance, paid breaks and lunches, holiday bonuses) are reduced.

The company does not make a schedule assigning specific employees neither to the products nor to the workstations. During the production process their decisions are made quickly by a foreman based only on his subjective opinions.

From the point of view of the company oriented to profit maximisation, the proper usage of human resources is an important activity. In order to maintain the competitive advantage, it is necessary to minimize all costs, especially those of production. Therefore, an employee scheduling management system should be flexible enough to make it able to adapt to the ambiguous and changing specificity of production and customer orders. Considering various conditions of orders, such as delivery deadlines and unit value as well as some unforeseen factors (sick leaves, birthdays, lateness as well as equipment failures, delivery delays, etc.), the system should offer various scenarios (which will be called further "options") of the production schedule. These variants are based on defining a specific production program (Fig. 2) for each part being in production in the observed moment of time.

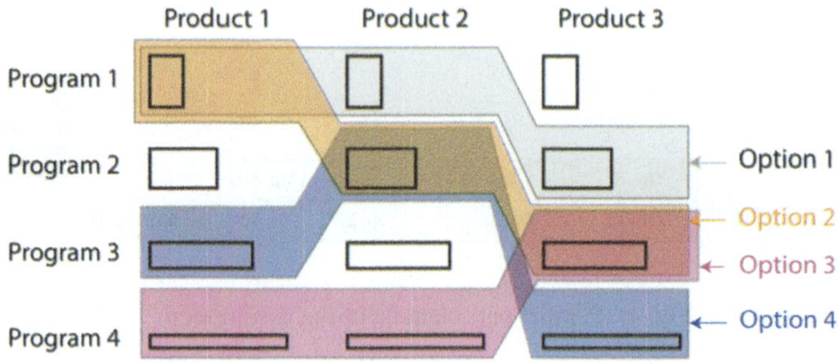


**Fig. 2.** Selecting a production plan for a different scheduling variant.

As step one, the programs were calculated - several for each specific product assuming a various number of employees and production time needed. An option is a set of the selected programs assuming the determined total demand of time and number of employees. Here, general analytical algorithms were used. In some particular situations, and for the further evolving of the idea as general, some empirical or statistical coefficients can be applied in order to achieve more precision.

## 2.2 Basic Example

In the paper one selected week was analysed. As a basis for the research, historical data were accepted. For the needs of the research, a group of 80 products was analysed. The machine park consists of 212 machines divided into 28 groups – one for each product. Each machine has to be operated by one of the two types of employees (type A or type B). The company works in 3-shift system. Based on the time of all operations and the volume of production, a set of employees is planned for shifts for the next week.

For example: 100 units of product 1, 750 units of product 2 → 20 employees type *A* and 15 employees type *B* are planned for this week. The number of workplaces is bigger than the number of employees, and the employees operate various machines and various products.

Considering the above, the problem of decision-making in production scheduling is complicated and it is impossible to find the optimal or near-optimal solution based only on a human factor.

### 2.3 Parameters, Technical Approach and Details

In the paper different approaches of scheduling process were proposed. Instead of the times of particular operations the thesis focuses on the performance of each product production process using various sets of employees at various time.

**Table 1.** Performances.

| Product | Set 1 | | | Set 2 | | |
|---|---|---|---|---|---|---|
| | Employee type 1 amount | Employee type 1 amount | Performance [pcs/h] | Employee type 1 amount | Employee type 1 amount | Performance [pcs/h] |
| 1 | 3 | 5 | 24,4 | 3 | 4 | 19,5 |
| 2 | 2 | 4 | 16,4 | 2 | 3 | 11,4 |
| 3 | 4 | 4 | 34,9 | 3 | 3 | 28,6 |
| 4 | 5 | 3 | 12,3 | 4 | 3 | 9,1 |
| 5 | 6 | 6 | 16,4 | 5 | 5 | 13,7 |
| 6 | 5 | 2 | 8,1 | 4 | 2 | 6,8 |
| 7 | 4 | 6 | 24,8 | 4 | 5 | 21,4 |
| 8 | 4 | 3 | 17,7 | 3 | 3 | 13,6 |
| 9 | 5 | 4 | 18,4 | 4 | 3 | 11,5 |
| 10 | 4 | 3 | 15,7 | 3 | 3 | 12,8 |

The performances of 10 products production taken under account are presented in Table 1. For example, product 1 can be produced with the performance/speed of 24,4 pc/h with the set of 3 employees type 1 and 5 employees type 2 or with the speed of 18,5 pc/h with the set of 3 employees type 1 and 4 employees type 2.

"Set 1" was calculated to give the maximum performance. Based on the unit times of operations, a set of employees was calculated in order to achieve the maximum production speed. Increasing of the number of employees does not speed up the process. "Set 2" onwards correspond to a lower number of employees what causes the lower production performance. Each of 80 products has 4 to 8 "sets" taken under consideration.

## 3 Optimization Methods

### 3.1 Selected Algorithms

Many kinds of decision-making problems can be solved by using meta-heuristic algorithms. Their application allows to find a near-optimal solution in a reasonable time without the transformation into mathematical formulations [9]. For the considered problem of decision-making process in the selection of suppliers, intelligent algorithms were chosen to use. For the examined case, Tabu Search and genetic algorithms were considered.

Both the Tabu Search and genetic algorithm can be used in solving NP (nondeterministic polynomial) optimization problems [8]. Although none of them gives a guarantee of finding the optimal solution, the solutions proposed by these two algorithms are fully acceptable [11–13].

**The Tabu search** algorithm solves optimization problems searching the solution space created by all possible solutions by means of a specific sequence of movements [12–14]. These movements are used to change the current solution to the new one. The algorithm checks the neighboring solutions trying to find a better one [9]. There are also taboo (forbidden) movements [12]. The algorithm avoids the oscillation around the locally optimal solution with the use of information stored in a Tabu List ($TABU_l(i_k)$). Thanks to the memory functions, the algorithm does not consider solutions that have been already visited or violated the rule [9].

For local search algorithms to work, it is necessary to determine the neighborhood of the solution. In our case, the neighborhood of the solution $i_k$ will be called the set of solutions $N(i_k)$, obtained by a single replacement of the production plan applied to one product in the current solution $i_k$. We also consider a randomized neighborhood $N_P(i_k) \subseteq N(i_k)$, where each element of the neighborhood $N(i_k)$ is included in the set NP ($i_k$) with probability $0 \leqslant P \leqslant 1$ independently of other elements. The Tabu Search algorithm presented in this section performs a probabilistic local search on a randomized neighborhood, taking steps both to improve the objective function and to deteriorate it. The general scheme of the algorithm is presented below.

1. Construct the initial solution $i_0$
2. Define $TABU_l(i_0) := \emptyset$, $i^* := i_0$, $L^* := L(i^*)$, $i^r := i_0$, $L^r := L(i^r)$, $k := 0$, $P := P_{min}$, flag:=+1.
3. Repeat until the break criterion is met:
    a. Execute the loop $N_{loop}$ times:
        i. Define neighbourhood $N_p(i_k)$,
        ii. Find $i_{k+1}$ thus, that $L(i_{k+1}) = \min_{j \in N_p(i_k)/TABU_l(i_k)} L(j)$
        iii. If $L^* > L(i_{k+1})$, then $L^* := L(i_{k+1})$, $i^* := i_{k+1}$.
        iv. Else if $L^r > L(i_{k+1})$, then $L^r := L(i_{k+1})$, $i^r := i_{k+1}$, flag += 1.
        v. Define k+=1, update $TABU_l(i_k)$.
    b. If flag = +1, then $i_k := i^r$.
    c. Define P:=P+ flag $\cdot \Delta P$
    d. If $P \geq P_{max}$, than flag -= 1
    e. If $P \leq P_{min}$, than flag +=, $L^r := L(i_k)$, $i^r := i_k$
4. Result $i^*$, $L^*$

The parameters $P_{min}$, $P_{max}$ are the upper and lower limits of the randomization parameter, $\Delta P$ is the value of incrementing randomization parameter, l is the length of the TABU list and $N_{loop}$ is the number of cycle iterations. These parameters are a set. k is the current iteration number, $i_k$ is the current solution, $TABU_l(i_k)$ is the list of prohibitions on the k-th iteration, P is the neighbourhood randomization parameter, flag $\in \{-1, +1\}$ indicates a decrease or increase in the randomization parameter, i*, L* is the best solution found and the value of the objective function for $i_t$ and correspondingly $i^r$, $L^r$ is the best solution found by a fixed value of the randomization parameter and the value of the objective function for it. The quality criterion for the current solution is the production costs for the selected variant – L.

**The Genetic algorithm** is an example of an algorithm analyzing many samples in each step, in contrast to Tabu Search based on a simple sample analysis. It has been chosen because of its successful application and good quality results in previous authors studies in production scheduling problems [8, 9, 14, 15]. The Genetic algorithm belongs to the class of evolution algorithms and has been used successfully in many areas [16–19]. A scheme of genetic algorithm is presented in Fig. 3.

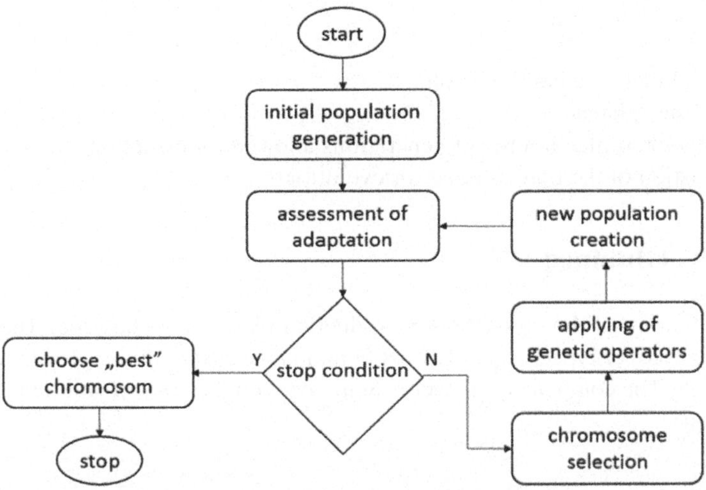


**Fig. 3.** A scheme of genetic algorithm.

The genetic algorithm consists of several main operations. The first – called initialization – generates an initial population. In the next step, the individual generated before is evaluated and selected. The chosen individuals can be reproduced or moved with no changes to the next generation. A reproduction process consists of 2 main activities: crossover and mutation.

In the scope of this particular case study, we define population as 10 different random generated production plans. All characteristics of the algorithm, such as the number of populations, generations, the number of individuals who have not undergone selection and randomized factor p, were obtained empirically. Result of testing the algorithm with various parameters shows that direction of increasing qualitatively volume of different permutation in one generation is more perspective than other options.

After each permutation population is sorted by the key of maximizing goal function, then first 60% of population passing selection. Farther, in order of descending, first a couple of the most promising individuals will give offspring, and along with them will start new generation, individual then will have a nonzero probability to mutate, remain unchanged or make a interbreeding with another individual, with what one permutation does not exclude another one, For probability of permutation, randomization operator P is responsible, which in this algorithm makes the same as in the Tabu Search. It depends on the speed of the algorithm and allows it to expand faster at the initial stages, and narrow at the end. Now, permutations in this scope are defined as follows: mutation is changing one program of one product randomly. It produces one new individual per one alteration. Crossover is basically mutation, although it changes program of one product not random, but as it is in second "parent" of couple individuals. It generates two individuals (each for autosomal dominant parent) and requires additional verification of mutated genome not to alter the one on the same.

The stop conditions are next:

- finding a global or suboptimal solution,
- access to the "plateau",
- the exhaustion of the number of generations allowed for evolution,
- the exhaustion of the time allotted for evolution.

## 4 Results Obtained

For the evaluation of the algorithms 8 production plans were analyzed. The company cost of the employee type 1 is 65 PLN per hour and the cost of the employee type 2 is 45 PLN per hour. The company costs were compared with the costs of solutions calculated by algorithms.

**Table 2.** Company production plans.

| Production plan | Number of shifts | Employee type 1 amount | Employee type 2 amount | Employee cost |
|---|---|---|---|---|
| 1 | 14 | 27 | 26 | 356720 |
| 2 | 15 | 25 | 25 | 360000 |
| 3 | 12 | 32 | 28 | 347520 |
| 4 | 12 | 19 | 20 | 224160 |
| 5 | 15 | 39 | 37 | 548400 |
| 6 | 17 | 41 | 40 | 661640 |
| 7 | 16 | 37 | 33 | 540160 |
| 8 | 15 | 26 | 29 | 394200 |

**Table 3.** Production plans generated by Tabu Search.

| Production plan | Number of shifts | Employee type 1 amount | Employee type 2 amount | Employee cost |
|---|---|---|---|---|
| 1 | 14 | 26 | 26 | 320320 |
| 2 | 14 | 27 | 25 | 322560 |
| 3 | 12 | 30 | 28 | 308160 |
| 4 | 12 | 19 | 21 | 209280 |
| 5 | 15 | 37 | 35 | 477600 |
| 6 | 16 | 42 | 41 | 585600 |
| 7 | 16 | 35 | 33 | 481280 |
| 8 | 15 | 26 | 28 | 354000 |

**Table 4.** Production plans generated by genetic algorithm.

| Production plan | Number of shifts | Employee type 1 amount | Employee type 2 amount | Employee cost |
|---|---|---|---|---|
| 1 | 14 | 26 | 26 | 320320 |
| 2 | 15 | 23 | 24 | 309000 |
| 3 | 12 | 30 | 28 | 308160 |
| 4 | 12 | 19 | 19 | 200640 |
| 5 | 15 | 36 | 37 | 480600 |
| 6 | 15 | 45 | 43 | 583200 |
| 7 | 16 | 35 | 34 | 487040 |
| 8 | 15 | 25 | 28 | 346200 |

Table 2 presents 8 company production plans including a number of shifts and a number of employees taking part in these. The employee cost is presented as well. Table 3 and Table 4 present the calculations of production plans for Tabu Search and genetic algorithm respectively.

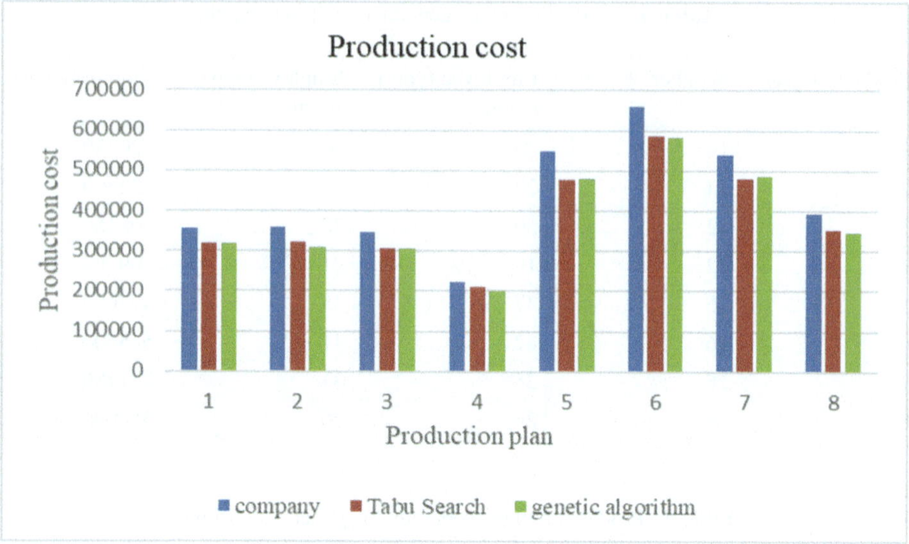


**Fig. 4.** Comparison of production costs.

Calculations have been proceeded in polish currency PLN. Accuracy of the calculation is equal to 1 PLN. As seen in Fig. 4, Tabu Search and genetic algorithms propose similar solutions that in all 8 cases are better than the production plan up to 12% (Tabu Search, plan no. 5).

## 5 Conclusion

Algorithms decreased the production costs by 6% to 12% for different production scheduling plans. It was an expected improvement considering that the company manually estimates the demand for employees. Received results will be verified in the production process soon. Nowadays, the production process is more advanced, and it becomes impossible to consider each factor from various possibilities manually. Scheduling algorithms have a long history of implementation in manufactures. This article aimed to underline their potential and ability to consider various factors. In addition to the deterministic case, algorithms can also be solved quickly in real-time while obtaining an optimal or a close to an optimal solution that will help to deal with unexpected changes. It is highly recommended to continue research at a bigger scale - where process complexity is way higher than the computing possibilities of current computers.

One of the future directions of improvement is defining and validating the empirical characteristics of algorithms, such as Tabu tenure, randomized factors, number of populations and generations, selection limit, number of possible iterations, and others.

## References


1. Tachizawa, E.M., Thomsen, C.G.: Drivers and sources of supply flexibility: an exploratory study. Int. J. Oper. Prod. Manag. **27**(10), 1115–1136 (2007)
2. Rojek, I.: Tooling selection in technological processes using neural networks. Arch. Mech. Technol. Mater. **35**, 1–15 (2015)
3. Olender, M., Kalinowski, K., Grabowik, C.: Practical approach of flexible job shop scheduling using costs and finishing times of operations. In: Burduk, A., Chlebus, E., Nowakowski, T., Tubis, A. (eds.) ISPEM 2018. AISC, vol. 835, pp. 391–400. Springer, Cham (2019). https://doi.org/10.1007/978-3-319-97490-3_38
4. Kłos, S., Patalas-Maliszewska, J.: An analysis of simulation models in a discrete manufacturing system using artificial neural network. In: Machado, J., Soares, F., Veiga, G. (eds.) HELIX 2018. LNEE, vol. 505, pp. 315–322. Springer, Cham (2019). https://doi.org/10.1007/978-3-319-91334-6_43
5. Antosz, K., Stadnicka, D.: The results of the study concerning the identification of the activities realized in the management of the technical infrastructure in large enterprises. Eksploat. Niezawodn. (Maintenance Reliab.) **16**(1), 112–119 (2014)
6. Rosienkiewicz, M., Helman, J., Cholewa, M., Molasy, M., Krause-Juettler, G.: Analysis and assessment of bottom-up models developed in central Europe for enhancing open innovation and technology transfer in advanced manufacturing. In: Scholz, S.G., Howlett, R.J., Setchi, R. (eds.) Sustainable Design and Manufacturing 2020. SIST, vol. 200, pp. 119–128. Springer, Singapore (2021). https://doi.org/10.1007/978-981-15-8131-1_11
7. Gola, A., Kłosowski, G.: Development of computer-controlled material handling model by means of fuzzy logic and genetic algorithms. Neurocomputing **338**, 381–392 (2019)
8. Bożejko, W., Uchroński, M., Wodecki, M.: Parallel Tabu search algorithm with uncertain data for the flexible job shop problem. In: Rutkowski, L., et al. (eds.) ICAISC 2016. LNCS (LNAI), vol. 9693, pp. 419–428. Springer, Cham (2016). https://doi.org/10.1007/978-3-319-39384-1_36
9. Burduk, A., Musiał, K.: Genetic algorithm adoption to transport task optimization. In: Graña, M., López-Guede, J.M., Etxaniz, O., Herrero, Á., Quintián, H., Corchado, E. (eds.) SOCO/CISIS/ICEUTE -2016. AISC, vol. 527, pp. 366–375. Springer, Cham (2017). https://doi.org/10.1007/978-3-319-47364-2_35
10. Rojek, I., Dostatni, E., Hamrol, A.: Ecodesign of technological processes with the use of decision trees method. In: Pérez García, H., Alfonso-Cendón, J., Sánchez González, L., Quintián, H., Corchado, E. (eds.) SOCO/CISIS/ICEUTE -2017. AISC, vol. 649, pp. 318–327. Springer, Cham (2018). https://doi.org/10.1007/978-3-319-67180-2_31
11. Lin, L., Gen, M.: Hybrid evolutionary optimisation with learning for production scheduling: state-of-the-art survey on algorithms and applications. Int. J. Prod. Res. **56**(1–2), 193–223 (2018)
12. Ojstersek, R., Brezocnik, M., Buchmeister, B.: Multi-objective optimization of production scheduling with evolutionary computation: a review. Int. J. Ind. Eng. Comput. **11**(3), 359–376 (2020)
13. Nurprihatin, F., Jayadi, E., Tannady, H.: Comparing heuristic methods' performance for pure flow shop scheduling under certainand uncertain demand. Manag. Prod. Eng. Rev. (2020)
14. Krenczyk, D., Skolud, B.: Transient states of cyclic production planning and control. Appl. Mech. Mater. **657**, 961–965 (2014)
15. Musiał, K., Kotowska, J., Górnicka, D., Burduk, A.: Tabu search and greedy algorithm adaptation to logistic task. In: Saeed, K., Homenda, W., Chaki, R. (eds.) CISIM 2017. LNCS, vol. 10244, pp. 39–49. Springer, Cham (2017). https://doi.org/10.1007/978-3-319-59105-6_4

16. Abdelsalam, A.M., El-Shorbagy, M.: Optimization of wind turbines siting in a wind farm using genetic algorithm based local search. Renew. Energy **123**, 748–755 (2018)
17. Alcantar, V., Ledesma, S., Aceves, S., Ledesma, E., Saldana, A.: Optimization of type III pressure vessels using genetic algorithm and simulated annealing. Int. J. Hydrogen Energy **42**(31), 20125–20132 (2017)
18. Didari, S., Wang, Y., Harris, T.A.: Modeling of gas diffusion layers with curved fibers using a genetic algorithm. Int. J. Hydrogen Energy **42**(36), 23130–23140 (2017)
19. Kammler, M., Janke, S.M., Kandratsenka, A., Wodtke, A.M.: Genetic algorithm approach to global optimization of the full-dimensional potential energy surface for hydrogen atom at fcc-metal surfaces. Chem. Phys. Lett. **683**, 286–290 (2017)

# Robust Adversarial Reinforcement Learning for Optimal Assembly Sequence Definition in a Cobot Workcell

Alessandro Alessio, Khurshid Aliev, and Dario Antonelli(✉)

DIGEP, Politecnico di Torino, Turin, Italy
dario.antonelli@polito.it

**Abstract.** The fourth industrial (I4.0) revolution encourages automatic online monitoring of all products to achieve zero-defect and high-quality production. In this scenario, collaborative robots, in which humans and robots share the same workspace, are a suitable solution that integrates the precision of a robot with the ability and flexibility of a human. To improve human-robot collaboration, human changeable choices or even non-significant mistakes should be allowed or corrected during work. This paper proposes a robust online optimization of the assembly sequence through Robust Adversarial Reinforcement Learning (RARL), where an artificial agent is deliberately trying to boycott the assembly completion. To demonstrate the applicability of robust human-robot collaborative assembly using adversarial RL, an environment composed of Markov Decision Process (MDP) like grid world is developed and a multi-agent RL approach is integrated. The results of the framework are promising: the robot observation on human activities has been successfully achieved thanks to a penalty-reward system adopted and the alternation of human to robot actions for the wrong terminal state is the one pursued by the human, but due to robot blockage wrong actions, the right terminal state is followed by human, which is the same as the robot target.

**Keywords:** Smart manufacturing · Machine learning · Human-robot collaboration · Industrial assembly

## 1 Introduction

I4.0 is promoting companies trying to find new solutions to achieve high-quality products by integrating new enabling technologies into the sector. Integration of collaborative robots and machine learning (ML) into the assembly workspaces can improve and optimize assembly products and processes and reduce human errors during production. During assembly processes, human operators tend to do wrong actions and human-robot reliability is important during collaboration with machines [1, 2]. However, machine learning tools can serve as thinking tools for machines, especially for cobots to monitor human actions. Cobots are designed to interact with humans directly and physically inside a shared workspace [3]. According to some publications, human


J. Trojanowska et al. (Eds.): MANUFACTURING 2022, LNME, pp. 25–34, 2022.
https://doi.org/10.1007/978-3-030-99310-8_3

errors generate around 50%–90% of quality problems in assembly manufacturing processes [4, 5]. For this reason, the authors of [6] proposed an algorithm to assess individual memory structures and evaluation methods of human errors in different assembly tasks. Another research proposed by [7] is a method of analyzing human errors caused by quality defects on automobile engine assembly lines. A proposed method integrates cognitive reliability and error analysis methods and fault tree analyses. Machine learning method to detect human error and recovery in assembly is presented by the integration of supervision architecture at different levels of abstractions, functions, actions, and execution monitoring [8]. Above mentioned studies lack online autonomous monitoring of human actions during assembly processes. Thus, this paper presents a robust human-robot collaboration approach based on reinforcement learning (RL) that monitors human errors during collaborative job execution.

The paper is organized as follows: first human-robot cooperative assembly formal definition through RL is presented in Sect. 2, RL based assembly framework is described and explained in Sect. 3, the results of the RL based framework for cooperative assembly is discussed in Sect. 4 and the conclusion is described in Sect. 5.

## 2 Human-Robot Cooperative Assembly Definition Through RL

Multi agent reinforcement learning (MARL) is an extension of single agent RL where multiple agents learn to maximize their individual cumulative rewards by collaborative interaction. Learning in single agent reinforcement learning is based on the Markov Decision Process (MDP), which is described by a 5-tuple $(S, A, P, r, \gamma, s_0)$ where, $S$ is a finite set of states of the environment, composed of agent's all possible sensing information about the environment; A is a finite set of actions, including the agent's all possible actions; P is the state transition matrix $P^a_{SS'} = P\left[S_{t+1} = s' | S_t = s, A_t = a\right]$, $R$ is the reward function $R^a_S = E\left[R_{t+1} | S_t = s, A_t = a\right]$; γ is the discount factor $\gamma \in [0, 1]$ for the future rewards and $s_0$ is the initial state distribution.

Our proposed MARL system for optimal collaborative assembly can be expressed as a stochastic Markov game [9], where cobot (supervisor) agent engages to learn the optimal assembly sequence and we consider human (adversary) as a second agent who is involved to learn optimal path and have a tendency to do an error in the system. Thus MDP in this paper can be reformulated as a tuple: $(S, A_h, A_r, P, r_h, r_r, \gamma, s_0)$ where $A_h$ actions of the human and $A_r$ robot actions that can be performed. $P : S \times A_h \times A_r \times S \to R$ is the transition reward and $r_h : S \times A_h \to R^2$ is the reward of human and $r_r : S \times A_r \to R^1$ is the reward of robot. If cobot is performing strategy $\mu$ and human is performing wrong strategy $v$, the reward function is $r_{\mu,v} = E_{a^1 \sim \mu(s), a^2 \sim v(.|s)}\left[r\left(s, a^1, a^2\right)\right]$. In this case robot is maximizing the $\gamma$ discounted reward while the human is minimizing it.

A multi-agents cooperative assembly framework has been developed in the next section to demonstrate the robustness of the proposed human-robot cooperative assembly through MARL.

## 3 Reinforcement Learning-Based Human-Robot Cooperative Assembly Framework

The assembly task planning is a longtime field of study that has a first practical and successful solution in the AND/OR graph proposed by de Mello and Sanderson [10]. To produce the optimal assembly sequence, the authors of [11] proposed a simple simulated annealing method. To find the optimal assembly sequence, several capability factors were examined. The authors of [12] used a genetic algorithm (GA) to produce optimal assembly sequences by combining factory information with the evaluation of assembly sequence plans. The performance of the GA method was enhanced further by the authors [13]. The latest implementations of task planning algorithms, in human-robot, machine-to-machine collaborative/cooperative assembly applications can be found in the following researches [14, 15]. Another key point of the human-robot cooperative/collaborative assembly applications is task assignment where authors [16] propose dynamic task classification and assignment approach for human and robot assembly in the collaborative work-cell.

The theoretical part of the proposed framework has been developed based on the studies described in [17, 18].

The algorithms discussed above operate in a deterministic assembly work cell in which the robot or even a human follows the planned job sequence. A degree of uncertainty exists in manual assembly because an operator may follow the wrong or alternative job sequence, either because he knows it is equal to the one intended, or because of a minor fault, which frequently has minimal implications on the completion time.

To reduce human faults during cooperative assembly, this research proposes a framework that monitors multiple agents' actions to reach optimal paths using RL during cooperative assembly jobs.

Interactions between the human, the robot, and the environment take place as represented in Fig. 1 in which four main parts are distinguished: the environment, the agents, the reinforcement learning algorithm, and the trained neural networks.

The environment is constituted by the workspace where assembling operations are performed. For the reason that assembly sequences might vary depending on the job, performance, workload on the agents, and other factors, all potential combinations are examined, and each intermediate step in the process represents a different state. Only certain transitions are admissible from one state because they are sequenced according to assembly logic and the MDP structure provided in Sect. 3.2.

The agents also must develop the knowledge about the admissible actions for each state and then find the best assembling sequence to accomplish the task, for this reason, they are trained following a specific reinforcement learning algorithm which is explained in detail in Sect. 3.1.

Agents trained independently can act according to different policies: in this case study, we used adversarial MARL directly compared to single agent approach. Since focus is on robot's behavior, this agent was trained to perform the best policy while human attempts to pursue a random objective, thus the robot must correct human's error during practice.

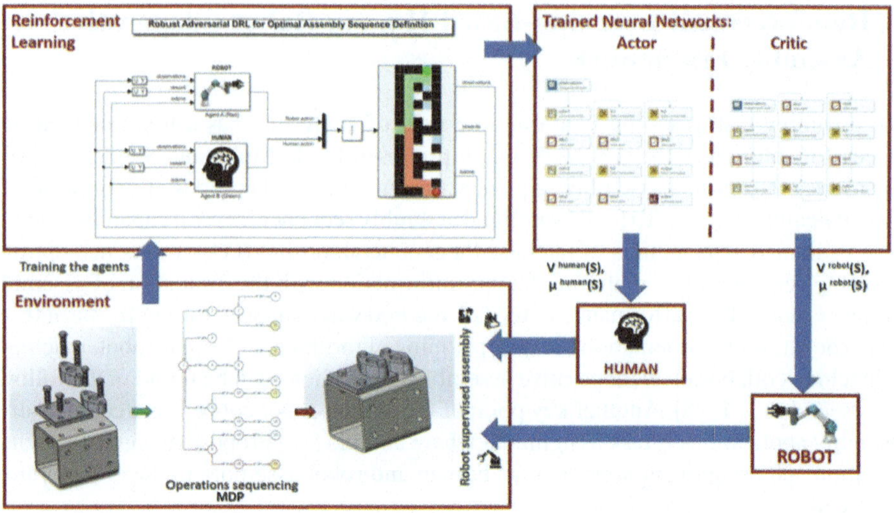


**Fig. 1.** Reinforcement learning-based framework for cooperative assembly.

### 3.1 Adversarial RL for Cooperative Assembly

In the adversarial environment assembly process, at every timestamp $t$ both agents (robot and human) observe the state $s_t$ and take actions $a_t^r \sim \mu(s_t)$ and $a_t^h \sim v(s_t)$. The state transitions $s_{t+1} = P\left(s_t, a_t^r, a_t^h\right)$ and a reward $r_t = r\left(s_t, a_t^r, a_t^h\right)$ is obtained from the environment. In the human robot assembly process robot gets a reward $r_t^r = r_t$ while a human is adversary receives a reward $r_t^h = r_t$. Thus, each step of the assembly MDP can be represented as $(s_t, a_t^r, a_t^h, r_t^r, r_t^h, s_{t+1})$. In the assembly robot protagonist is attempting to optimize the following reward function,

$$R^i = E_{s_0 \sim p, a^1 \sim \mu(s), a^2 \sim v(s)}\left[\sum_{t=0}^{\infty} \gamma^t r^i\left(s, a^1, a^2\right)\right] \tag{1}$$

because policies $\mu$ and $v$ are learnable elements, $R^1 \equiv R^1(\mu, v)$. Likewise, the human seeks to do an error action and maximize its own reward: $R^2 \equiv R^2(\mu, v)$. In our example, the assembly path is optimized first using a robot agent, and then with the involvement of a second human agent. In this case, the human operator's objective is to pursue a terminal condition which is not necessary the same of the robot, for this reason if it would be any mismatch between the agents' assembly sequences, the robot would correct the human driving him performing the right action.

### 3.2 Environment (MDP and GridWorld)

The physical environment is the workspace where the assembly is executed but to train the agents also a virtual one is needed. As the same as the real world with constrains and feasible actions, the agents can perform only one operation at time and can only move to certain states: they are allowed to advance to the next state, to regress to a previous step or to wait without doing nothing. This kind of behavior has been simulated by means of a grid world which has the same structure of the MDP schema represented in Fig. 2. In the MDP chart arrows indicate the admissible transitions, blue labels indicate whether that action should be done by the human ($a^h$) or the robot ($a^r$), yellow highlighted states are terminals and red-colored path is the one that must be learnt by agents thanks to RL algorithm.

Matlab software was used to create the grid world, which shows the sequence of potential assembly steps. Black cells indicate constraints that force agents to follow only approved trajectories by limiting their movement, each white cell in the grid represents a single elementary operation in the assembly. The same designation was used to indicate states in MDP and grid world, and the terminals' background is distinguished with a light blue tone. Agents in the grid are represented by circles, with red indicating the robot and green indicating the human; each path covered is marked with the color of the corresponding agent.

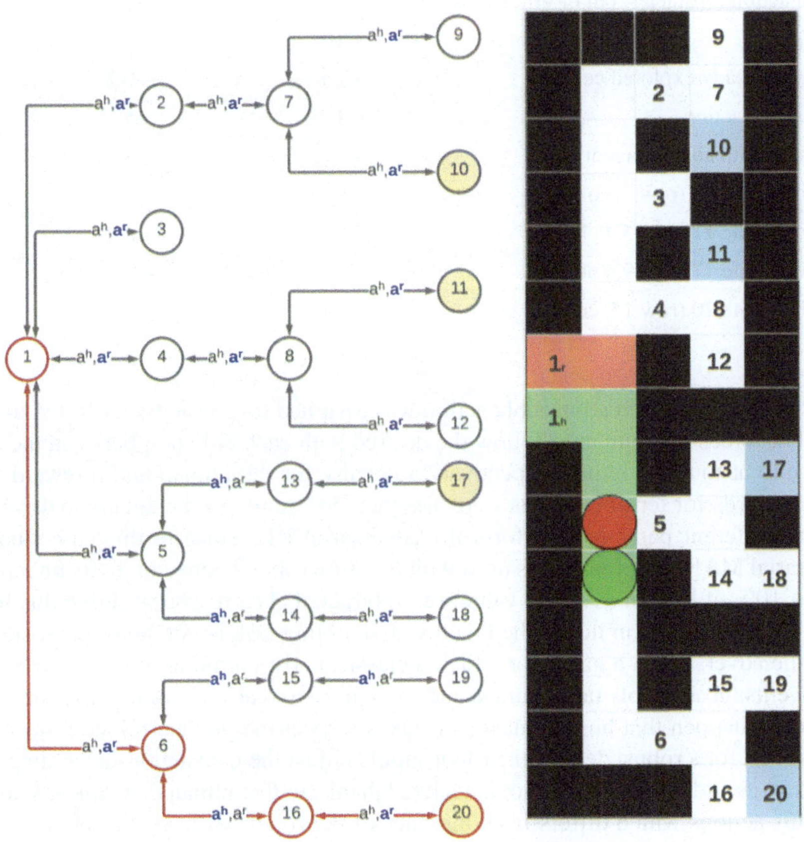

**Fig. 2.** Human robot collaborative assembly MDP structure and grid world frame during training.

The training was performed as a multiple simulation sequence during which the PPO agent's neural networks of actor and critic were updated in weights and biases.

An observation is taken at each time step by taking a photograph of the current simulation. Four channels are supplied as input to the agents for a single observation: The first channel is for obstacles, which defines the grid world's structure; the second is the "self-channel", which defines the agent's path; the third is the "other-agent-channel", which describes the path covered by the other agent; the fourth is for terminal states. Agents explored the grid world looking for terminal states according to the environment's constraints. Rewards were assigned differently in relation to the reached cells and regarding the agents, Table 1 shows the rewards in detail for each agent.

**Table 1.** Rewards and penalties for each agent.

| Action | Robot reward | Human reward |
|---|---|---|
| Illegal action (obstacles, out of grid world) | −10 | −10 |
| Idle | −10 | −10 |
| Move to already explored cell | +0.5 | −0.5 |
| Admissible action | −1 | −1 |
| Collision with another agent | +1 | −1 |
| Terminal state 10 (row 3, col 4) | +1 | +5 |
| Terminal state 11 (row 5, col 4) | +1 | +1 |
| Terminal state 17 (row 9, col 5) | +1 | +1 |
| Terminal state 20 (row 15, col 5) | +5 | +1 |

The penalty to each admissible action was assigned to get the target faster, to make the robot forces the human to follow the desired path each collision between them was taken into account to return respectively a penalty for the human and a reward to the robot, rewards for terminal states were assigned differently to the agents to drive them through different paths and perform the adversarial RL. Since in this case study the adversarial MARL approach was used with focus on robot's behavior, to its final reward value a 10% of human's reward value was subtracted. The reason of doing this for the robot but not for human lies in the fact that assembling task is not the same of standing up to the adversary in a game for which a classical adversarial approach is needed for both agents: in assembly tasks human and robot have to reach a common goal but in real case it can happen that human makes mistakes respect the predetermined sequence; in those cases, for a robust design, the robot should adjust the next actions according to the best strategy. Adversarial behavior is realized thanks to the human that "unconsciously" (with his actions which differs from the ones of the robot) reduces the robot's reward. This behavior pushes the robot to hamper human actions when they are wrong. If both agents were totally adversarial the assembly wouldn't be possible.

In the single-agent approach rewards assignment doesn't change except for the penalty inflicted to the robot in relation to human's reward which is obviously absent.

### 3.3 Actor-Critic Agents' Network

Proximal Policy Optimization (PPO) agent is an online, model-free, policy gradient reinforcement learning method [19, 20]. This algorithm alternates sampling data from interaction with the environment and optimizing surrogate objective function: PPO agent estimates the probability to take each action in a specific state and acts with respect to probability distribution; the current policy is implemented for a determined number of epochs and then both actor and critic are updated using a minibatch. Using PPO agents either the observations or the actions can be both discrete and continuous.

Policy and value function are estimated thanks to two function approximators: actor $\mu(S)$ and critic $V(S)$. The actor takes the observations S and returns the probabilities of taking each action in that state. The critic, from observations S, returns the expectation of discounted long-term reward. At the end of the training, the optimal policy is stored in the actor.

Policy gradient methods estimate the weights of the policy using the gradient ascent algorithm. According to Schulman [19], the loss policy for PPO agents is:

$$L_t^{CLIP+VF+S}(\theta) = \hat{E}_t\left[L_t^{CLIP}(\theta) - c_1 L_t^{VF}(\theta) + c_2 S[\pi_\theta](s_t)\right] \tag{2}$$

where $c_1$, $c_2$ are coefficients, $L_t^{CLIP}(\theta)$ is the clipped policy gradient objective ("surrogate"), $S$ is the entropy bonus to promote the exploration of the agent and $L_t^{VF}(\theta)$ is the squared-error loss $\left(V_\theta(s_t) - V_t^{targ}\right)^2$.

The training algorithm, after initialization of both actor $\mu(S)$ and critic $V(S)$ with random parameter values $\theta_\mu$, $\theta_V$ respectively, is in this way executed:

1. N experiences are generated by following the current policy: $S_{ts}, A_{ts}, R_{ts}, S_{ts+1}, A_{ts+1}, R_{ts+1}, \ldots S_{ts+N}, A_{ts+N}, R_{ts+N}$ where S are the states, A are the actions and R the Rewards; N corresponds to a terminal state or at maximum to the Experience Horizon value.
2. For each episode step compute the return and advantage function.
3. Learn from mini-batches of experience over K epochs:

   a. Sample random mini-batch data set from the current set of experience.
   b. Update the critic parameters by minimizing the loss $L_{critic}$ across all sampled mini-batch data.
   c. Update the actor parameters by minimizing the loss $L_{actor}$ across all sampled mini-batch data and additional entropy loss is added to this term, which encourages policy exploration.

Steps are repeated until the training episode reaches a terminal state.

## 4 Results and Discussions

In this example, the grid world environment was used to visualize the training results. The grid world was structured to look like the MDP graph for a better interpretation.

Since they have the same structure, one could expect to have a terminal condition for each training episode corresponding to each terminal state but in this way, in multi-agent scenario, no convergence was achieved. For this reason, during multi-agent training only the desired end condition was set as terminal among all the possibilities.

The robot surveillance on human activities has been successfully achieved thanks to the penalty-reward system adopted and a step-by-step alternation of human to robot actions. Terminal state 10 is the one pursued by the human during the very first training episodes; on the contrary, robot's target is terminal state 20. Opposite direction between these terminal states have been exploited to cause agents collision and allow them to understand how to react on these occurrences to maximize their own rewards.

Assuming that the robot's aim is right because it leads to a correctly assembled item, and since the robot's actions are free of decision-making autonomy compared to humans', the robot blocks human's incorrect movements. Because the single episode does not terminate if agents are not in the planned final position or the single episode reaches the maximum number of steps, even if the human reaches a wrong terminal state without being blocked by the robot, the human's return value is heavily affected due to the numerous collisions between agents.

Figure 3 depicts the agents' learning progress: red marks and lines denote robot behavior, whereas green marks and lines denote human behavior. The early episodes are required for each agent to explore, as seen in the graph; nevertheless, rewards are low in value since they do numerous illegal actions that result in high penalties. Agents begin to understand the path after a first phase of random actions, and a second phase of training is visible in the plot by the first "horizontal" trend: agents begin to recognize the goodness of terminal states that are regularly achieved, but human and robot objectives are still different. Collisions appear to be considered in the latter stages of training, just before the convergence asymptote: there are remarkable spikes in rewards values that alternate between the agents. Given all the above-mentioned rewards and penalties, maximizing the reward value for each agent causes the human to change its objective and not deviate from the robot's optimal path in order to avoid more penalties. Training came to an end when the agents reached the terminal designated condition in the shortest time possible at episode 1363.

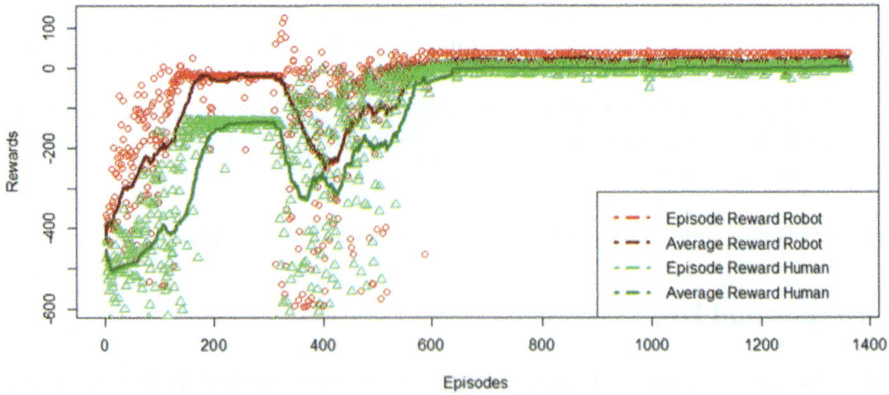


**Fig. 3.** Training process diagram: dots represent single episode reward for each agent, lines represent corresponding average reward computed on a set of 50 episodes.

## 5 Conclusions

Flexibility and adaptability in tasks execution are unrivalled characteristics proper of human nature, anyway, this property can bring the human to execute actions in a different way from the planned sequence. In collaborative work-cells is then necessary to perform robust programming to allow the robot to manage every situation to avoid production slowdowns and stops. In this case study was demonstrated that the robot can successfully force the human to follow the assembling sequence thanks to the proper implementation of MARL.

The use of a grid world environment to simulate agents' state transitions is both a strength as well as a limit of the approach: even if it allows to show clearly the achieved results, respect the real scenario in which one agent's activity changes the state of both, in this example agents have their own states, so they can't be in the same state at the same time.

For the sake of simplicity, we tested the algorithm on a limited set of operations in a virtual environment only. This work can be further improved by modifying the training environment and applying the created algorithm to a real assembly process: extending the present method to a full assembly derived from an industrial case study will consistently increase the number of states in the MDP, and it will be interesting to test the real-time correction of the robot to human errors in the execution of elementary operations.

## References

1. Park, K.S.: Human Reliability: Analysis, Prediction, and Prevention of Human Errors. Elsevier (2014)
2. Aliev, K., Antonelli, D.: Proposal of a monitoring system for collaborative robots to predict outages and to assess reliability factors exploiting machine learning. Appl. Sci. **11**(4), 1621 (2021)
3. Galin, R., Meshcheryakov, R.: Review on human–robot interaction during collaboration in a shared workspace. In: Ronzhin, A., Rigoll, G., Meshcheryakov, R. (eds.) ICR 2019. LNCS (LNAI), vol. 11659, pp. 63–74. Springer, Cham (2019). https://doi.org/10.1007/978-3-030-26118-4_7
4. Liu, G.Z., Zhang, Y.J., Li, Z., Ying, Y., Cai, Z.X.: Human errors analysis in substation operation based on CREAM. Electr. Power **40**(5), 85–89 (2007)
5. Di Pasquale, V., Miranda, S., Neumann, W.P., Setayesh, A.: Human reliability in manual assembly systems: a systematic literature review. IFAC-Papersonline **51**(11), 675–680 (2018)
6. Strenge, B., Schack, T.: Empirical relationships between algorithmic SDA-M-based memory assessments and human errors in manual assembly tasks. Sci. Rep. **11**(1), 1–12 (2021)
7. Le, Y., Qiang, S., Liangfa, S.: A novel method of analyzing quality defects due to human errors in engine assembly line. In: 2012 International Conference on Information Management, Innovation Management and Industrial Engineering, vol. 3, pp. 154–157. IEEE (2012)
8. Lopes, L.S., Camarinha-Matos, L.M.: A machine learning approach to error detection and recovery in assembly. In: Proceedings 1995 IEEE/RSJ International Conference on Intelligent Robots and Systems. Human Robot Interaction and Cooperative Robots, vol. 3, pp. 197–203. IEEE (1995)
9. Perolat, J., Scherrer, B., Piot, B., Pietquin, O.: Approximate dynamic programming for two-player zero-sum Markov games. In: International Conference on Machine Learning, pp. 1321–1329. PMLR (2015)

10. De Mello, L.H., Sanderson, A.C.: AND/OR graph representation of assembly plans. IEEE Trans. Robot. Autom. **6**(2), 188–199 (1990)
11. Milner, J.M., Graves, S.C., Whitney, D.E.: Using simulated annealing to select least-cost assembly sequences. In: IEEE International Conference on Robotics & Automation, pp. 2058–2063. IEEE (1994)
12. Tseng, Y.J., Chen, J.Y., Huang, F.Y.: A multi-plant assembly sequence planning model with integrated assembly sequence planning and plant assignment using GA. Int. J. Adv. Manuf. Technol. **48**(1–4), 333–345 (2010)
13. Lu, C., Yang, Z.: Integrated assembly sequence planning and assembly line balancing with ant colony optimization approach. Int. J. Adv. Manuf. Technol. **83**(1–4), 243–256 (2015). https://doi.org/10.1007/s00170-015-7547-7
14. Aliev, K., Antonelli, D., Bruno, G.: Task-based programming and sequence planning for human-robot collaborative assembly. IFAC-PapersOnLine **52**(13), 1638–1643 (2019)
15. Aliev, K., Antonelli, D.: Analysis of cooperative industrial task execution by mobile and manipulator robots. In: Trojanowska, J., Ciszak, O., Machado, J.M., Pavlenko, I. (eds.) MANUFACTURING 2019. LNME, pp. 248–260. Springer, Cham (2019). https://doi.org/10.1007/978-3-030-18715-6_21
16. Bruno, G., Antonelli, D.: Dynamic task classification and assignment for the management of human-robot collaborative teams in workcells. Int. J. Adv. Manuf. Technol. **98**(9–12), 2415–2427 (2018). https://doi.org/10.1007/s00170-018-2400-4
17. Morales, M.: Grokking Deep Reinforcement Learning. Simon and Schuster (2020)
18. Pérolat, J., et al.: Learning Nash equilibrium for general-sum Markov games from batch data. In: Artificial Intelligence and Statistics. PMLR (2017)
19. Schulman, J., Wolski, F., Dhariwal, P., Radford, A., Klimov, O.: Proximal Policy Optimization Algorithms. ArXiv:1707.06347 [Cs] (2017)
20. Schulman, J., Levine, S., Moritz, P., Jordan, M.I., Abbeel, P.: Trust Region Policy Optimization. ArXiv:1502.05477 [cs.LG] (2015)

# Designing an Automated Assembly Workplace in a Simulation Environment

Kateryna Kovbasiuk(✉), Michal Balog, and Kamil Židek

Faculty of Manufacturing Technologies With a Seat in Prešov, Department of Industrial Engineering and Informatics, The Technical University of Košice, Bayerova 1, 080 01 Prešov, Slovak Republic
kateryna.kovbasiuk@tuke.sk

**Abstract.** Assembly is the most complex stage of the production process. This stage not only affects the quality and reliability of products, but also the ongoing production time, labor productivity and efficiency of the entire system. Therefore, any changes or improvements to the assembly phase need to be thoroughly tested before their implementation. This paper suggests a solution for the transition from manual to automated assembly using simulation. For this purpose, the manual assembly of cam switches is chosen. The focus is then on designing the automated workplace for the cam switches assembly. One of the key points of this paper is the use of simulation as a tool for production solutions. Using the method of quantifying assembly tasks complexity, the tasks which are suitable for further automation are defined, and the appropriate level of automation is defined too. Then the automated workplace is modelled in Tecnomatix Plant Simulation. The solution presented in this paper is relevant for process and production managers involved in planning a new assembly line or redesigning an existing one.

**Keywords:** Assembly · Automation · Tecnomatix Plant Simulation

## 1 Introduction

The assembly process very often plays a key role in a production system [1]. Its improvement and optimization are essential for the competitiveness of production: about 50% of the cost of the product should be attributed to the assembly phase [2]. Flexibility and agility are the key factors in developing efficient and competitive production systems. Nowadays global competition is forcing companies to reduce production costs and production time while improving quality. The solution can be the automation of assembly processes which results in increased accuracy and repeatability, reduced production time, workload, staff errors and labor costs [3].

The increasing degree of automation in industrial enterprises is a modern trend in production. By choosing the right level of automation, a company can still maintain a high level of flexibility with fewer manual tasks. Identification of the right level of automation for the best system performance is based on a balanced and holistic approach to automation. By using simulation software tools, organizations gain opportunities to


J. Trojanowska et al. (Eds.): MANUFACTURING 2022, LNME, pp. 35–49, 2022.
https://doi.org/10.1007/978-3-030-99310-8_4

explore how their assembly processes and systems will work before automation implementation. Due to the fact that the commissioning of new production lines, processes and production facilities is often costly and capital-intensive, the use of simulation in production can bring huge benefits. The data obtained through the simulation allow to test different combinations and scenarios in the virtual environment.

### 1.1 Simulation

The desire to implement the Industry 4.0 concept leads to the need to use the capabilities of modern simulation, since it is an effective means of researching new processes and testing new products, devices, technologies and systems. One of the most prominent technologies that drives the fourth industrial revolution is simulation [4, 5]. First computer simulation was used to solve complex problems in aerospace and steel corporations, however, today it has a great number of application domains, especially in manufacturing [6, 7].

Simulation can be integrated with almost any other Industry 4.0 technology. It can be used for pro-active purposes such as designing and evaluating a vertical integration level before machine integration, as well as it can be used for re-active purposes to evaluate various "what-if" scenarios resp. alternative solutions [6]. Simulation models may help improve the performance of manufacturing systems, assess the expected outcomes of different systems, processes or machines implementation before the actual implementation and can eliminate planning failures. Moreover, simulation is a cost-effective method for testing decisions before their realization [8].

### 1.2 Simulation in Assembly System Automation

Despite the fact that Industry 4.0 has introduced high levels of digitization and automation, assembly systems still remain manual in most cases, as automation cannot completely replace human cognitive and problem-solving abilities [9]. However, even though manual assembly systems (MAS) provide flexibility, they do not offer the same high level of productivity as automated systems do. In countries with high labour costs, this aspect affects the cost of the product. Also, automated assembly systems make it possible to continuously perform the same assembly tasks, taking the same time and using the same tools [10].

Productivity and revenues can be increased by the reduction in costs. This can be achieved by implementing robots and autonomous machines in assembly systems. Productivity then can be achieved since robots can perform recurrent tasks better than humans. Apart from common industrial robots, there are also collaborative ones, which can cooperate with humans with no risks connected to it [11]. This is an optimal alternative for assembly system automation since fully automated robotic assembly links are not always beneficial or even possible. Thus, using a specific simulation software package, a manufacturing company can prototype and evaluate automated or semi-automated assembly system to decide whether the automation is needed. The range and variety of such software packages is constantly growing. As the dominant basic concepts in modern simulation modelling are used [11]:

- discrete-event simulation systems (AnyLogic, Arena, SIMUL8, FlexSim, Tecnomatix Plant Simulation, WITNESS etc.),
- dynamic systems (MATLAB),
- systems based on network paradigms (ARIS),
- systems based on continuous modelling (AnyLogic, iThink, Powersim, Vensim),
- other.

Discrete-event simulation software packages are the most commonly used for simulating processes and systems in manufacturing such as assembly ones [12].

## 2 Methodology

Several steps were taken to model an automated assembly workstation in a simulation environment: choosing an object which is assembled manually, describing its assembly sequence, identifying the automation potential of each task in the assembly process, determining the optimal automation level and modelling the automated assembly workplace in a simulation environment.

### 2.1 Manual Assembly Process

The selected device is a cam switch – a device that is designed to switch electrical circuits (see Fig. 1). Cam switches are usually assembled manually with the occasional use of electromechanical tools.

**Fig. 1.** Cam switch.

**Table 1.** Components of a cam switch.

| | Components | Number of components |
|---|---|---|
| 1 | Switching chamber I | 1 |
| 2 | Switching chamber II | 1 |
| 3 | Cam switch cover | 1 |
| 4 | Shaft | 1 |
| 5 | Stud | 2 |
| 6 | Latching chamber | 1 |
| 7 | Bolt nut | 4 |
| 8 | Latching star 60$^{o}$ | 1 |
| 9 | Latching slide | 2 |
| 10 | Latching spring | 2 |
| 11 | Flat stop | 1 |
| 12 | Stopper I | 1 |
| 13 | Stopper II | 1 |
| 14 | Screw bolt washer | 2 |
| 15 | Cam | 4 |
| 16 | Circlip | 1 |
| 17 | Insulating tube | 2 |
| 18 | Flat washer | 1 |
| 19 | Plastic button | 3 |
| 20 | Contact spring | 3 |
| 21 | Electrical contact bridge (with 2 precious clad metals stamped into it) | 3 |
| 22 | Terminal (with a contact riveted on its right side) | 3 |
| 23 | Terminal (with a contact riveted on its left side) | 3 |

Table 1 presents components and number of components needed for one cam switch assembly:

Each assembly task consists of three or four basic subtasks [14] (Fig. 2):

1. Identification and gripping of the part.
2. Bringing the part to the placement/insertion position.
3. Insertion/fitting.
4. Optionally securing the part.

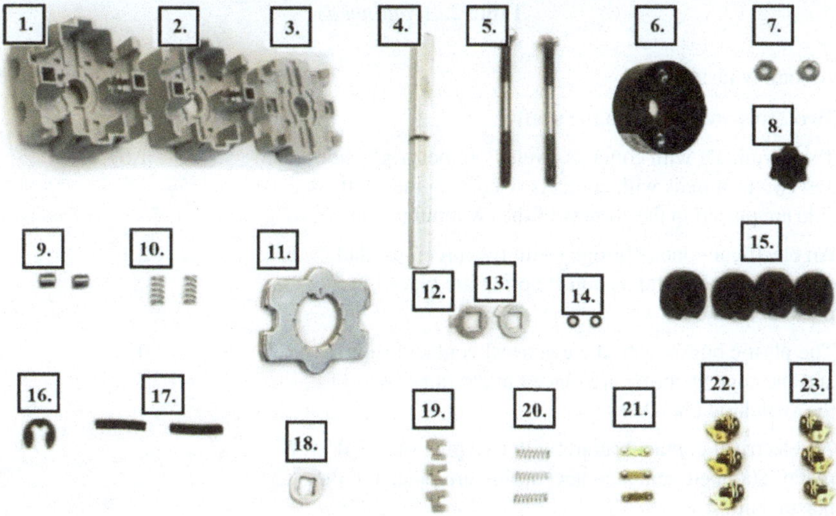


**Fig. 2.** Cam switch components [13].

Table 2 presents the sequence of the cam switch assembly tasks and the average time needed to perform each task separately. Time is divided into main (insertion and securing of the part) and additional (identification and gripping of the part).

**Table 2.** Sequence of the cam switch assembly tasks.

| № | Assembly tasks | Main time [s] | Additional time [s] |
|---|---|---|---|
| 1 | Two studs are inserted into a latching chamber with two bolt nuts already inserted into it under the press | 1,3 | 2,3 |
| 2 | The latching chamber is placed on a position fixing base | 0,5 | 0,1 |
| 3 | A latching star is inserted into the centre of the latching chamber | 1,6 | 2,7 |
| 4 | A shaft with the circlip on it is inserted into the centre of the latching chamber | 2,5 | 1,2 |
| 5 | Two latching slides are placed in parallel on both sides of the latching star | 3,3 | 2,7 |
| 6 | Two latching springs are placed in the latching chamber perpendicular to the latching rollers | 4,1 | 6,5 |
| 7 | A stop plate is placed on the top | 0,9 | 2 |
| 8 | Two stoppers and a flat washer are placed on the shaft | 3,9 | 4,6 |
| 9 | A switching chamber is placed on the top | 0,5 | 3,3 |

(*continued*)

**Table 2.** (*continued*)

| № | Assembly tasks | Main time [s] | Additional time [s] |
|---|---|---|---|
| 10 | Two cams are placed on the shaft | 6,6 | 3,5 |
| 11 | Two terminals with contacts riveted on their right side and two terminals with contacts riveted on their left side are placed in the corners of the switching chamber | 18,6 | 1,6 |
| 12 | An electrical contact bridge (with two precious clad metals stamped) and a contact spring are placed in the plastic button | 5 | 2 |
| 13 | The plastic button with the electrical contact bridge and the contact spring are placed in the upper part of the switching chamber | 1,5 | 0 |
| 14 | An electrical contact bridge (with two precious clad metals stamped) and a contact spring are placed in the plastic button | 5 | 2 |
| 15 | The plastic button with the electrical contact bridge and the contact spring are placed in the lower part of the switching chamber | 1,5 | 0 |
| 16 | A switching chamber is placed on the top | 3,1 | 10 |
| 17 | Two cams are placed on the shaft | 3,1 | 0,2 |
| 18 | A terminal with a contact riveted on its right side and a terminal with a contact riveted on its left side are placed in the upper part of the switching chamber | 12 | 1,1 |
| 19 | Two insulating tubes are put on the studs | 1,9 | 2,7 |
| 20 | An electrical contact bridge (with two precious clad metals stamped) and a contact spring are placed in the plastic button | 5 | 2 |
| 21 | The plastic button with the electrical contact bridge and a contact spring are inserted between the terminals | 3 | 0 |
| 22 | The switching chamber is closed with a plastic cam switch cover | 4,2 | 0,2 |
| 23 | Two screw bolt washers and two bolt nuts are inserted | 2,4 | 2,8 |
| 24 | The bolt nuts are screwed in | 3 | 4,8 |
| 25 | The type plate is affixed to the latching chamber | 1,6 | 1,7 |
| 26 | Contacts are tighten | 4,3 | 2,9 |
| | | **100,4** | **62,9** |

The average time of manual assembly of one cam switch is 2 min 43 s. In percentage correlation, 40% of the manual assembly time is spent identifying, taking, and placing the part in the insertion position; 60% of the manual assembly time is spent inserting and fixing the part.

## 2.2 Assessment of Task Potential for Human-Robot Collaboration

When the assembly operation is decomposed into separate tasks, by using the method of tasks-distribution in human-robot collaboration (HRC) the automation potential of each task can be identified. The HRC automation potential is calculated based on product complexity model [15].

Table 3 presents the scores which represent the ease of automation. Using this potential score sheet for robotic assembly in HRC, each cam switch components are evaluated based on their quantitative and qualitative characteristics to identify the tasks for robotic automation.

**Table 3.** Potential score sheet for robotic assembly in HRC.

| | | | | | |
|---|---|---|---|---|---|
| Component | Weight | | | | |
| | >12 kg | 8–12 kg | 3–8 kg | 1–3 kg | <1 kg |
| | 0 Points | 1 Point | 2 Points | 3 Points | 4 Points |
| | Sensitivity | | | | |
| | Highly sensitive | Damage in careless handling | Damage in light force | Damage at high force | Robust |
| | 0 Points | 1 Point | 2 Points | 3 Points | 4 Points |
| | Dimensional stability | | | | |
| | Shapeless | Deformation possible | Deformation under force | Deformation at high force | Rigid |
| | 0 Points | 1 Point | 2 Points | 3 Points | 4 Points |
| | Handling (ease of gripping of the component) | | | | |
| | Human hand | Custom gripper | 2 fingers gripper | | |
| | 0 Points | 2 Points | 4 Points | | |
| Mounting | Direction of mounting of components | | | | |
| | Not a straight line | Straight line from side | Straight line from above | | |
| | 0 Points | 2 Points | 4 Points | | |
| | Tolerance limits | | | | |
| | <0,5 mm | ≥ 0,5 < 1 mm | >1 mm | | |
| | 0 Points | 2 Points | 4 Points | | |

*(continued)*

**Table 3.** (*continued*)

| | | | |
|---|---|---|---|
| Safety | Tools increasing safety risks (sharp edged and pointed tools) | | |
| | The use would increase the danger to humans | No consequences | The use would reduce the danger to humans |
| | 0 Points | 2 Points | 4 Points |
| | Risk of collision in the head/neck area | | |
| | The use of HRC involves the risk of collision injury | No work at head height | The use of HRC would reduce the risk of collision injury |
| | 0 Points | 2 Points | 4 Points |
| Miscellaneous | Alignment needed for the part features | | |
| | Human dexterity is required | Can be handled by the robot | No alignment is needed |
| | 0 Points | 2 Points | 4 Points |
| | Components attachment | | |
| | Force required | Easy to do | No attachment task |
| | 0 Points | 2 Points | 4 Points |
| | Equipment/machine triggering for assembly | | |
| | Required | Not required | |
| | 0 Points | 4 Points | |

The evaluation process then proceeds with the calculations [15]:

1. HRC potential influenced by the physical properties of the component being assembled:

$$\mathrm{HRC}_{\mathrm{CP}} = \frac{\sum_{1}^{J} CP}{4J} \cdot 100 \tag{1}$$

CP – a potential score for physical characteristics of a selected part;
J – number of part related attributes.

2. HRC potential based on the assembly characteristics:

$$\mathrm{HRC}_{\mathrm{MT}} = \frac{\sum_{1}^{K} MT}{4K} \cdot 100 \tag{2}$$

MT – potential score for mounting characteristics for assembling of a part;
K – number of mounting related attributes.

3. HRC safety potential:

$$\mathrm{HRC}_{\mathrm{SF}} = \frac{\sum_{1}^{M} SF}{4M} \cdot 100 \tag{3}$$

SF – potential score for safety related attributes;
M – number of safety related attributes.

4. HRC potential due to miscellaneous characteristics (attachment, alignment, equipment triggering) for the assembly task:

$$HRC_{Misc} = \frac{\sum_1^N Misc}{4N} \cdot 100 \quad (4)$$

Misc – potential score for different assembly characteristics;
N – number of attributes considered in this category.

5. Percentage share of the score for each component:

$$HRC_{Part} = \frac{\sum AS}{\sum AF} \cdot 100 \quad (5)$$

AS – scores achieved for each factor considered for HRC task evaluation;
AF – the number of all factors considered for HRC potential.

If the automation potential is above 70%, then the task can be assigned to a robot. Some tasks such as inserting studs, placing an electrical contact bridge and a contact spring in the plastic button, placing insulating tubes, and affixing a type plate, have to be performed manually and are not evaluated because they require precision. Ten cam switch components were evaluated to identify the automation potential for the assembly tasks in which they are used. The results are presented in Table 4.

**Table 4.** Results of evaluating the automation potential of the cam switch tasks.

| № | Component | Potential |
|---|---|---|
| **1** | **Latching star** | **70,4%** |
| **2** | **Shaft** | **72,7%** |
| **3** | **Latching slide** | **86,3%** |
| 4 | Latching spring | 31,8% |
| 5 | Stop plate | 63,6% |
| 6 | Stopper | 63,6% |
| 7 | Flat washer | 63,6% |
| 8 | Switching chamber | 59% |
| **9** | **Cam** | **70,4%** |
| **10** | **Terminal (with a contact riveted on it)** | **72,7%** |

As it can be seen from the results represented in Table 4, assembly tasks in which a latching star, a shaft, a latching slide, a cam, and a terminal are used, can be automated, thus, they can be performed by robots. Thuswise the most appropriate level of automation is semi-automation.

## 2.3 Modelling Assembly Workplace in a Simulation Environment

To test the effectiveness of manual cam switch assembly automation, assembly workplace is modelled in a simulation environment. For this purpose, a discrete-event simulation software package, Tecnomatix Plant Simulation, is selected. This software is considered the most appropriate tool for the computer simulation [16, 17] and one of the best solved software in the area of production solutions [18, 19].

A semi-automated workplace for cam switches assembly, where collaborative robots can work with people and share a common workspace without physical barriers, was modelled in Tecnomatix Plant Simulation program (see Fig. 3).

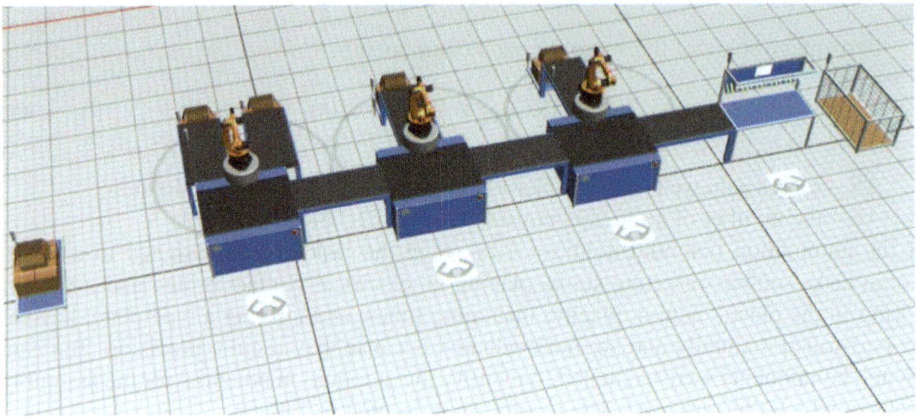

**Fig. 3.** A semi-automated workplace for cam switches assembly.

The proposed workplace consists of three stations, where workers assemble cam switches, while cooperating with collaborative robots, and one assembly station with one worker. In Fig. 4, each assembly stage is marked with a number, which indicates the order in which the cam switch is assembled. The components are fed to the robots with the help of vibratory bowl feeders which are represented by the turnplates.

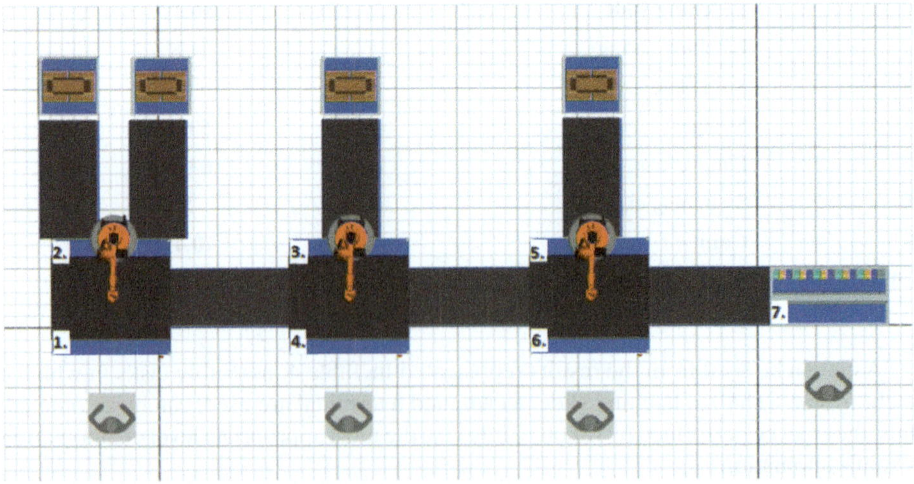


**Fig. 4.** The order of the cam switch assembly tasks which are distributed between workers and collaborative robots.

A worker at the first parallel assembly station (1) inserts two studs into the latching chamber, and then at the same parallel assembly station the robot (2) inserts a latching star and a shaft into a latching chamber. Next, this part of the cam switch is moved by a conveyor to another parallel assembly station, where the second robot (3) inserts two latching slides into the latching chamber and the worker (4) inserts two latching springs, puts a stop plate, stoppers, a flat washer, and places the switching chamber and cams. Then the cam switch is moved by the conveyor to the last parallel station, where the robot (5) inserts four terminals with riveted contacts and the worker inserts two plastic buttons with electric contact bridges and contact springs, places the switching chamber, inserts two cams, and two terminals. Then the cam switch is moved to the last assembly station (6), where the worker inserts two insulating tubes, a plastic button with an electric contact bridge and a contact spring, two screw bolt washers and two bolt nuts. Then the worker screws in the bolt nuts and then affixes a type plate on the locking chamber and tightens the contacts.

The processing time is set before starting a simulation. The processing time on each station is based on the time needed to perform each assembly task on a given station manually, assuming that the tasks performed by robots will take relatively less time (Fig. 5).

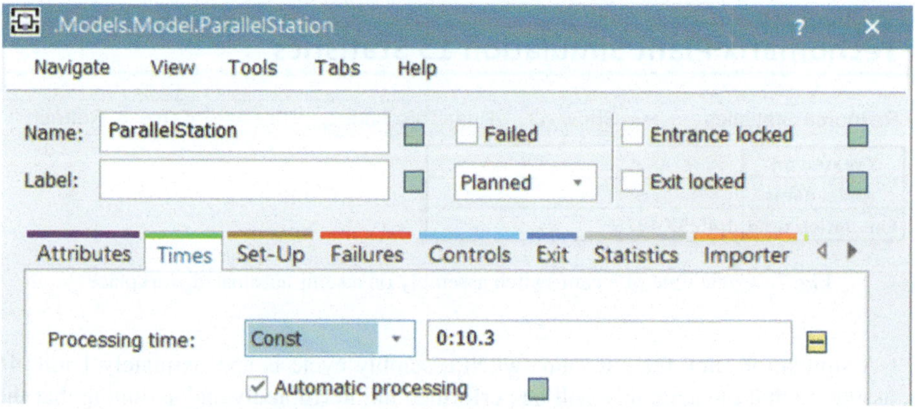


**Fig. 5.** Setting the processing time on each assembly station.

After setting the processing time on each station, the simulation of one cam switch assembly is started (Fig. 6).

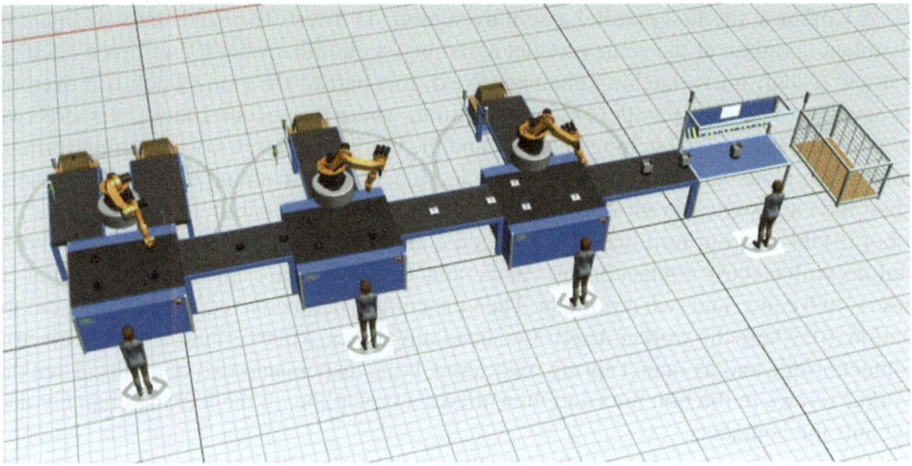

**Fig. 6.** Ongoing simulation of the cam switches assembly.

After the simulation is finished, the cycle time is found in the **Statistics Report.**

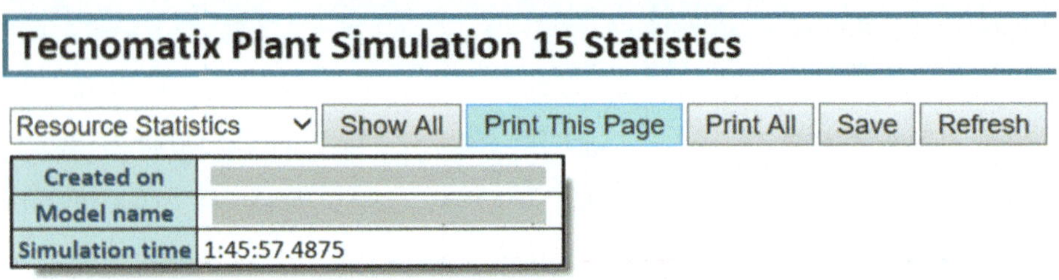


**Fig. 7.** Cycle time of a cam switch assembly on a semi-automated workplace.

A simulation time for one cam switch assembly cycle is approximately 1 min 46 s, assuming that the assembly will be performed uninterruptedly and assuming that the robots will perform assembly tasks 1–4 s quicker than the same tasks are performed manually (Fig. 7).

## 3 Results

With the proposed workplace, assembly speed and process quality can be improved without the risk of injury associated with working in close proximity to machines. The approximate cycle time is 1 min 46 s, although the workers are still assigned many tasks (Table 5).

**Table 5.** Cycle time on the manual and semi-automated workplaces.

| | Cycle time |
|---|---|
| Manual assembly | 2 min 43 s |
| Assembly on a semi-automated workplace | 1 min 46 s |

Thus, the cycle time of a cam switch assembly at the proposed semi-automated workplace is shorter than the cycle time of a cam switch assembly performed manually. Having these results provided, it may be assumed that:

– the company with the semi-automated workplace for cam switch assembly can assemble more cam switches at the same period of time that is needed to assemble the planned number of these electrical devices,
– the labor cost can be reduced since the planned amount of cam switches can be assembled in a shorter period of time.

Semi-automated assembly workplace's biggest advantage over the fully automated one is the possibility to implement new assembly methods or the possibility of changing them, thus, with the automation being adopted, the flexibility is not excluded.

According to the specifications of the product based on which an automated assembly workplace was proposed, the following technical components and their required charateristics are suggested in Table 6.

**Table 6.** Technical components of automated assembly workplace and their required characteristics.

| Technical components | Number | Required characteristics |
|---|---|---|
| Collaborative robots (e.g. IRB 14050 Single-arm YuMi collaborative robot) | 3 | Repeatability: ≤ 0.02 mm<br>Payload: ≤ 0,5 kg |
| Robot gripper | 3 | Since the proposed automated assembly requires high flexibility, robot grippers should be adaptive. In fact, these grippers have to adapt to the shape of the parts |
| Conveyor belt | 3 | Conveyor belts have to have sensors for detecting objects on it, which are then conveyed to the next station for further assembly operations |
| Vibratory bowl feeder | 4 | It is necessary to evaluate how quickly the parts will be sorted and how fast the vibratory feeder will need to convey them to maintain the required material flow |

## 4 Conclusion

Assembly systems have to handle a large number of product variations in order to keep their competitiveness, so in most cases they are manual or hybrid. Automated assembly systems may offer lower level of flexibility and there are certain complications concerning the assembly tasks assignment between humans and robots. That's why it is crucial to find a proper approach to automation implementation. Discrete-event simulation software programs allow to test various production systems and lines before their real implementation and realization.

Firstly, the focus was on the manual assembly of cam switches. The starting point was to define each assembly task and average time needed to perform each task. This information was needed for further automation potential identification of each task based on the quantitative and qualitative characteristics of the components used in these assembly tasks. After the evaluation and calculations, the assembly tasks which could be performed by the robots were defined. Semi-automated workplace for cam switches assembly was modelled in Tecnomatix Plant Simulation. The workplace involved three collaborative robots, four vibratory bowl feeders represented by turnplates, and four workers. After setting the processing time, the simulation was started. As a result of simulation, it appeared that the cycle time for one cam switch assembly on the semi-automated workplace is 35% shorter than during the manual assembly. Thus, semi-automated cam switches assembly can be cost-effective for the company which wants to implement it. Such assembly workplace still provides the flexibility.

As for the future directions, the next research can be done on the transforming a semi-automated workplace into a fully automated one, making the emphasis on technical solutions. Another research can be focused on the digitalization of the assembly process by using digital assistance systems. It may include provision of personnel with interactive and digital assembly instructions and installation of sensors and industrial controllers on equipment. Even though there exist multimodal human-machine interfaces, they all require specific digital assistance system workflow, which is demanding, and thus new approach can be worked out. Another future research can be aimed at providing the framework to create and implement a digital twin of the assembly system. This can appear to be a highly beneficial step in a company with an automated assembly workplace.

**Acknowledgements.** This work was supported by the Slovak Research and Development Agency under the contract No. APVV-19-0590, by the projects KEGA 055TUKE-4/2020 granted by the Ministry of Education, Science, Research and Sport of the Slovak Republic.

## References

1. Rewers, P., Trojanowska, J., Diakun, J., et al.: A study of priority rules for a levelled production plan. In: Advances in Manufacturing, Manufacturing 2017, Lecture Notes in Mechanical Engineering, pp. 111–120. Springer, Cham (2018). https://doi.org/10.1007/978-3-319-68619-6
2. Varela, M.L.R., et al.: Collaborative paradigm for single-machine scheduling under just-in-time principles: total holding-tardiness cost problem. Manag. Prod. Eng. Rev. **9**(1), 90–103 (2018). https://doi.org/10.24425/119404

3. Bevilacqua, M., Ciarapica, F.E., De Sanctis, I., Mazzuto, G., Paciarotti, C.: The automation of an assembly system: a business process re-engineering (BPR) perspective. In: 6th International Conference on Industrial Engineering and Systems Management, pp. 371–377, Seville, Spain (2015)
4. Rüssmann, M., et al.: Industry 4.0: the future of productivity and growth in manufacturing industries. Boston Consulting Group (2015)
5. Kuric, I., Kandera, M., Klarák, J., Ivanov, V., Więcek, D.: Visual product inspection based on deep learning methods. In: Tonkonogyi, V., et al. (eds.) InterPartner 2019. LNME, pp. 148–156. Springer, Cham (2020). https://doi.org/10.1007/978-3-030-40724-7_15
6. Gunal, M.: Simulation and the fourth industrial revolution. In: Gunal, M.M. (ed.) Simulation for Industry 4.0. SSAM, pp. 1–17. Springer, Cham (2019). https://doi.org/10.1007/978-3-030-04137-3_1
7. Zidek, K., Pitel, J., Adamek, M., Lazorik, P., Hosovsky, A.: Digital Twin of experimental smart manufacturing assembly system for Industry 4.0 concept. Sustainability **12**(9), 3658 (2020)
8. Gunal, M., Karatas, M.: Industry 4.0, digitisation in manufacturing, and simulation: a review of the literature. In: Gunal, M.M. (ed.) Simulation for Industry 4.0. SSAM, pp. 19–37. Springer, Cham (2019). https://doi.org/10.1007/978-3-030-04137-3_2
9. Bortolini, M., Faccio, M., Galizia, F.G., Gamberi, M., Pilati, F.: Adaptive automation assembly systems in the Industry 4.0 era: a reference framework and full–scale prototype. Appl. Sci. **11**(3), 1256 (2021)
10. Rosati, G., Faccio, M., Carli, A., Rossi, A.: Fully flexible assembly systems (F-FAS): a new concept in flexible automation. Assem. Autom. **33**(1), 8–21 (2013)
11. Gunal, M.: Simulation for the better: the future in industry 4.0. In: Gunal, M.M. (ed.) Simulation for Industry 4.0. SSAM, pp. 275–283. Springer, Cham (2019). https://doi.org/10.1007/978-3-030-04137-3_16
12. Abramova, I.G., Pronichev, N.D., Abramov, D.A., Korotenkova, T.N.: Simulation modelling of the organization of production processes of machine-building enterprises in the instrumental environment Tecnomatix Plant Simulation: laboratory practice. Samara State Aerospace University, Samara (2014). (Original in Russian)
13. Zidek, K., et al.: CNN training using 3D virtual models for assisted assembly with mixed reality and collaborative robots. Appl. Sci. Basel **11**(9), 4269 (2021)
14. Lien, T.K.: Manual assembly. In: Laperrière, L., Reinhart, G. (eds.) The International Academy for Production Engineering. CIRP Encyclopedia of Production Engineering. Springer, Heidelberg (2014)
15. Malik, A.A., Bilberg, A.: Collaborative robots in assembly: a practical approach for tasks distribution. In: 52nd CIRP Conference on Manufacturing Systems, Procedia CIRP, vol. 81, pp. 665–670. Elsevier B.V. (2019)
16. Fedorko, G., Molnar, V., Vasil, M., Hanzl, J.: Application of the Tecnomatix Plant Simulation program to modelling the handling of ocean containers using the AGV system. Naše more **65**(4), 230–236 (2018)
17. Lazar, I., Husar, J.: Validation of the serviceability of the manufacturing system using simulation. J. Effi. Responsib. Educ. Sci. **5**(4), 252–261 (2012). https://doi.org/10.7160/eriesj.2012.050407
18. Daneshjo, N., Malega, P.: Proposal of the reworking station model using Plant Simulation. TEM J. Technol. **10**(1), 197–203 (2021)
19. Kaščak, J., et al.: Design of an atypical construction of equipment for additive manufacturing with a conceptual solution of a printhead intended for the use of recycled plastic materials. Appl. Sci. **11**, 2928 (2021). https://doi.org/10.3390/app11072928

# Integrating the Assessment of Sustainability and an ERP System in Small and Medium Manufacturing Enterprise - A Case Study

Justyna Patalas-Maliszewska[1](✉), Sławomir Kłos[1], and Ewa Dostatni[2]

[1] University of Zielona Góra, Licealna 9, 65-417 Zielona Góra, Poland
j.patalas-maliszewska@iim.uz.zgora.pl
[2] Poznań University of Technology, M. Skłodowskiej-Curie 5, 60-965 Poznań, Poland

**Abstract.** Nowadays, the managers of manufacturing companies from the small and medium-sized enterprises sector, are looking for some kind of a business model that would allow the use of information technologies to support processes while, at the same time, enabling the implementation of actions to increase the level of sustainable production. The purpose of this paper is to focus on such business model. The innovativeness of our work is defining the SBM-ERP, combined with a methodology for implementing the ERP system and an approach to the assessment of sustainability (SA). We have also highlighted the practicality, for managers, of providing parallel, sustainable manufacturing activities using information technologies that are already used based on the real-life case study.

**Keywords:** The assessment of sustainability · The ERP system · Polish small and medium manufacturing enterprise · A case study

## 1 Introduction

To be competitive on the market, manufacturing companies need not only to be flexible, in terms of offering high quality products, in order to satisfy customer needs, but they must also operate according to sustainable production. Sustainable Manufacturing (SM) is an integration of the economic, environmental, and social approach of the manufacturing enterprise [1]. Research aimed at providing solutions towards more sustainable development in manufacturing is of the utmost importance today and more urgent than ever [2]. The driving force in under-taking this research work were the expectations of managers of manufacturing companies from the small and medium-sized enterprises sector, to find a such business model that would allow the use of information technologies, to support processes while, at the same time, enable the implementation of actions to increase the level of sustainable production.

In the literature, interest in the concept of sustainable business models (SBM) has increased greatly, in recent years [3–8]. The main objectives of SBM is to create a competitive advantage by improving value as perceived by the customer while contributing


J. Trojanowska et al. (Eds.): MANUFACTURING 2022, LNME, pp. 50–60, 2022.
https://doi.org/10.1007/978-3-030-99310-8_5

to the sustainable development of the company [3]. Moreover, the SBM enables a company's ethos, vis-à-vis sustainable value, to be described, analysed and managed and also to be communicated to its customers [6]. The Sustainable Business Model (SBM) research topic is a relatively new, yet major field of research interest in sustainable production [9]. There is now increased research in the role of an enterprise as to how sustainable production impacts on sustainable consumption. New business models become now a milestone marking the advent of SM [10]. According to Bocken [7], Boons [8] and Lüdeke-Freund [3] companies need to keep experimenting with their BM to find new ways to drive sustainable consumption.

We are looking for a new approach to building SBM for manufacturing companies; this includes integrating the application, using the ERP and assessing the sustainability (SA) of a manufacturing company; it may also be incorporated into SBM literature and practice. The ERP system helps employees collect data and information across business functions. There is also a need to develop a new approach to the assessment of sustainability (SA), integrated with the processes carried out and supported by the ERP system within a company [11], namely the new SBM-ERP. The authors focus their research on the small and medium enterprises sector (SMEs), due to the fact that the implementation of sustainable development in SMEs is still at a lower level than is the case in larger enterprises [12].

## 2 Materials and Methods

The research was carried based on the analysis of the literature of SM and SA and on the results of empirical research [13, 14] from implementation of the ERP system in a manufacturing company in order to define the sustainable business model (SBM) integrated with the ERP system (SBM-ERP). The SBM-ERP (Fig. 1) includes elements of the business model in line with the approach of Osterwalder and Pigneur [15], namely: key partners (KP), key activities (KA), key resources (KR), value proposition (VP), customer segments (CS), channels (CH), customer relations (CR), cost structure (CST) and revenue streams (RS). Each of SBM-ERP element was detailed described in our previous work [16].

A detailed description of the SBM-ERP model (Fig. 1) is presented on the example of a real case study.

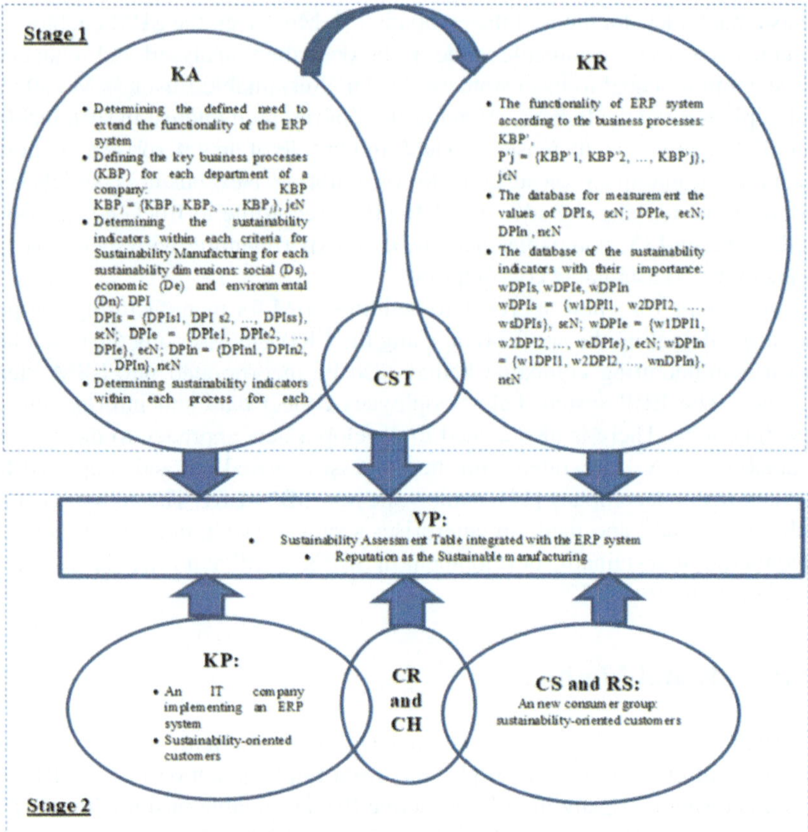


**Fig. 1.** Overlap of a sustainable business model integrated with ERP (SBM-ERP).

## 3 A Case Study

The central element of the proposed SBM-ERP is VP. We agree that it should be compatible with ethical purchasing [17] and has, therefore, been defined as achieving the Assessment of Sustainability table integrated with the ERP system and as achieving, also, the reputation of Sustainable Manufacturing (VP2).

In order to verifying the new approach SBM-ERP the data were received from the case study - medium sized, Polish, metal manufacturing enterprise. Therefore, each element of SBM-ERP was detailed defined and implemented.

- KA1. Extension of the functionality of the ERP system by a module supporting the assessment of sustainability.

Managers have decided to extend the functionality of the ERP system by a module supporting the assessment of sustainability.

- KA2: Defining Key Business Processes (KBP).

The business processes realised within an analysed company analysed and supported by the ERP system were defined:

- KBP1 (Production and Technology Management): analysis of customers' technical requirements (design documentation), material specification of products and semi-finished products (generally MS Excel files), preparing the technical documentation of products, specification of the technology of products (generally Excel files), estimation of the costs of products (based on the weight of the products), description of the simplified technology and design of the product.
- The business process of Production (KBP2): production planning (generally Excel files), production scheduling (generally MS Excel files), preparation and completion of work cards, distribution and completion of production tasks.
- The business process of Sales (KBP4): registration of customers' enquiries (emails), preparing of quotations (generally Excel or Word files), preparation of orders (generally Excel or Word files, sending by emails), preparation of invoices (generally Excel or Word files, sending by emails),
- The business process of Logistics (KBP6): material orders, service orders, preparation of documents, regarding the turnover of materials (receipts for materials, the issue of materials, the movement of materials, etc.), stocktaking (inventory of materials, products and semi-finished products), preparation of transport orders.

- KA3: Determining the Sustainability Indicators (SI) within the criteria for Sustainable Manufacturing.

As proposed in our approach, the sustainability indicators in each process were determined in [16].

- KR1: The Functionality of the ERP System, according to the KBP

The functionality of the ERP System, according to KBP analysed and supported by the ERP system were defined. Business processes defined by the manager of the company that supported ERP system unfortunately do not correspond to sustainability dimensions: social. In our approach we state, that in order to obtain SI values in the sustainability dimensions: economic and social, the current database of the implemented ERP system should be used. Therefore, in further stages of designing and implementing the SBM-ERP model, this area is not analysed.

- KR2: Database for Measuring the Values of DPI.

The values of the quantitative indicators (Table 1) could be obtained with the help of the ERP system. The company's managers decided that the additional functionalities of the ERP system would include the following indicators: $DPI_{e1}$, $DPI_{e2}$, $DPI_{e3}$, $DPI_{n2}$ and $DPI_{n3}$. Therefore, the additional functionalities of ERP the rules for additional functionality of the ERP and have been defined (Table 1).

**Table 1.** Additional functionality of ERP for the company analysed.

| Sustainability dimensions | Processes | Sustainability indicators (SI) | Rules for additional functionality of the ERP |
|---|---|---|---|
| Economic: $DP_e$, $e \in N$ | $KBP_1$, $KBP_2$ | $DPI_{e1}$ total monthly production costs | Move the values of $DPI_{e1}$, $DPI_{e2}$, $DPI_{e3}$ from the ERP database to the sustainability table |
| | $KBP_6$ | $DPI_{e2}$ total monthly logistic costs | |
| | $KBP_2$ | $DPI_{e3}$ productivity | |
| Environmental $DP_n$, $n \in N$ | $KBP_1$, $KBP_2$, $KBP_4$, $KBP_6$ | $DPI_{n2}$ energy usage | Add:<br>Share routing of the product<br>Define the power consumption for each technological operation<br>Calculate the power consumption for semi-finished products<br>Calculate the power consumption for finished products<br>Move the values of $DPI_{n2}$ to the sustainability table, to $DPI_n$ |
| | $KBP_1$, $KBP_2$, $KBP_4$, $KBP_6$ | $DPI_{n3}$ greenhouse gas emission (monthly) | Add:<br>Share bill of material for each product<br>Define the dioxide emission level for each material<br>Calculate the carbon dioxide emission for semi-finished products<br>Calculate the carbon dioxide emission for finished products<br>Move the values of $DPI_{n3}$ to the sustainability table, to $DPI_n$ |

The following values of the SI are obtained: $DPI_{e1}$: 820 920 euro, $DPI_{e2}$: 17 240 euro, $DPI_{e3}$: moving average productivity (volume EUR/costs EUR for 4 month): 72.3, $DPI_{n1}$: 48 020 KWh per month, $DPI_{n1}$: 354.2 kg per month.

- KR3: The Database of Sustainability Indicators with their Importance

In order to receive the weighted value of the criteria for Sustainable Manufacturing, the FAHP method is used [18–20]. Linguistic variables describe the rules for each SI criteria, i.e.: $DPI_{e1}$, $DPI_{e2}$, $DPI_{e3}$ (U), $DPI_{n2}$ (D), $DPI_{n3}$ (I), according to Nydick and Hill [21], a fuzzy number can be assigned $\tilde{a} = (l, m, u)$ o with a triangular membership function. This number is defined in the range [*l*, *u*] with its primary function taking the value of 1 in point m. The owner of the company assessed the importance of each individual indicator according to the above-mentioned rules. For the SI criteria calculated, the following results were obtained, based on the marked responses of the business owner and using the FAHP method (Table 2). Thanks to the FAHP method, each SI criteria receives different weights based on the table.

**Table 2.** Importance of SI criteria according to the scale of preferences used in the FAHP method.

| | $DPIe_1$ | $DPIe_2$ | $DPIe_3$ | $DPIn_2$ | $DPIn_3$ |
|---|---|---|---|---|---|
| $DPIe_1$ | $\tilde{a}11 = (1, 1, 1)$ | $\tilde{a}12 = (1/3, 1, 1)$ | $\tilde{a}13 = (3, 5, 7)$ | (3, 5, 7) | (3, 5, 7) |
| $DPIe_2$ | (1, 1, 3) | (1, 1, 1) | (3, 5, 7) | (3, 5, 7) | (3, 5, 7) |
| $DPIe_3$ | (1/7, 1/5, 1/3) | (1/7, 1/5, 1/3) | (1, 1, 1) | (1/5, 1/3, 1) | (1/5, 1/3, 1) |
| $DPIn_2$ | (1/7, 1/5, 1/3) | (1/7, 1/5, 1/3) | (1, 3, 5) | (1, 1, 1) | (1/3, 1, 1) |
| $DPIn_3$ | (1/7, 1/5, 1/3) | (1/7, 1/5, 1/3) | (1, 3, 5) | (1, 1, 3) | (1, 1, 1) |

As a result, the weight values for SI criteria: $wDPI_{e1}$, $wDPI_{e2}$, $wDPI_{e3}$, $wDPI_{n2}$, $wDPI_{n3}$ and the weight values after standardization: $swDPI_{e1}$, $swDPI_{e2}$, $swDPI_{e3}$, $swDPI_{n2}$, $swDPI_{n3}$ (Table 3) were obtained.

**Table 3.** Weight values for each SI criteria.

| $wDPIe_1$ | $wDPIe_2$ | $wDPIe_3$ | $wDPIn_2$ | $wDPIn_3$ |
|---|---|---|---|---|
| 0.4146 | 0.4853 | 0.0717 | 0.1092 | 0.1282 |
| $swDPIe_1$ | $swDPIe_2$ | s $wDPIe_3$ | $swDPIn_2$ | $swDPIn_3$ |
| 0.3429 | 0.4014 | 0.0593 | 0.0903 | 0.1060 |

Thanks to the FAHP method, each sustainability indicator receives different parameter weights, which means that the most important indicators are DPIe1 and DPIe2 (Table 4).

**Table 4.** The importance of each SI criterion in the company analysed.

| Sustainability dimensions | Processes | Sustainability indicators (SI) | The value of IS | The ranking of SI according to the weight of each SI |
|---|---|---|---|---|
| Economic: $DP_e$, $e \in N$ | KBP1, KBP2 | $DPIe_1$ | | 2 |
| | $KBP_6$<br>$KBP_2$ | $DPIe_2$<br>$DPIe_3$ | 72.3 | 1<br>5 |
| Environmental $DP_n$, $n \in N$ | KBP1, KBP2<br>KBP4, KBP6<br>KBP1, KBP2<br>KBP4 KBP6 | $DPIn_2$<br>$DPIn_3$ | 48020 kWH<br>354.2 kg | 4<br>3 |

Therefore, the most important indicators for obtaining the status of SM are the indicators: production costs and logistic costs, followed by CO2 emission. Productivity is the least significant indicator. By using the FAHP method (element of KR of the model) in our approach, it is possible to determine which SI are the most important indicators for obtaining the status of SM and, thanks to this, define the order of implementation the actions needed for improving sustainability levels of the company analysed, with regard to the specific business processes within an enterprise. Based on the data about the implementation costs of SBM-ERP they are about 14 000 EUR higher than the traditional ERP system. The cost of adapting the ERP system to a sustainable business model requires additional analyses, regarding, for example, the environmental impact of products and certificates. Examples of the costs of additional development and software design are: data acquisition (water or energy consumption, level of pollutant emissions), the on-line calculation of worker and machinery productivity, etc., the on-line registration costs of logistics including the daily distance of forklifts and daily gantry operation times, etc. implementation of the FAHP analytical module.

- KP: Key partners

It is very important that SBM-ERP should be implemented by a partner experienced in implementing ERP systems and who also has the know-how and experience of the development of sustainability. The current customers of the company investigated are mainly the construction companies which order such products as balustrades and balcony platforms, etc. The company produces about 70 constructions, monthly, with each construction weighing several tons. Due to implementation of the SBM-ERP model, the company will be able to increase its client base with clients for whom co-operation with a company with the status of Sustainable Manufacturer is very important; such new clients will invariably have a bearing on future ordering patterns.

- CS and RS: Customer Segments and Revenue Streams, CR and CH: Customer Relations and Channels

The SBM-ERP model is currently being implemented in the company under analysis; for this reason, it cannot be stated how any increase in revenue streams in the company will be affected, as yet, by the implementation of this model. However, it is assumed that if a company receives the status of Sustainable Manufacturer, it will gain new sustainability-oriented clients. Such a company image will also help to develop good relations with existing clients and the acquisition of new ones.

- VP1: Value Proposition: The Assessment of Sustainability Table integrated with the ERP System

Finally, an Assessment of Sustainability table has been designed, integrated with the ERP system (Table 5). The setting of recommendations for entrepreneurs is possible on the basis of setting a reference value for the SI indicators obtained. The following data from the Polish Central Statistical Office in Poland (Statistical Yearbook of Industry – Poland, 2018) was obtained with these values:

- 32% of manufacturing companies in Poland are in western Poland, the company surveyed is from western Poland. There are 69 347 manufacturing companies in western Poland.
- for $DPI_{e1}$ - production costs - annual costs in PLN million for manufacturing companies from western Poland: 354 527,8 million PLN per year, monthly: PLN 29 543,98 million = 7 385,995 million EUR, per enterprise: 0,106508 million EUR.
- for $DPI_{e2}$ - logistics costs no data available.
- $DPI_{n2}$ - energy usage (monthly). Based on the data from the Statistical Yearbook of Industry – Poland, 2018: energy usage in Polish manufacturing companies: 8392 GWh, manufacturing enterprises in western Poland constitute 32% of all enterprises in Poland, therefore 32% from the given indicator is assumed, so: 2685,44 per year, monthly: 223,7867 per enterprise: 0,003227 Gwh.
- for $DPI_{n3}$ - CO2 emission (monthly). Based on the data from the Statistical Yearbook of Industry – Poland, 2018: CO2 emissions in Polish manufacturing companies: 339,8 thousand tonnes, manufacturing enterprises in Western Poland constitute 32% of all enterprises in Poland, therefore 32% from the given indicator is assumed, so: 108,736 annually, monthly: 9.06 per single enterprise in western Poland; 0,000131 thousand tonnes.

Thanks to the implementation of the SBM-ERP model in the analysed company, it is possible to define the needed corrective actions that the company must take to increase its SI level. So, we received the results of the assessment of sustainability, the table integrated with the ERP system, so-called, (Table 5), which expresses the validity of SI and at the same time compares them with reference values of SI obtained from the Polish Central Statistical Office in Poland (Statistical Yearbook of Industry – Poland, 2018). The most important SI are: production costs and logistics costs, followed by CO2 emission. These indicators have been appropriately assigned to processes: KBP1: Production and Technology Management, KBP2: Production, KBP4: Sales, KBP6: Logistics.

For the business process: KBP6: Logistics, where the indicator: DPIe2 (logistics cost) is located, no actions for the improvement of the sustainable level of the company

**Table 5.** The assessment of sustainability table integrated with the ERP system.

| The importance of criteria in Sustainable Manufacturing | Processes | The value of criteria in Sustainable Manufacturing according to their importance | The reference value of criteria in Sustainable Manufacturing | SI assessment |
|---|---|---|---|---|
| 1 | $KBP_1$, $KBP_2$ | $vDPIe_2$=346,05 euro/month | - | |
| 2 | $KBP_6$ | $vDPIe_1$= 14074,67 euro/month= 0,01407 mln euro/month | 0,1065 mln euro/month | |
| 3 | $KBP_2$ | $vDPIn_3$= 3 542 KG of CO2 = 0,0035 thousand tonnes | 0,0001 thousand tonnes | |
| 4 | $KBP_1$, $KBP_2$, $KBP_4$, $KBP_6$ | $vDPIn_2$= 48 020 KWh /month] = 0,0048 GWh | 0,0032 GWh | |
| 5 | $KBP_1$, $KBP_2$, $KBP_4$, $KBP_6$ | $vDPIe_3$= 72,3 | 100 | |

- green - level of SI recommended to be maintained.
- red - level of SI recommended to be improved.
- yellow - level of SI recommended to be monitored.

analysed are required, because the sustainable level has been specified as green, which means that this level is recommended to be maintained. However, within the processes KBP1: Production and Technology Management, KBP2: Production, KBP4: Sales and-still in process- KBP6: Logistics where the indicator: $DPI_{n3}$ (greenhouse gas emission) is located, defining the actions for improving the sustainable level of the company analysed is needed. Knowing that the main technological processes in the metal industry company analysed, are welding and plasma cutting steel components and also that welding and plasma cutting processes are energy-consuming and cause high CO2 emissions into the atmosphere, the following activities are proposed:

- the proper preparation of the material prior to welding, in order to reduce deficiencies and unnecessary corrections.
- the introduction of controls into the parameters for welding and plasma cutting (voltage and current, proper selection of shielding gases, etc.) in order to shorten these processes, reduce energy consumption and reduce the number of shortages (e.g. by using the metaheuristic algorithms to optimize the cycle time.
- changes to the batch sizes of orders in order to reduce CO2 through external and internal transport.

## 4 Discussion and Conclusions

The added value of our research to the recent state of the research field is our innovative approach to integrating the assessment of sustainability in an ERP system within small and medium manufacturing enterprises: SBM-ERP.

This study allowed the state of knowledge about the measurement of SI, integrated with the ERP system, to be identified as well as to identify the need to implement the SBM-ERP model in manufacturing companies. The practical significance of our work is determined in the form of a complex business model, the implementation of which will allow managers to obtain the status of Sustainable Manufacturer in the context of obtaining possible benefits for their manufacturing companies. Our approach assumes that manufacturing SMEs have and use the ERP system, but the need for its expansion is related to the need to implement activities in the company, leading to "Sustainable Manufacturer" status.

The proposed approach makes it possible to continuously assess the sustainability level of an enterprise that uses the ERP system, along with monitoring the effects of the corrective actions implemented, according to the results including in the Assessment of Sustainability table. The concept, here presented, of the approach to integrating the assessment of sustainability with an ERP system, in SME manufacturing enterprises: SBM-ERP, is part of the broader work of the authors. Each stage is appropriately designed and refined depending on the specificity of the enterprise in which it is implemented. Therefore, the future works will present an example of the practical implementation of the proposed model SBM-ERP in a medium-sized, Polish, metal manufacturing enterprise.

**Acknowledgements.** This work is supported by program of the Polish Minister of Science and Higher Education under the name "Regional Initiative of Excellence" in 2019–2022, project no. 003/RID/2018/19, funding amount 11 936 596.10 PLN.

## References

1. Malek, J., Desai, T.N.: Prioritization of sustainable manufacturing barriers using Best Worst Method. J. Clean. Prod. **226**, 589–600 (2019)
2. Stark, R., Seliger, G., Bonvoisin J.: Sustainable Manufacturing, Life Cycle Engineering and Management. Springer, Heidelberg (2017)
3. Lüdeke-Freund, F., Dembek, K.: Sustainable business model research and practice: emerging field or passing fancy? J. Clean. Prod. **168**, 1668–1678 (2017)
4. Geissdoerfer, M., Vladimirova, D., Evans, S.: Sustainable business model innovation: a review. J. Clean. Prod. **198**, 401–416 (2018)
5. Evans, S., et al.: Business model innovation for sustainability: towards a unified perspective for creation of sustainable business models. Bus. Strateg. Environ. **26**, 597–608 (2011)
6. Schaltegger, S., Hansen, E.G., Lüdeke-Freund, F.: Business models for sustainability: origins, present research, and future avenues. Organ. Environ. **29**(1), 3–10 (2016)
7. Bocken, N.: Business-led sustainable consumption initiatives: impacts and lessons learned. J. Manag. Dev. **36**(1), 81–96 (2017)
8. Boons, F., Lüdeke-Freund, F.: Business models for sustainable innovation: state-of-the-art and steps towards a research agenda. J. Clean. Prod. **45**, 9–19 (2013)

9. Pal, R., Gander, J.: Modelling environmental value: an examination of sustainable business models within the fashion industry. J. Clean. Prod. **184**, 251–263 (2018)
10. Rojek, I., Mikołajewski, D., Dostatni, E.: Digital twins in product lifecycle for sustainability in manufacturing and maintenance. Appl. Sci. **11**, 31 (2021)
11. Moldavska, A., Welo, T.: A holistic approach to corporate sustainability assessment: incorporating sustainable development goals into sustainable manufacturing performance evaluation. J. Manuf. Syst. **50**, 53–68 (2019)
12. Chang, A.Y., Cheng, Y.T.: Analysis model of the sustainability development of manufacturing small and medium-sized enterprises in Taiwan. J. Clean. Prod. **207**, 458–473 (2019)
13. Patalas-Maliszewska, J.: Assessing the impact of ERP implementation in the small enterprises. Found. Manag. **4**(2), 51–62 (2012)
14. Kłos, S.: A model of an ERP-based knowledge management system for engineer-to-order enterprises. In: Dregvaite, G., Damasevicius, R. (eds.) ICIST 2016. CCIS, vol. 639, pp. 42–52. Springer, Cham (2016). https://doi.org/10.1007/978-3-319-46254-7_4
15. Osterwalder, A., Pigneur, Y.: Business Model Generation: A Handbook for Visionaries, Game Changers, and Challengers. Wiley (2010)
16. Patalas-Maliszewska, J., Kłos, S.: The methodology of the S-ERP system employment for small and medium manufacturing companies. IFAC-PapersOnLine **52**(10), 85–90 (2019)
17. Viciunaite, V., Alfnes, F.: Informing sustainable business models with a consumer preference perspective. J. Cleaner Prod. **242**, 118417 (2020)
18. Pislaru, M., Herghiligiu, I.V., Robu, I.B.: Corporate sustainable performance assessment based on fuzzy logic. J. Clean. Prod. **223**, 998–1013 (2019)
19. Bocewicz, G., Nielsen, I.E., Banaszak, Z.A.: Production flows scheduling subject to fuzzy processing time constraints. Int. J. Comput. Integr. Manuf. **29**(10), 1105–1127 (2016)
20. Kłosowski, G., Gola, A., Świć, A.: Application of fuzzy logic in assigning workers to production tasks. In: Distributed Computing and Artificial Intelligence, Conference 2016 Advances in Intelligent Systems and Computing, pp. 505–513 (2016)
21. Nydick, R.L., Hill, R.P.: Using the analytic-hierarchy process to structure the supplier-selection procedure. Int. J. Purch. Mater. Manag. **28**, 31–36 (1992)

# Comparative Analysis of the Environmental Assessment of an Exemplary Product in Selected IT Tools

Marcin Radzki and Jacek Diakun(✉)

Poznań University of Technology, M. Skłodowskiej-Curie 5, 60-965 Poznań, Poland
jacek.diakun@put.poznan.pl

**Abstract.** The article presents the environmental assessment of an exemplary product (plastic box). The article outlines the concepts of eco-innovation and eco-design. The evolution of the product life cycle concept is also presented. The LCA method, its structure and applied metrics were characterized. The analysis was carried out in two IT tools that are modules of 3D CAD systems: Eco Materials Adviser (Autodesk Inventor module) and SolidWorks Sustainability.

**Keywords:** Eco-design · Environmental assessment · LCA

## 1 Introduction

The aspects of durability and sustainable product design are more and more often the main topic of discussion in the field of design. The World Commission on Environment and Development defines sustainable development as development that meets the present needs for the development of future generations to meet their own needs. The concept of sustainable development covers social, environmental and economic aspects and therefore accompanies the entire product life cycle. In practice, sustainable design is the process of creating a product that allows you to generate profits for the enterprise. In addition, it must use as little energy as possible, and therefore the minimum amount of raw material, which is associated with a smaller amount of generated waste, which can potentially have a negative impact on the environment.

The concept of eco-innovation is associated with any form of innovation, both technical and non-technical. Its purpose is primarily to create new opportunities for companies and bring benefits to the environment by preventing or limiting the negative impacts of enterprises on the ecosystem. Eco-innovation is closely related to the methods of using natural resources and their production and consumption. Eco-innovations make it possible to minimize the flow of materials and energy going beyond the enterprise as a result of changes in production methods and materials used, which gives companies the opportunity to gain a competitive advantage on the market.

The subject of ecodesign has been widely used in economic practice relatively recently. Currently, enterprises consider it necessary to monitor the environmental impact


J. Trojanowska et al. (Eds.): MANUFACTURING 2022, LNME, pp. 61–75, 2022.
https://doi.org/10.1007/978-3-030-99310-8_6

of their products or services. Ecodesign can be used for existing products, services or processes, but it is also used for new products.

All kinds of provided services or manufactured products, to some extent, affect the ecosystem. The scale of this impact depends on several variables, such as: the means used to manufacture the product, the materials used, or the product life time. Currently, the product is required to minimize its impact on the environment throughout its life cycle, with the greatest attention being paid to the phases with the greatest negative impact on the ecosystem. In addition to environmental aspects, important are also those related to the reduction of costs for each stage of the life cycle, which allows a given company to improve the issue of competitiveness on the market. There are several methods that allow a product to be assessed in terms of its environmental impact. The best known method is the Life Cycle Assessment (LCA). It is believed that this method, due to its complexity, is one of the most accurate and objective in terms of environmental assessment. It was developed and popularized at the beginning of the nineties of the twentieth century. It increasingly influences the ecological assessment of services, technologies or products. It is supported by various laws and procedures included in the ISO 1404x series.

## 2 Product Life Cycle

The product life cycle (PLC) should be understood as the period of time from the creation of the idea of a certain product, then the development of its concept, and then design, execution, distribution, sale, operation and final scrapping [1]. The concept of the product life cycle is determined by the following principles [1]:

- products have a limited life span.
- the sale of a product consists of several stages. Each of them generates challenges and problems for the manufacturer.
- depending on the stage of the product life cycle, profits increase or decrease.
- products require the application of human resources strategies and various financial, marketing, production and purchasing activities at every stage of their life cycle.

With the development of tools for design, manufacturing and computer aided engineering (CAD/CAM/CAE), a new era of introducing the product to the market has begun. After the introduction of Computer Integrated Manufacturing (CIM) in the early 1980s, the commonly used method of proceeding was the cooperation of design and manufacturing processes. As a result, terms such as design for production or designed for production were created [2]. However, the stages of use of the product and its disposal were still not addressed.

In the mid-1980s, there was a breakthrough in terms of designing the product life cycle with regard to environmental protection. Many European countries have introduced regulations relating to packaging and packaging waste management. Directive 85/339/EEC was introduced, which made the LCC and LCA subject to tests in order to modernize the methods of environmental assessment of the life cycle of the product, so that they meet the latest requirements and conditions of competitiveness, and at the same time could meet the growing trend of environmental protection [3].

The E-PLC concept focuses on the whole product life cycle "from cradle to grave", i.e. product conception, design, production, sale, use and end-of-life. Due to the fact that there is currently no specific E-PLC model imposed by standards, scientists continue to study individual PLC components independently of one another. The result of these activities has been the emergence of many new perspectives for E-PLC over the last twenty years. The common element of individual studies was relying on the same scientific publications. Analyzing the product in terms of design, Alting proposes a six-phase life cycle: market needs research, product concept development, production, distribution, use and disposal or recycling [2]. In addition, he believed that each of the phases should be taken into account and thoroughly analyzed at the stage of creating the product concept. Based on the work of Alting, the concept of the material life cycle was introduced, which is an extension of the product life cycle. The sample material analysis of a product was based on the residual amount of material that was recycled at the end of its life cycle.

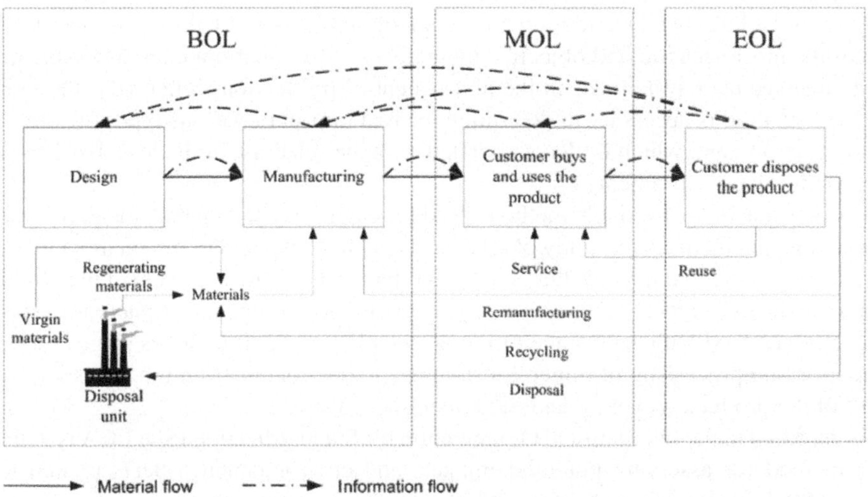


**Fig. 1.** Closed loop of the engineering product life cycle [4].

Currently, the product life cycle model uses the flow of both information and materials. It is divided into three phases. The first phase of the cycle, called BOL (Beginning of life) consists of the stages of product design and production. Then the product life cycle goes to the second phase, in which the customer purchases the finished product. In this part of the cycle, called the Middle of Life (MOL) phase, the item is used and possibly repaired. This phase is characterized by the separation of the flow of information and materials, as well as the return of data to BOL. The final phase is related to the end of life of the product that is subject to the recycling process. Product decommissioning is defined as the end of life (EOL). In this phase, the flow of information and materials is finally separated. Materials and components are passed back to BOL and MOL, and all information related to the product and its design to BOL. This form of the product life cycle model enables the free flow of data between BOL, MOL and EOL. The closed loop enables designers to constantly improve the product at every stage of its life. This type of concept is currently the most advanced known model of the engineering product life cycle [4]. The diagram of the engineering life of the product is shown in Fig. 1.

## 3 Environmental Assessment of the Product Life Cycle

Among the many available tools and methods for environmental management, the Environmental Life Cycle Assessment (LCA) deserves special attention. It is a product analysis method that covers the environmental aspects and potential environmental impacts throughout the product life cycle ("cradle to grave"), starting from the raw material extraction stage, then through production, operation and decommissioning of the product. It has been used since the end of the nineties of the twentieth century. Since then, it has been constantly popularized and developed, which means that it plays an increasingly important role in the ecological evaluation of products, services and technologies. Additionally, the method was supported by specific procedures described in ISO 14040 [5] and ISO 14044 [6]. Due to the wide scope of application, LCA is considered to be one of the most accurate and objective methods used for environmental assessment. In waste management systems or in the assessment of production technology, the use of the LCA method is necessary to determine the real impact of various types of solutions on the environment, which is also associated with the selection of the least burdensome solution for the environment [7].

LCA is one of the available methods used for environmental management, the application of which includes the study of all aspects related to the environment as well as the estimation of potential impacts that may occur during the entire life cycle of the product. Thanks to the LCA method, it is possible to assess the environmental impacts and aspects that are associated with each stage of the product life cycle. It includes stages such as: extraction and processing of mineral resources, production, distribution, transport, use, reuse of the product, recycling and final disposal of waste.

According to the International Organization for Standardization ISO, LCA is defined as a method for assessing potential impacts and environmental aspects related to a product [8]. It includes four phases: defining the purpose and scope of research, reviewing the product system in terms of a set of key inputs and outputs, evaluation of possible environmental impacts related to the system inputs and outputs, analysis of the obtained results of the set assessment and impact assessment phases related to research objectives [5, 6].

LCA is currently treated as a tool supporting decision-making related to the selection of the most convenient way to design a new product or technology. It can also greatly contribute to their development. The environmental assessment of the product life cycle is related to the model system, which consists of unit processes that fulfill one or more specific functions. These processes are a resource used for material and energy flows between processes. Due to this fact, it is necessary to collect data related to energy consumption or raw materials in each phase of the life cycle of a given product [9].

The basic tasks assigned to the LCA method are [9]:

- reporting of possible environmental impacts of the product in each phase of its life,

- review of the available opportunities for the emergence of related environmental impacts, so that the implemented remedial measures do not cause the creation of further environmental problems,
- outlining priorities in improving the production of products,
- a compilation of the different ways in which a given process can be performed or a comparison of all available solutions to a given problem.

Based on the results obtained as a result of LCA analyzes, the product production system with the most favorable impact on the ecosystem is selected. The analysis of a product's environmental impact starts at the design stage, which ensures that any possible impacts on the product are anticipated at each stage of its life cycle. This is the stage with the highest risk for the product or any project, because at this point you need to define such aspects as: raw materials, base materials, product production process, transport or disposal. In addition, a preliminary estimate of the life of the product and the possibility of its repair should be made.

The LCA method additionally determines the so-called the transfer of the environmental impact of pollutants transferred from one phase of the cycle to another, or one environmental component to another. An example of such a transfer may be, for example, processing and re-use of a product instead of obtaining a raw material for the production of a new product [7].

The structure of the LCA consists of several important steps in the assessment. Figure 2 shows the phases of the life cycle assessment.

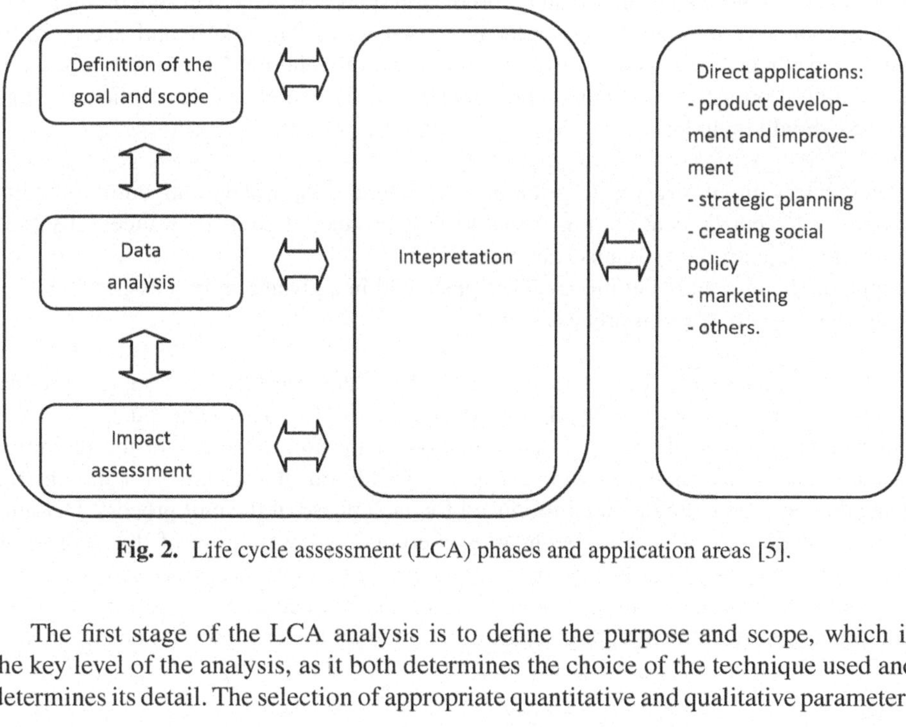


**Fig. 2.** Life cycle assessment (LCA) phases and application areas [5].

The first stage of the LCA analysis is to define the purpose and scope, which is the key level of the analysis, as it both determines the choice of the technique used and determines its detail. The selection of appropriate quantitative and qualitative parameters and the determination of the limits of a given model depend on the adopted goal and the anticipated manner of using the obtained results. A very important element of each study using the LCA method is to set the goal and target group of the study for which the results will be presented. The key in this case is the fact that LCA is only a decision support tool, and the so-called interested parties. According to the ISO 14040 standard, they are defined as "units or groups associated with or affected by the environmental performance of the product system or the results of the life cycle assessment" [5]. The comment to the standard explains that "the purpose should clearly define the intended use, the reasons for conducting the research and the intended recipient, ie to whom the test results are to be communicated" [10].

In the first step of the LCA life cycle assessment, the purpose, the envisaged use of the study results and the description of the principal, the person carrying out the research and the target audience should be established and adequately justified. Additionally, the type of analysis used should also be declared, as two types of analysis can be performed: non-comparative and comparative.

The scope of research should be understood as a specific type of collected data and their characterized scope, as well as system boundaries. In particular, the level of sophistication of the system and the exact stages of the product life cycle to be tested should be determined. Within the examined stage, the time, geographic and technological scope of the LCA study to be conducted should be specified. In addition, it is required to indicate the type of environmental impacts and the methodology used to estimate them, which is the basis for the correct characterization and classification [9].

Defining the scope of the LCA test involves the characterization of three important, closely related issues: product system, product system boundary, and functional unit.

A product system should be understood as a set of "materially and energetically connected unit processes that fulfill one or more specific functions" [5, 6, 9]. It is possible to link unit processes when using a product stream. An exemplary description of a unit process should include:

- elementary input streams (e.g. crude oil, natural gas, water) and output streams (e.g. water eutrophication, emissions to air), product streams (e.g. electricity) and intermediate product streams (raw materials),
- type of changes and operations taking place within a given unit process,
- the place where the unit process begins.

The product system boundary is described as "the interface between the product system and the environment or other product systems" [9]. The main functions of this stage of the LCA study are the determination of the time period and the definition of the technological and geographical area. In addition, the accuracy of the entered data and their completeness are introduced for each phase of the unit process. Defining the area of the system and its boundaries is crucial as the purpose of this stage is to determine the energy used in each phase of the process and the necessary sources of raw materials. When determining the boundaries of a unit process, it is necessary to define unit processes. Additionally, unit stages of the product life cycle should be considered, such as [9]:

- input and output streams in the product production process,
- transport and distribution,
- production and consumption of raw materials and energy,
- exploitation of products,
- consumption of secondary raw materials,
- aspects related to the installation of the product or its additional components.

By the term functional unit it is meant "the quantitative effect of the product system used as a reference unit in LCA studies". A functional unit is associated with the entire product system and is considered in terms of functional properties or functions of the product. When analyzing a given product in relation to this unit (as a parameter), one should refer to it quantitatively, ie 1 MW of consumed or received energy, 1 kg of CO2 emissions, etc. [7].

The report prepared by the US Environmental Protection Agency (EPA) prepared a list of LCA metrics and their characteristics. It complements the general methodology of the product life cycle, consisting of the three pillars of sustainable development: environmental, social and economic. These metrics include [11, 12]:

- Cumulative Energy Demand - is the total amount of energy that is consumed during the entire life cycle of the product.
- Cumulative Fossil Energy Demand - a CED subcategory that describes the amount of energy over the entire life cycle of a product. This energy comes from the combustion of raw materials such as oil, natural gas and coal.
- Cumulative Renewable Energy Demand - is a subset of CED, which characterizes the amount of renewable energy in the entire life cycle of the product. Renewable energy consists of such energy sources as: hydro, solar, geothermal and wind energy.
- Global Warming Potential - often also referred to as the carbon footprint. It shows the impact on climate change over time. It is usually presented for the next 100 years. It shows the emissions of all greenhouse gases into the air. Examples of this type of gas are: methane (CH4), carbon dioxide (CO2) or nitrous oxide (N2O)
- Ozone Depletion Potential - describes the total impact of emissions of all gases that negatively affect ozone in the stratospheric ozone layer. The analysis of this metric is applied throughout the product life cycle and is determined based on the functional unit delivered to the customer (including product end-of-life management). Stratospheric ozone occurs as a layer of natural gas to protect living cells, against excessive exposure to ultraviolet (UV) radiation. Overexposure to radiation can cause, inter alia, cancer or has a negative effect on agriculture, which results, among others, in lower yields.
- Acidificaton Potential - shows the total impact of all acid gas emissions, such as nitrogen oxides (NOx), hydrochloric acid (HCl), hydrofluoric acid (HF), sulfur oxides (SOx) or ammonia (NH4). Excessive emission of these substances causes acidification of soil and water reservoirs and accelerates the corrosion of building structures.
- Eutrophication Potential - This is a category that illustrates algae overgrowth as a result of the excessive emission of limiting nutrients such as nitrogen and phosphorus. Emissions can occur directly or indirectly and target water bodies.

- Photochemical Ozone Creation Potential - Shows the relative total impact of nitrogen oxides and VOC emissions to the atmosphere over the entire life cycle of a product. In the presence of nitrogen oxides and sunlight, when volatile organic compounds are emitted, for example non-metallic hydrocarbons, chemical reactions take place, the product of which is ozone (O3). It is produced at ground level, which causes the phenomenon of the so-called photochemical smog.
- Waste water footprint and water emission - a category that describes the total water demand at each stage of the product life cycle necessary to provide the customer with a functional unit. Most often, the category is divided into fresh and salt water.
- Environmental and human toxicity assessment - quantifies the ecosystem fate of emissions of all kinds of chemicals and their impact on human health and the environment on the basis of possible effects.
- Direct Land UseChange - an indicator relating to the conversion of natural land, such as forests, pastures or farmland, to a changed state. The aim of this type of activity is the possibility of producing forest and agricultural products, e.g. raw materials for the production of biofuels. By-products that arise as a result of this are, inter alia, greenhouse gases.
- Indirect Land Use Change - an indicator describing the phenomenon of land use change in the event of a change in the location of the crop and its relocation to another location, which is associated with a change in the condition of the land. The effect of this phenomenon is, inter alia, a change in the carbon stock in a given area or an increase in greenhouse gas emissions.

Examples of environmental analysis can be found in the literature. The paper [13] presents a comparative environmental analysis of acoustic barriers made of five types of materials using the LCA method. In the article [14] two approaches to concrete structures service life modeling in LCA were tested. LCA was performed for 94 CC and HVFAC mix designs. A number of publications are also devoted to the environmental analysis of packaging. In the report [15], the environmental analysis of plastic and paper straws for portion-sized carton packages was presented using the LCA method. Another report [16] presents a comparative LCA analysis of carton packages and alternative packaging systems for liquid food on the Nordic market. The subject of research is also the packaging recycling system. In [17], two scenarios were compared using the LCA method: no packaging recycling system and was compared with two hypothetical scenarios where all the packaging waste that was selectively collected. The researchers are also interested in the differences in the results of the analysis performed with the use of different software. The article [18] presents a comparative analysis of the processes of exploration and production of oil and gas performed in the SimaPro and OpenLCA programs.

# 4 Environmental Product Assessment in Selected Tools

## 4.1 Research Methodology

The purpose of this study is to conduct an environmental analysis of the same product in specialized modules of two popular 3D CAD systems. The research methodology consists in modeling a real product, conducting an analysis in two 3D CAD systems and then comparing the results. The choice of the product was dictated by two reasons: the possibility of performing the analysis in the evaluation versions of both systems and the material homogeneity of the product.

## 4.2 Product Being Analyzed

The subject of the analysis will be a 1.5 L storage box (Fig. 3a). It is a product consisting of four parts, all of them are made of polypropylene (PP). For the purposes of the environmental analysis, the product was modeled in the 3D CAD environment (Fig. 3b). The analyzes were carried out in the following tools: Eco Materials Adviser [19] and SolidWorks Sustainability [20], both in evaluation versions.

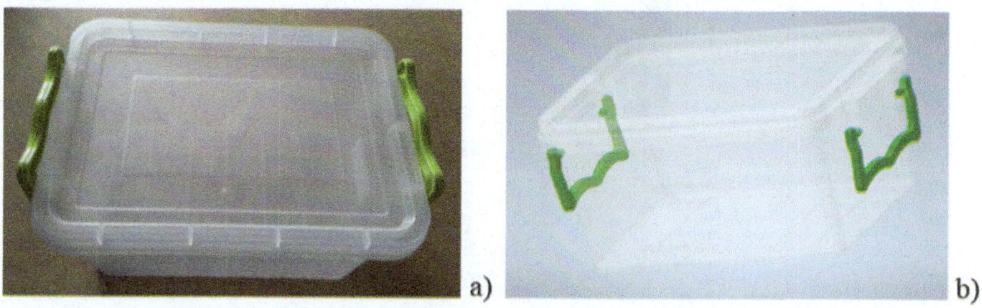

**Fig. 3.** Analyzed product: a) real product, b) 3D CAD model.

## 4.3 Analysis in Eco Materials Adviser

Using the Eco Materials Adviser tool, operating in the CAD 3D Autodesk Inventor 2020 environment, the LCA analysis of the product was carried out using the CML method. As part of the analysis, the product manufacturing process was defined as injection molding. Data on the distribution and installation of the product were not included in the analysis.

The product was analyzed in terms of energy consumption throughout its life cycle. Particularly important in this case is the part called Pudelko_1: 1, which according to the program exceeds the acceptable standard at every stage of the life cycle. Parts such as the handles of the container exceed the permissible value of energy consumption at the end of life stage of the product. The remaining analyzes are shown in Figs. 4, 5, 6, 7, 8 and 9.

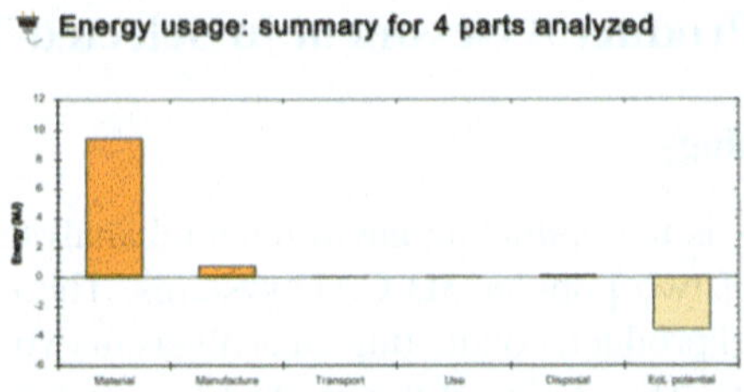


| | Energy (MJ) | Percentage |
|---|---|---|
| Material | 9.4 | 92 % |
| Manufacture | 0.73 | 7 % |
| Transport | Available in Full Version of Eco Materials Adviser | |
| Use | Available in Full Version of Eco Materials Adviser | |
| Disposal | 0.083 | 0.8 % |
| Total (for first life) | 10 | 100 % |
| End of life potential | -3.6 | |
| Total | 6.7 | |

**Fig. 4.** Product analysis using the eco materials adviser - values of energy consumption.

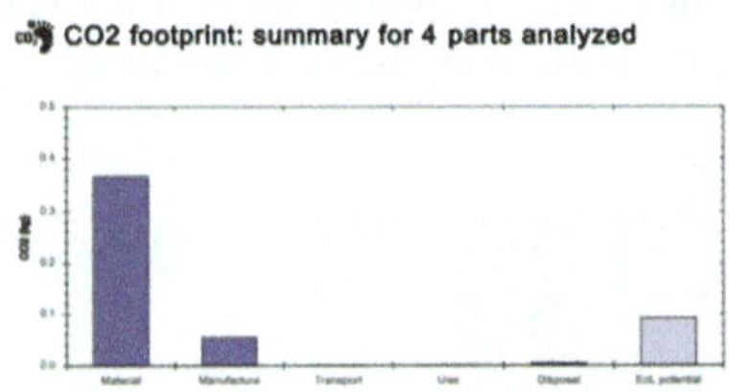


| | CO2 (kg) | Percentage |
|---|---|---|
| Material | 0.37 | 86 % |
| Manufacture | 0.055 | 13 % |
| Transport | Available in Full Version of Eco Materials Adviser | |
| Use | Available in Full Version of Eco Materials Adviser | |
| Disposal | 0.0050 | 1 % |
| Total (for first life) | 0.43 | 100 % |
| End of life potential | 0.092 | |
| Total | 0.52 | |

**Fig. 5.** Product analysis using the eco materials adviser - values of carbon footprint.

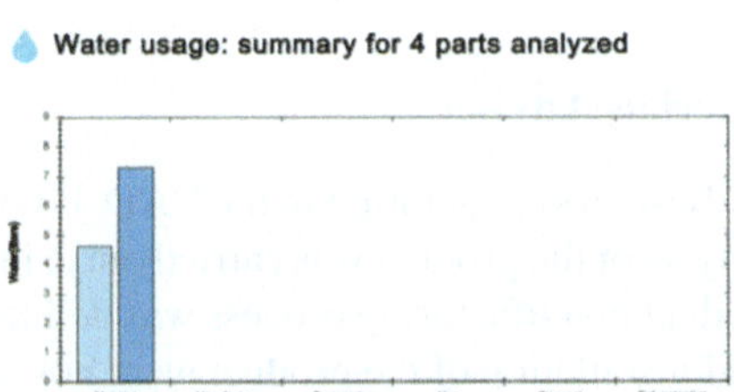


| | Baseline | New concept | Change | Percentage |
|---|---|---|---|---|
| Material (liters) | 4.6 | 7.3 | 2.7 | 58 % increase |
| Manufacture (liters) | Not included in analysis | | | |
| Transport (liters) | Not included in analysis | | | |
| Use (liters) | Not included in analysis | | | |
| Disposal (liters) | Not included in analysis | | | |
| End of life potential (liters) | Not included in analysis | | | |

**Fig. 6.** Product analysis using the eco materials adviser - values of variants in terms of water consumption.

Cost: summary for 4 parts analyzed

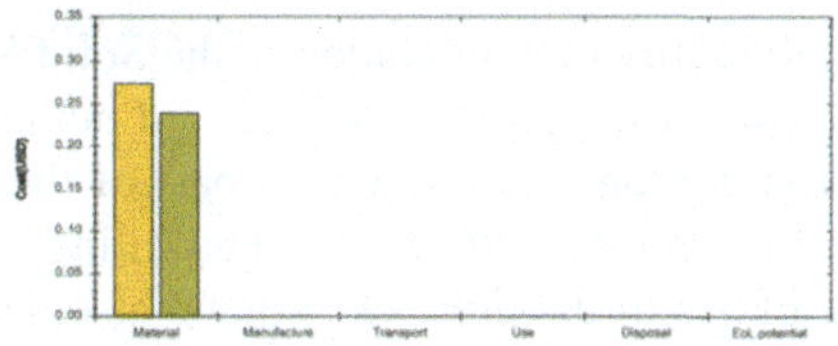

| | Baseline | New concept | Change | Percentage |
|---|---|---|---|---|
| Material (USD) | 0.27 | 0.24 | -0.034 | 13 % reduction |
| Manufacture (USD) | | Not included in analysis | | |
| Transport (USD) | | Not included in analysis | | |
| Use (USD) | | Not included in analysis | | |
| Disposal (USD) | | Not included in analysis | | |
| End of life potential (USD) | | Not included in analysis | | |

**Fig. 7.** Product analysis using the eco materials adviser - values of variants in terms of production cost.

RoHS compliance and Food-contact compatibility: summary for 4 parts analyzed

RoHS compliance

| | Baseline | New concept | Change |
|---|---|---|---|
| Compliant Parts | 4 | 4 | 0 |
| Non-Compliant Parts | 0 | 0 | 0 |
| Conditions apply, status unknown or no material assigned | 0 | 0 | 0 |
| Total | 4 | 4 | 0 |

Food-contact compatibility

| | Baseline | New concept | Change |
|---|---|---|---|
| Compatible Parts | 4 | 4 | 0 |
| Non-Compatible Parts | 0 | 0 | 0 |
| Conditions apply, status unknown or no material assigned | 0 | 0 | 0 |
| Total | 4 | 4 | 0 |

**Fig. 8.** Product analysis using the eco materials adviser - values of product compliance with the EU RoHS directive and compatibility in use in contact with food.

End of life: summary for 4 parts analyzed

| | Baseline | New concept | Change |
|---|---|---|---|
| Reuse | 0 | 0 | 0 |
| Recycle | 4 | 4 | 0 |
| Downcycle - comminution | 0 | 0 | 0 |
| Downcycle - reprocessing | 0 | 0 | 0 |
| Downcycle - metal recovery | 0 | 0 | 0 |
| Combustion | 0 | 0 | 0 |
| Landfill | 0 | 0 | 0 |
| Total (for first life) | 4 | 4 | 0 |

**Fig. 9.** Product analysis using the eco materials adviser - values of the end of the product life cycle.

## 4.4 Analysis in SolidWorks Sustainability

Using the SolidWorks Sustainability tool, operating in the SolidWorks 3D CAD system environment, the LCA analysis of the product was carried out using the CML method (Fig. 10). The following assumptions were made as part of the analysis: product life cycle length 5 years, end of the product life cycle - landfilling, place of production of the product - Poland, place of use of the product - Poland also. Data on the distribution and installation of the product were not included in the analysis.

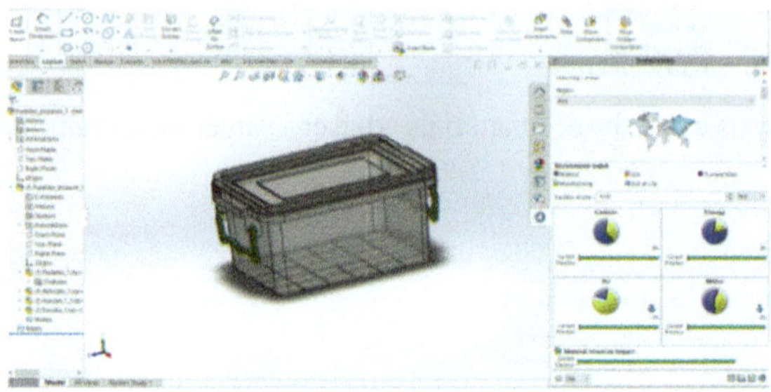

**Fig. 10.** Product analysis in SolidWorks Sustainability.

The obtained results were obtained in the form of the product environmental impact window. The analysis broken down into four main environmental categories: carbon footprint, energy consumption, air acidification and water eutrophication. The obtained results show that the stages of product production and transport have the greatest impact on the ecosystem (Fig. 11). The remaining analyzes are presented in Figs. 12–13. The main indicators are compared in Table 1.

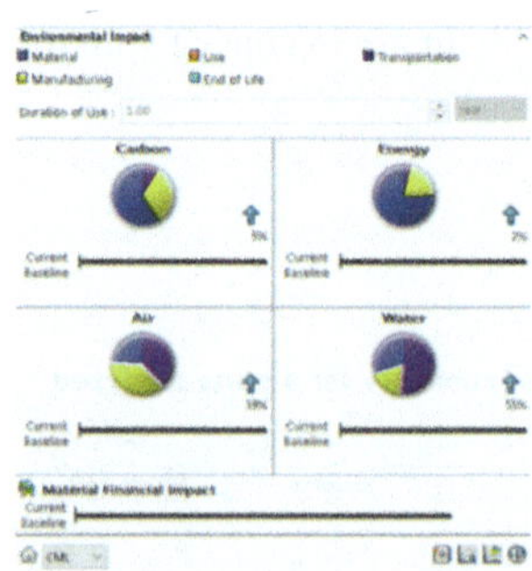


**Fig. 11.** Product environmental impact in the SolidWorks sustainability program.

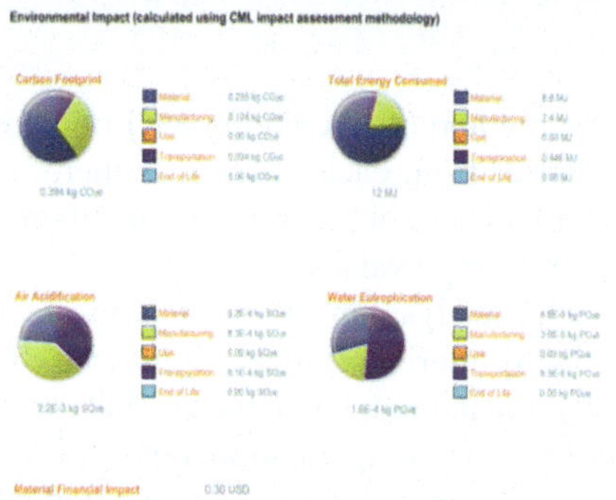


**Fig. 12.** List of environmental indicators in terms of the negative impact of the product on the environment with a breakdown into individual stages of its life cycle in SolidWorks sustainability.

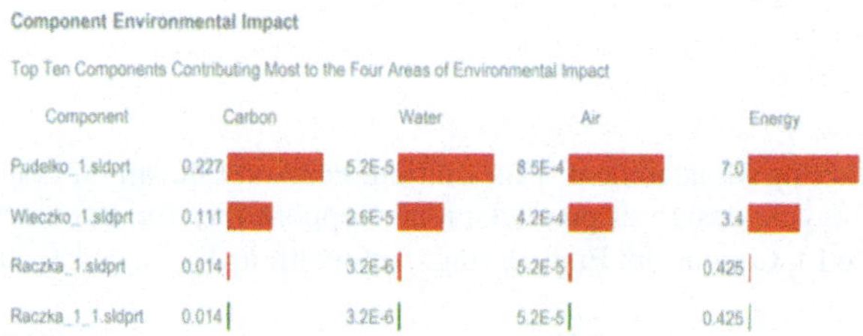


**Fig. 13.** Summary of the parts of the tested product in terms of negative environmental impact in SolidWorks sustainability.

**Table 1.** Comparison of indicators from eco materials adviser and SolidWorks sustainability.

| Indicator | Eco materials adviser | SolidWorks sustainability |
|---|---|---|
| Energy consumption | | |
| Material | 9.400 MJ | 08.800 MJ |
| Manufacture | 0.730 MJ | 02.400 MJ |
| Transport | Not available | 00.446 MJ |
| Use | Not available | 00.000 MJ |
| Disposal | 0.083 MJ | 00.000 MJ |
| Total | 6.700 MJ | 12.000 MJ |
| CO2 footprint | | |
| Material | 0.370 kg | 0.235 kg |
| Manufacture | 0.055 kg | 0.124 kg |
| Transport | Not available | 0.034 kg |
| Use | Not available | 0.000 kg |
| Disposal | 0.0050 kg | 0.000 kg |
| Total | 0.4300 kg | 0.394 kg |

## 5 Conclusions

Despite the use of the same method of calculating environmental indicators, the values obtained from both tools differ from each other. It is therefore advisable to calculate environmental indicators with the use of several tools and then aggregate the results, e.g. in the form of calculation of average values.

The use of environmental analysis in the evaluation of production processes and resulting products is becoming more and more significant. The range of information processed by it is constantly expanding, which means that the assessment of the product life cycle is extended over time to new application areas. It is anticipated that in the near future the environmental assessment tools will be integrated with other tools to support decisions in situations where environmental aspects will be an important factor. It seems, however, to mention that in very few cases environmental analysis can be used as the main factor supporting decision-making.

## References

1. Kotler, P.: Marketing Management. 12th Edition, Pearson Education, London (2006)
2. Alting, L.: Life-cycle design of products: a new opportunity for manufacturing enterprises. In: Kusiak A. (ed.), Concurrent Engineering: Automation, Tools, and Techniques, New York (1993)
3. Kao, Y.-C., Cheng, H.-Y., She, C.-H.: Development of an integrated CAD/CAE/CAM system on taper-tipped thread-rolling die-plates. J. Mater. Process. Technol. **177**(1–3), 98–103 (2006). https://doi.org/10.1016/j.jmatprotec.2006.04.082
4. Cao, H., Folan, P.: Product lifecycle: the evolution of a paradigm and literature review from 1950–2009. Prod. Planning Control (2011). https://doi.org/10.1080/09537287.2011.577460
5. EN ISO 14040:2006: Environmental management — Life cycle assessment — Principles and framework
6. PN-EN ISO 14044:2006: Environmental management — Life cycle assessment — Requirements and guidelines
7. Lesiuk, A., Oleszczuk, P., Kuśmierz, M.: Zastosowanie techniki LCA w ekologicznej ocenie produktów, technologii i gospodarce odpadami, [w:] Ryczkowski, J., Adsorbenty i katalizatory, Wybrane technologie a środowisko, Uniwersytet Rzeszowski, Rzeszów (2012)
8. PKN-ISO/TR 14062:2004 Environmental management - Integrating environmental aspects into product design and development, PKN, Warszawa (2004)
9. Kowalski, Z., Kulczycka, J., Góralczyk, M.: Ekologiczna ocena cyklu życia procesów wytwórczych (LCA), Wydawnictwo Naukowe PWN, Warszawa (2007)
10. The Polish Committee for Standardization, Environmental Management. Commentary on the ISO series standards 14000,Warszawa (2005)
11. Hunkeler, D., Lichtenvort, K., Rebitzer, G.: Environmental Life Cycle Costing, SETAC (2008). ISBN: 978-1-880611-83-8
12. EPA, Life CycleAssessment: Principles and Practice, EPA/600/R-06/060, US EnvironmentalProtection Agency, Office of Research & Development, Cincinnati, Ohio (2006)
13. Abdulkareem, M., Havukainen, J., Nuortila-Jokinen, J., Horttanainen, M.: Life cycle assessment of a low-height noise barrier for railway traffic noise. J. Cleaner Prod. **232**, 129169 (2021). https://doi.org/10.1016/j.jclepro.2021.129169
14. Marinković, S., Carević, V., Dragaš, J., The role of service life in life cycle assessment of concrete structures. J. Cleaner Prod **290**, 125610 (2020) https://www.sciencedirect.com/science/article/abs/pii/S0959652620356560

15. LCA of plastic & paper straws for portion-sized carton packages (2019). https://www.tetrapak.com/content/dam/tetrapak/publicweb/gb/en/sustainability/documents/LCA-paper-plastic-straws.pdf
16. Comparative Life Cycle Assessment of Tetra Pak® carton packages and al-ternative packaging systems for liquid food on the Nordic market (2020). https://www.tetrapak.com/content/dam/tetrapak/publicweb/uk/en/sustainability/2020-lca-tetra-pak-european-market.pdf
17. Ferreira, S., Cabral, M., da Cruz, N.F., Simões, P., Marques, R.C.: Life cycle assessment of a packaging waste recycling system in Portugal. Waste Manage. **34**(9), 1725–1735 (2014). https://doi.org/10.1016/j.wasman.2014.05.007
18. Iswara, A.P., et al.: A comparative study of life cycle impact assess-ment using different software programs. In: IOP Conference Series: Earth and Environ-mental Science, Joint International Conference on Civil, Environmental, and Geo Engineering, vol. 506 (2019)
19. Eco Materials Adviser
20. SolidWorks Sustainability

# The Environmental Analysis of a Product Manufactured with the Use of an Additive Technology

Anna Dudkowiak[1](✉), Ewa Dostatni[1], Izabela Rojek[2], and Dariusz Mikołajewski[2]

[1] Poznan University of Technology, Pl. Marii Skłodowskiej-Curie 5, 60-965 Poznan, Poland
anna.dudkowiak@put.poznan.pl

[2] Kazimierz Wielki University, ul. Chodkiewicza 30, 85-064 Bydgoszcz, Poland

**Abstract.** The use of additive technologies is one of the key elements of the Industry 4.0 structure. An unquestionable advantage of additive manufacturing is the speed of introducing changes in computer models, and thus the ease of product customization. Examples of such products are orthoses, prostheses and exoskeletons, which are personalized depending on the requirements of the person for whom they are intended. Despite the widespread use of additive manufacturing, mainly for rapid prototyping, there is relatively little information about the environmental impact of this process. This impact depends on the choice of 3D printer and the filament used in production. This paper attempts to conduct a comparative life cycle environmental analysis of two alternative versions of a product that was manufactured with the use of additive technologies. The structure of the product was identical and the research experiments consisted in changing the materials used in the additive manufacturing (from PLA to ABS). The effects of these changes on the environmental factors were observed and a direct comparison of the effects in the different factors was made. SimaPro software with implemented databases was used for the analysis. Missing information on the environmental impact of additive manufacturing of PLA and ABS parts was taken from the literature for the purpose of the study. The results of the research are presented in the paper.

**Keywords:** Additive manufacturing · Eco-design · Life Cycle Assessment (LCA)

## 1 Introduction

Environmental protection is becoming one of the most important assets in the world. Air pollution contributes to loss of health of many people. Reduction of air pollution can save millions of lives [1, 2].

Innovative technologies, such as 3D scanning, additive printing, and reverse engineering, integrated into the Industry 4.0 paradigm, can negatively affect the environment. Innovative technologies are increasingly used in medicine. An example is the use of additive technologies in the design and manufacturing of assistive devices, such as,


J. Trojanowska et al. (Eds.): MANUFACTURING 2022, LNME, pp. 76–89, 2022.
https://doi.org/10.1007/978-3-030-99310-8_7

e.g., exoskeletons. About 15% of the world's population suffer from various types of disabilities, of which 110–190 million are persons with reduced mobility, who require the support of equipment in daily activities [2, 3]. Each assistive device must be adapted to individual needs of its user. To meet this demand, there is a strong need for flexible manufacturing processes. Additive manufacturing provides this level of flexibility - each product can have a unique design. Considering some undeniable advantages, such as short and simple manufacturing process and relatively low manufacturing costs depending primarily on the quantity of material used, additive technologies are expected to keep gaining in popularity.

A CAD model developed in an additive technology becomes a sufficient basis for the end product. The time- and labour-consuming preparations required to manufacture a product in a traditional technology are circumvented here. The soaring popularity of additive technologies is bound to increase their environmental footprint. Key issues here are manufacturing waste, end of life product recycling, use of non-renewable resources, emissions in the manufacturing process, and energy consumption. In order to curb the environmental impact, the size of the problem must be examined and preventive measures must follow. The aim of this paper was to discuss the environmental assessment of a product manufactured with the use of an additive technology. The authors investigate how the material used in the manufacturing process affects the product's environmental parameters. The SimaPro software has been used for the purpose of the study.

The second and the third chapter of this article provides an extensive analysis of literature. Subsequently, IT tools supporting environmental analysis were presented. The fifth part describes the steps and the results of the environmental impact study. In the end, the conclusions were made.

## 2 Sustainable Manufacturing vs Additive Technologies

Accelerated development of Industry 4.0 brings a number of environmental threats, such as increasing demand for electrical energy [4] and emissions of harmful compounds, to mention just a few. Electrical energy is required to operate machines and meet the growing demand for hardware computing power. Innovative solutions introduced under the Industry 4.0 framework, such as, e.g., 3D printing, generate emissions of harmful substances. 3D printers use electrical energy to melt the working material. The environmental impact depends on the energy source. It is negligible with solar energy, but increases with the use of energy from the mains supply, depending on the combination of energy sources engaged by the supplier. In order to minimise the negative environmental impact of, inter alia, new technologies, in 2015 the World Commission for Environment and Development [5] defined 17 sustainable development goals and 169 targets. Aimed to help implement the sustainable development policy, they are accompanied by the concept of sustainable manufacturing, defined as the creation of manufactured products that use processes that minimise negative environmental impacts, conserve energy and natural resources, are safe for employees, communities, and consumers and are economically sound [6].

Sustainable manufacturing encompasses not only manufacturing, but also preparatory and post-manufacturing processes, such as designing, procurement of raw materials, distribution, use and reuse of the product. One cannot overestimate the importance of the design process, when decisions are made concerning recyclability of the product, minimization of dangerous substances, and use of environmentally friendly materials. At the stage of procurement of raw materials, the supply chain should be considered and eco-efficient supplies secured. Manufacturing processes should generate zero emissions and provide for efficient use of resources, including the electrical energy. In compliance with the sustainable manufacturing policy, distribution should include product returns, reuse and recycling. While in use by the end consumer, a sustainable product should be reliable as well as generate low operating expenses and zero emissions. The last stage is end of life disposal, with various possibilities, such as reuse, disassembly, and landfill disposal at a low cost [7–10].

For an additive technology to be considered environmentally friendly, it should feature all the above-mentioned qualities of sustainable manufacturing. No product has been produced so far which would fully comply with the requirements of sustainable manufacturing. The key issue are emissions of contaminants, depending primarily on the type of material used. Gases and particles emitted during 3D printing contaminate the air [11, 12]. Some of them, such as those emitted by Bisphenol A (BPA), used as a plasticizer and antioxidant, are presumed to have carcinogenic potential for humans [13]. Various measures are recommended to mitigate the hazards, such as use of low-emission materials and low temperatures, installing shields with filters around the printer, and monitoring emissions on an ongoing basis. Wojtyła [14] has established that when used in 3D printing, PLA is much less toxic than ABS. In an environmental analysis of ankle foot orthoses, Górski [15] has found out, on the basis of the calculated carbon footprint, that a 3D-printed orthosis is much more environmentally friendly than one made in the conventional technology of plaster cast covered manually with layers of resin and fiberglass fabric [16]. In $CO^2$eg, the carbon footprint of an orthosis manufactured traditionally is three times that of a 3D-printed one.

A study of a 3D printer working in the fused deposition modelling (FDM) technology has shown that when diverse materials and various extrusion temperatures are used, emissions of super-micron particles go down to zero and give way to emissions of ultrafine (10–30 nm) particles. Emissions increase in line with the increase in the temperature of extrusion. What is more, even a relatively short, 40-min 3D printing cycle generates up to 200 $mm^2$ of emissions [17]. Steinle [18] has found out that emissions of ultrafine aerosol (UFA) are much higher when printing in PLA than ABS. A longer duty cycle of the 3D printer causes the emission rating to rise. What is important, 3D printing in spacious, well-ventilated rooms does not increase the UFA concentration significantly, as compared to printing in a confined, non-ventilated room. This is an important hint when choosing a room for a 3D printing laboratory. Other materials used in 3D printing are thermosetting photopolymers, which can be reused as materials reforming 3D printed objects into a new shape. The technology is used for repairing damaged parts [19].

As is the case with many other manufacturing technologies, the environmental impact of 3D printing depends largely on choices made by the user. Model design, energy source, materials in use, recyclability - all these factors determine the resulting environmental impact of the product.

Life Cycle Assessment (LCA) is just one of many tools facilitating the assessment of the environmental impact. Owing to a standardized analysis process and incorporation of numerous standards, the method ensures comparable results of analyses conducted in enterprises and research organizations, and facilitates the creation of a hierarchy of many issues related to the environmental impact of manufacturing processes and product life cycle.

## 3 Life Cycle Assessment (LCA)

Life Cycle Assessment (LCA) is defined in the ISO 14040:2006 standard as an environmental management technique aimed at the assessment of products, materials, processes, services and systems in terms of their impact on the natural environment. LCA covers the possible impact on the ecosystem throughout the product life cycle ('from cradle to grave'), i.e., from the procurement of raw materials to the disposal of materials at the end-of-life. Aggregated impact on the natural environment at all the stages of the product life cycle is assessed, based on the assumption that processes are interdependent, or that particular manufacturing stages affect one another [20]. LCA ensures a comprehensive overview of the product's environmental impact, taking into account processes that are otherwise excluded, such as extraction of raw materials, transport, etc. [21]. The LCA method is classified as a quantitative tool not only facilitating the classification of certain groups of impact, but also determining the impact quantitatively for each of the measures in use [22]. Owing to the complexity of calculations and sequentiality which can be easily expressed by means of algorithms, the LCA is widely implemented in software tools supporting product environmental assessment [23]. According to ISO 14040:2006, the LCA comprises four stages [24, 25]:

1. Goal & scope definition - definition of the product under analysis and its service life, scope of study, data source, target group and intended purpose of the study.
2. Inventory analysis - determination of system inputs, outputs and processes, determination of the raw material and energy balance, creation of the product's life cycle.
3. Impact assessment - classification of environmental impacts by a selected method and determination of their size (categorisation and quantitative analysis).
4. Interpretation - presentation and critical evaluation of results (this stage is in progress simultaneously with the three other stages).

There are defined methods by which an LCA can be performed, such as, e.g., Eco-Indicator 99, IMPACT 2002+, or ReCiPe 2016 Endpoint [26]. In this study, the ReCiPe 2016 Endpoint method, described in detail in chapter *Methodology of the study*, has been used.

## 4 Software Tools Supporting Environmental Analysis

There is a wide choice of software tools which facilitate the environmental analysis of manufactured products. One group of them are autonomous software tools which require manual and usually time-consuming implementation of the product life cycle. Not integrated into any 3D CAD environment, they do not support any data transfer, such as, e.g., product structure. Their ample databases retrieve proprietary data on processes. Typically, the analysis is carried out in accordance with the LCA environmental management technique (discussed in detail in chapter *LCA*) by a method (e.g., Eco-Indicator 99, ReCiPe 2016 Endpoint) implemented into the software. They support extensive analyses of emissions (positive and negative), and provide numerical values of emitted substances, gases, etc. Some examples of such software tools are GaBi, SimaPro, Umberto, OpenLCA. SimaPro is one the most commonly used tools. Data on the product can be retrieved from the implemented databases, and the calculation methods correspond to the LCA environmental management technique. Environmental impacts can be imaged for one product assembly, and alternative product assemblies can be compared [27].

Another group are autonomous tools integrated into a 3D CAD environment or one of its modules. Some examples are the Eco Materials Adviser environmental analysis module of Autodesk Inventor and SOLIDWORKS Sustainability of SOLIDWORKS. Solutions of this type streamline work through automated data interchange between a 3D product model and the environmental analysis module, thus saving the designer's time spent on entering the product structure data. Additionally, other data on the manufacturing process can be entered, such as methods of transport, place of production and use, service life, etc., which - as a standard - is not assigned to a 3D model. Analyses performed in such systems are trimmed down to the examination of water and energy consumption, carbon footprint of the manufacturing process, etc. Compared to the analyses performed by the autonomous systems referred to above, these tools are intended for management purposes. They do not support a thorough analysis of the environmental impact of designed products throughout the life cycle using the LCA environmental management technique [28].

Many enterprises develop proprietary environmental analysis software for particular products, such as, e.g., the Ecodesign Manual by Philips, the Handbook of Volvo, the Environmental Guidelines by Electrolux, etc. [29].

Changes in manufacturing processes aimed at reducing the environmental impact of products throughout the life cycle are driven by the pressure put by industry beneficiaries on environmental protection. Accordingly, the number of software programs supporting environmental analysis, equally customized (adapted for certain products and enterprises), autonomous and integrated into CAD 3D systems, available on the market, is growing rapidly.

# 5 Environmental Analysis of a Product Manufactured in an Additive Technology

## 5.1 Methodology of the Study

A comparative LCA environmental analysis has been performed for a hand exoskeleton in two alternative versions:

- the base assembly - featuring PLA elements manufactured in an additive technology,
- the alternative assembly - featuring ABS elements manufactured in an additive technology.

The study has been conducted in accordance with the LCA four-stage environmental management technique (in compliance with ISO 14040:2006):

**Goal & Scope Definition.** A hand exoskeleton developed by scientists of the Kazimierz Wielki University in Bydgoszcz, Poland has been examined. The study is aimed to determine the environmental impact of the product throughout its life cycle. Two assemblies of the product, with elements made of two different materials (both manufactured in an additive technology), have been examined, to find how the change of material affects the software output, i.e., information about the environmental impact of the assembly. The final output is a comparative analysis of the alternative product assemblies, shown as a compilation of graphs generated in the software. The study can support scientists in the selection of material for the exoskeleton.

The product has been assigned a service life of 5 years - at the end of that period, all its parts should be replaced with new ones.

The scope of the study has been defined as follows: the manufacturing processes of equipment, tools and vehicles used throughout the product's life cycle, such as a lorry, a drill-driver, a 3D printer and a laptop, have been excluded from the analysis; however, emissions to the environment in the processes related directly to the manufacturing of the exoskeleton with the use of the above mentioned equipment, tools and vehicles (e.g., transport of sub-assemblies in the lorry) have been included.

The analysis has been performed in the SimaPro software, with data sourced from the literature and the software databases. Universal substitutes available in databases have been used in place of the missing data required to create the life cycle of the product (e.g., linear servo controller - electronics, for control units).

**Inventory (inputs and outputs) Analysis.** At stage one, processes required to develop the product structure, which were missing in the software database, were developed. The focus was put on the additive manufacturing process. All other elements of the two assemblies under analysis were purchased and identical for both assemblies, so they had no impact on the outcome of the comparative analysis. Based on data sourced from [11], a universal additive manufacturing process was entered for both PLA and ABS, with the respective emissions generated by each material during a one-hour cycle of the

3D printer. Next, the process of additive manufacturing of the assembly parts, namely, finger phalanxes, a metacarpus and a carpal joint (connected by means of supports), a housing for the electronic circuit and fourteen mounting pegs, has been developed. Energy consumption by the 3D printer and the laptop (necessary to adapt a universal exoskeleton design to individual needs of the patient) has also been taken into account. The additive manufacturing process for PLA, implemented into the system, is shown in the screenshot in Fig. 1. The printing process was of the same duration for both assemblies, but the energy consumption by the printer was declared higher for ABS than for PLA.

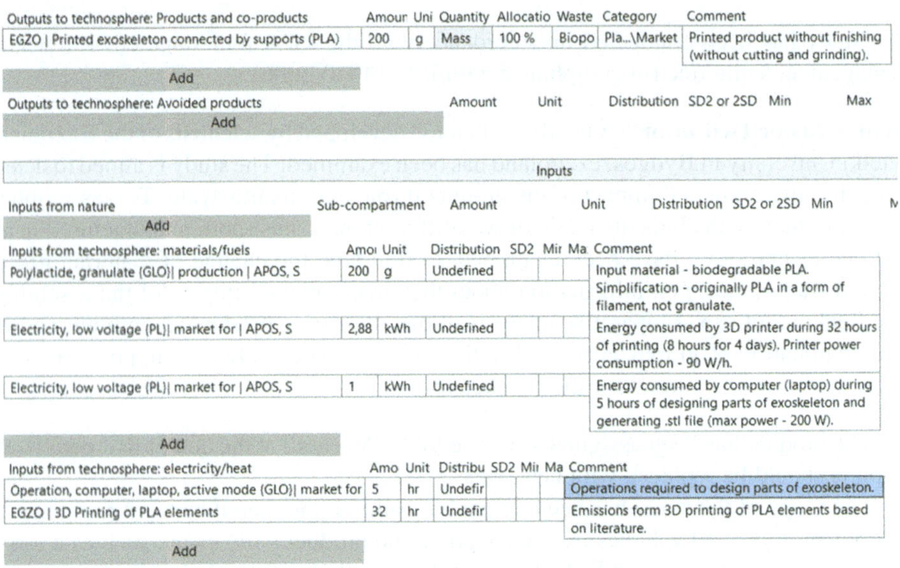


**Fig. 1.** Additive manufacturing of elements in PLA.

Similarly, the manufacturing process for the polyester elements - a wrist orthosis and a LiIon battery case - was developed. The process was identical for both assemblies. Other subassemblies were simplified (as they did not affect the result of the comparative analysis) and based on the data retrieved from the SimaPro database. As mentioned in section a), substitutes (similar processes for parts manufactured in similar technologies) were used in some assemblies for the data which was missing in the database.

At stage two, the product structure, identical for both assemblies, was developed. Modification of the alternative assembly relative to the base one consisted in using a different material (ABS in place of PLA). The product structure is shown in Fig. 2.

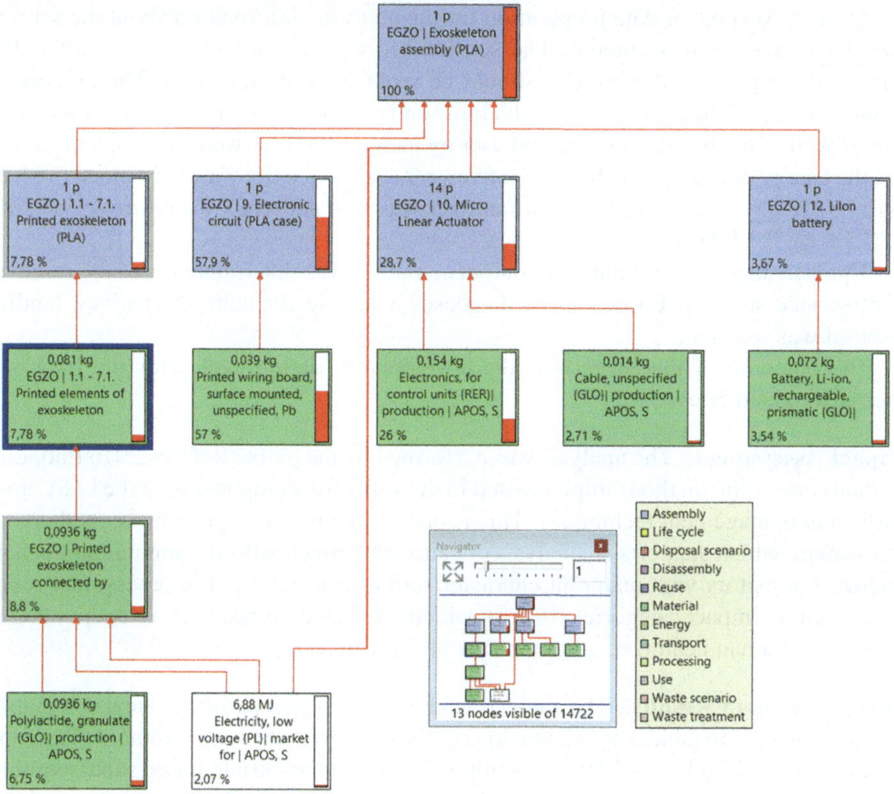


**Fig. 2.** The most important elements of the product structure.

According to section *Goal & scope definition*, considering the job production of the exoskeleton, the manufacturing processes of equipment, tools and vehicles used in the entire manufacturing process of the product under analysis were excluded from the study scope. However, certain processes related directly to the production of the exoskeleton, in which the above mentioned equipment, tools and vehicles were used, were identified, namely:

- delivery of the purchased subassemblies and the end product to the customer - transport by a lorry (underlying assumption: each subassembly is transported over a distance of 100 km),
- customization of the exoskeleton - performed on a computer in the active mode; duration: 5 h; mean energy consumption,
- additive manufacturing of the exoskeleton elements - with the use of a 3D printer; mean energy consumption,
- assembly of the exoskeleton - energy consumption by a drill driver for 3 h of operation.

At the next stage of data preparation for the analysis, information about the service life of the product was entered. The service life was defined as 5 years - after that period all the parts of the product should be replaced with new ones. The process of manufacturing of the carton box in which the end product would be delivered to the end user as well as the transport to the end user by means of a lorry were taken into account, and the energy consumption during the five years of service life was determined (164.25 kWh, 90 W/day). Consumption of a sanitizer was also added for the entire service life (182.5 l, ca. 0.1 l/day).

The last stage of the inputs and outputs analysis was determination of the removal from service strategy. Based on the databases available for both alternatives, landfill disposal was selected.

The outcome of this stage was the development of life cycles for the base and alternative assemblies.

**Impact Assessment.** The analysis was performed by the global ReCiPe 2016 Endpoint method (one of the methods implemented in the software, compliant with the LCA environmental management technique). This is one of the most comprehensive methods of assessment, which supports an analysis of cause and effect paths linking midpoint characterization factors with endpoint characterization factors [30]. The assessment relies on 22 indirect impact categories (Fig. 3), which are then assigned to three endpoint area categories (human health, ecosystems, resources) (Fig. 4).

**Interpretation of Results.** The analysis leads to a direct comparison (visual - on graphs and quantitative - in tables) of two versions of the assembly (the base assembly - PLA and the alternative assembly - ABS). As mentioned above, the same purchased subassemblies were used for both assemblies of the product; the only difference between the assemblies was the material used for the parts produced in the additive technology. The differences of the estimated environmental impact (midpoint characterisation factors) between the two assemblies throughout their life cycles are shown in the screenshot below (Fig. 3).

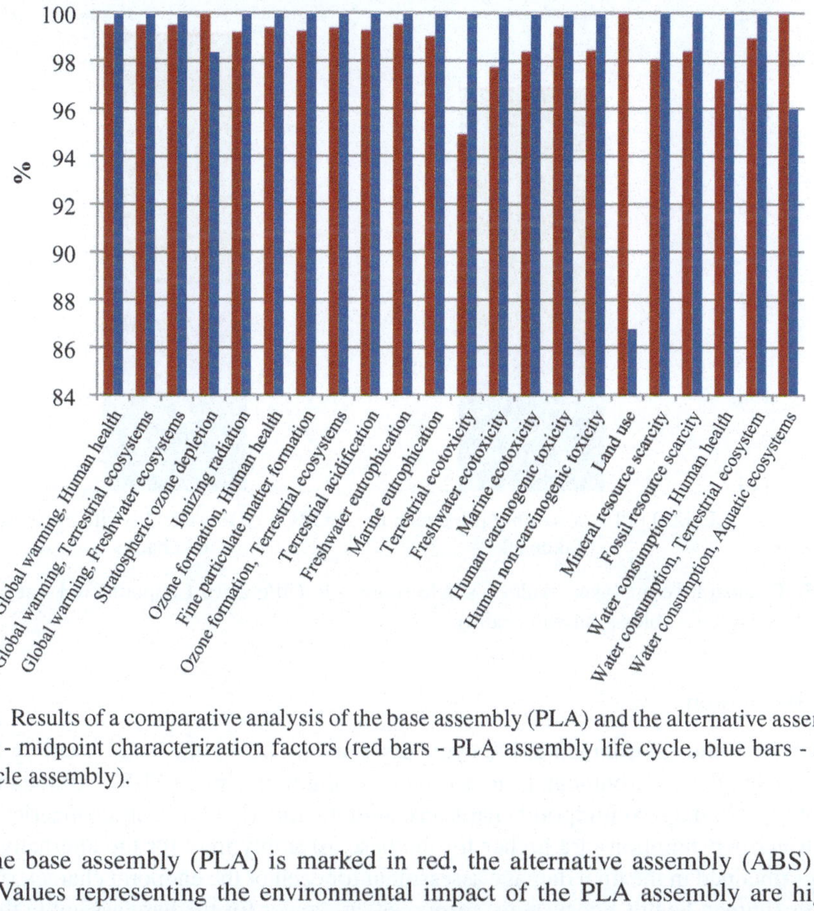


**Fig. 3.** Results of a comparative analysis of the base assembly (PLA) and the alternative assembly (ABS) - midpoint characterization factors (red bars - PLA assembly life cycle, blue bars - ABS life cycle assembly).

The base assembly (PLA) is marked in red, the alternative assembly (ABS) - in blue. Values representing the environmental impact of the PLA assembly are higher than those for the ABS assembly only in three of the 22 categories under analysis. Similarly, it follows from an analysis of the endpoint characterization factors - normalization (by the ReCiPe 2016 Endpoint method) that for each of the categories (human health, ecosystems, resources), values for the life cycle of the ABS assembly are higher than those for the PLA assembly (Fig. 4). All in all, the alternative assembly has a greater environmental impact than the base assembly (single score - Fig. 5).

| Damage category | Unit | EGZO \| PLA assembly life cycle, no takeback | EGZO \| ABS assembly life cycle, no takeback |
|---|---|---|---|
| Resources | | 0,00027 | 0,000275 |
| Ecosystems | | 0,0257 | 0,026 |
| Human health | | 0,402 | 0,407 |

**Fig. 4.** Comparative analysis results (normalization - ReCiPe 2016 Endpoint).

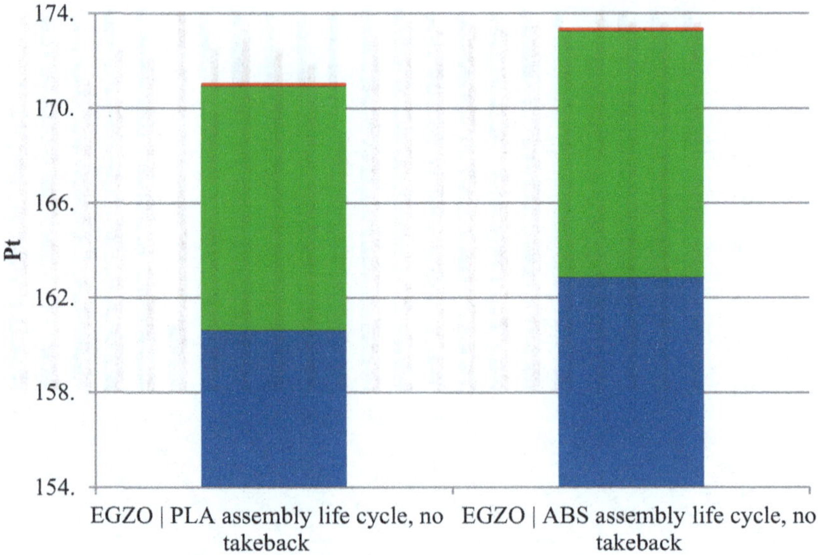


**Fig. 5.** Comparative analysis results – single score – ReCiPe 2016 Endpoint (red – resources, green – ecosystems, blue – human health).

### 5.2 Discussion

Subjected to the environmental life cycle assessment (LCA), the base assembly (PLA) shows a smaller environmental impact than the alternative one (ABS). Although three out of the 22 analyzed midpoint characterization factors (land use, stratospheric ozone, and water consumption) are higher for the base assembly than for the alternative one, values obtained in the final damage assessment for each of the endpoint characterization factors (human health, ecosystems, resources) are lower for the base assembly than for the alternative one.

A comparative analysis of the manufacturing process exclusively (excluding the service life and the end of life scenario) shows greater differences between the midpoint characterization factors for the two assemblies than an analysis of the entire life cycle (i.e., smaller percentage values for one assembly relative to the other). For instance, the land use factor for the entire life cycle of the ABS assembly (manufacturing, service life, and end of life disposal) is equivalent to 86% of the same factor for the PLA assembly; whereas in the analysis of the manufacturing process only, the land use factor for the ABS assembly is equivalent to 72% of the same factor for the PLA assembly. Greater differences between midpoint characterization factors for the manufacturing process only than for the entire life cycle result from the fact that the share of the additive manufacturing process (different for each assembly) in the manufacturing process of the assembly is greater than its share in the entire life cycle of the assembly.

Based on the performed analyses, it can be stated that throughout the life cycle, the base assembly (PLA) has a smaller environmental impact (as indicated by the endpoint characterization factors) than the alternative assembly (ABS). Due to a lack of precise data in the database and the literature, the PLA assembly has not been assigned recycling as the end of life scenario. The environmental impact of the difference between PLA and ABS in terms of the end of life disposal will be examined in another study.

The development of a product concept is such a complex process that it cannot be stated unequivocally which of the assemblies is optimal. Consideration should also be given to other properties of the product, such as its functionality, endurance, aesthetics, ergonomics, etc. Therefore, when performing the environmental analysis, it should be assessed whether differences in the environmental impact of the life cycle of each of the assemblies (in quantitative terms) are significant compared to, e.g., endurance of the compared materials. PLA shows less fragility than ABS. ABS is manufactured based on crude oil and is more demanding in the process than PLA, which is based on natural raw materials [31]. When printing in ABS, the 3D printer consumes more energy. ABS has a higher melting point and shrinkability, and is more harmful, especially to human health. Considering the concentration of emissions during 3D printing in an enclosed space, it is recommended that additive manufacturing, especially from ABS, be performed in ventilated spaces, ideally when nobody is in the room [31].

It follows from the analysis of any assembly that the process of service (energy consumption necessary for the operation of the exoskeleton) and the processes of manufacturing the Arduino Uno-based electronic circuit and the LiIon battery are of key importance in terms of the environmental impact throughout the product's life cycle. Therefore, particular focus should be given to possible modifications leading to optimization and reduction of the environmental impact without compromising the product's functionality. The study discussed in this paper has not only shown a more optimal assembly in terms of its environmental impact, but also helped to outline processes whose environmental impact should be minimized in the first place.

## 6 Conclusions

The use of software for ecological analysis enabled the selection of more environmentally friendly material from among alternative versions, improving the decision-making process to a large extent. The analysis also made it possible to identify the negligible impact of the part manufactured with the use of additive technology in relation to the production process of the whole assembly. This means that the choice of material from among the analyzed alternatives does not matter much from the comprehensive point of view of the environmental impact of the whole product.

The SimaPro software used in the study is a tool supporting a thorough analysis of the environmental impact of a product throughout its life cycle. The global methods implemented in the software (ReCiPe 2016 Endpoint, Eco-Indicator 99) facilitate calculation of the environmental impact and streamline clear representation of data. Limitations are put by databases, which lack some data on particular (often specific) processes necessary to build a precise representation of the product's life cycle. Another difficulty is posed by the fact that manufacturing processes of some subassemblies produced by third parties are unknown. Meticulous data gathering is required to build a product life cycle in the software which will truly and accurately represent the real life cycle of the product. Further study will focus on minimizing the simplifications made to the product, especially with regard to the disposal of particular assemblies at the end of life.

**Acknowledgements.** The work presented in the paper has been co-financed under grant no. 0613/SBAD/4710 and grant to maintain the research potential of Kazimierz Wielki University.

## References

1. WHO health an climate change survey report. https://www.who.int/phe/en. Accessed 14 Oct 2021
2. Rojek, I., Macko, M., Kopowski, J., Mikołajewski, D.: Reverse engineering as a way to save environment with-in patient-tailored production of assistive technology devices – based on own hand exoskeleton case study. In: Machado, J., Soares, F., Trojanowska, J., Ottaviano, E. (eds.) ICIENG 2021. LNME, pp. 82–91. Springer, Cham (2022). https://doi.org/10.1007/978-3-030-79165-0_8
3. Rojek, I., Mikołajewski, D., Dostatni, E., Macko, M.: AI-optimized technological aspects of the material used in 3D printing processes for selected medical applications. Materials **13**(23), 5437 (2020)
4. Stężowski, J.: Sustainability as an integral part of the fourth industrial revolution. https://mamstartup.pl/zrownowazony-rozwoj-jako-integralna-czesc-czwartej-rewolucji-przemyslowej/. Accessed 14 Oct 2021
5. Sustainability and Goals of Sustainable Development. www.unic.un.org.pl/strony-2011-2015/zrownowazony-rozwoj-icelezrownowazonego-rozwoju/2860. Accessed 12 Sept 2021
6. Feng, S.C., Joung, C.: A measurement infrastructure for sustainable manufacturing. Int. J. Sustain. Manuf. **2**(2/3), 204–221 (2011)
7. Czaplicka-Kolarz, K., Kruczek, M.: The concept of eco-efficiency in sustainable production management. In: Scientific Papers of Silesian University of Technology. Organization and Management Series, vol. 63, pp. 59–71 (2013)
8. Valdivia, S., Ugaya, C.M.L., Sonnemann, G., Hildenbrand, J.: Towards a Life Cycle Sustainability Assessment: Making Informed Choices on Products. UNEP, France (2011)
9. Karwasz, A., Dostatni, E., Diakun, J., Grajewski, D., Wichniarek, R., Stachura, M.: Estimating the cost of product recycling with the use of ecodesign support system. Manag. Prod. Eng. Rev. **7**(1), 33–39 (2016)
10. Dostatni, E., Rojek, I., Hamrol, A.: The use of machine learning method in concurrent ecodesign of products and technological processes. In: Hamrol, A., Ciszak, O., Legutko, S., Jurczyk, M. (eds.) Advances in Manufacturing. LNME, pp. 321–330. Springer, Cham (2018). https://doi.org/10.1007/978-3-319-68619-6_31
11. Karwasz, A., Osiński, F.: Literature review on emissions from additive manufacturing by FDM method and their impact on human health. Manag. Prod. Eng. Rev. **11**(3), 65–73 (2020)
12. Żukowska, M., Górski, F., Bromiński, G.: Rapid manufacturing and virtual prototyping of pre-surgery aids. In: Lhotska, L., Sukupova, L., Lacković, I., Ibbott, G.S. (eds.) World Congress on Medical Physics and Biomedical Engineering 2018. IP, vol. 68/3, pp. 399–403. Springer, Singapore (2019). https://doi.org/10.1007/978-981-10-9023-3_72
13. Nomiri, S., Hoshyar, R., Ambrosino, C., et al.: A minireview of bisphenol A (BPA) effects on cancerrelated cellular signaling pathways. Environ. Sci. Pollut. Res. **26**(9), 8459–8467 (2019)
14. Wojtyła, S., Klama, P., Baran, T.: Is 3D printing safe? Analysis of the thermal treatment of thermoplastics: ABS, PLA, PET and nylon. J. Occup. Environ. Hyg. **14**(6), 80–85 (2017)
15. Górski, F., Osiński, F., Wierzbicka, N., Żukowska, M.: Environmental impact of additive manufacturing of individualized orthopaedic supplies. In: Tonkonogyi, V., et al. (eds.) Advanced Manufacturing Processes II, pp. 384–393. Springer, Ukraine (2021)

16. Cha, H.Y., et al.: Ankle-foot orthosis made by 3D printing technique and automated design software. Appl. Bionics Biomech. **2017**(1–3), 1–6 (2017)
17. Stabile, L., Scungio, M., Buonanno, G., Arpino, F., Ficco, G.: Airborne particle emission of a commercial 3D printer: the effect of filament material and printing temperature. Indoor Air **27**(2), 398–408 (2017)
18. Steinle, P.: Characterization of emissions from a desktop 3D printer and indoor air measurements in office settings. J. Occup. Environ. Hyg. **13**(2), 121–132 (2016)
19. Zhang, B., Kowsari, K., Serjouei, A., Dunn, M.L., Ge, Q.: Reprocessable thermosets for sustainable three-dimensional printing. Nat. Commun. **9**(1), 1–7 (2018)
20. Cecchi, T., de Carolis, C.: Life cycle assessment. In: Biobased Products from Food Sector Waste, pp. 365–377. Springer, Cham (2021). https://doi.org/10.1007/978-3-030-63436-0_11
21. Curran, M.A.: Life Cycle Assessment: Principles and Practice. National Risk Management Research Laboratory, Cincinnati (2006)
22. Kronenberg, J., Bergier, T.: Challenges of Sustainable Development in Poland. Fundacja Sendzimira, Cracow (2010)
23. Wach, A.K.: Life Cycle Assessment (LCA) as the basis for computer-assisted product evaluation. In: Conference Materials of the 2nd National Science and Technology Conference "Ecology in Electronics", Warsaw (2002)
24. Guinee, J.B., et al.: Life cycle assessment: past, present, and future. Environ. Sci. Technol. **45**(1), 90–96 (2011)
25. ISO 14040:2006 Environmental management—Life cycle assessment—Principles and framework
26. Dostatni, E.: Ecological product design in 3D CAD environment with the use of agent technology. Wydawnictwo Politechniki Poznańskiej, Poznan (2014)
27. Silva, D., et al.: How important is the LCA software tool you choose? Comparative results from GaBi, openLCA, SimaPro and Umberto. In: Proceedings of the VII Conferencia Internacional de Análisis de Ciclo de Vida en Latinoamérica, Medellin, Colombia, pp. 10–15 (2017)
28. Dudkowiak, A., Grajewski, D., Dostatni, E.: Analysis of selected IT tools supporting ecodesign in the 3D CAD environment. IEEE Access **9**, 134945–134956 (2021)
29. Fargnoli, M., Kimura, F.: The optimization of the design process for an effective use in ecodesign. In: Takata, S., Umeda, Y. (eds.) Advances in Life Cycle Engineering for Sustainable Manufacturing Businesses, pp. 59–64. Springer, Tokyo (2007)
30. Huijbregts, M.A.J., et al.: ReCiPe2016: a harmonised life cycle impact assessment method at midpoint and endpoint level. Int. J. Life Cycle Assess. **22**(2), 138–147 (2016). https://doi.org/10.1007/s11367-016-1246-y
31. SYGNIS: The 5 most important differences between PLA and ABS. https://shop.sygnis.pl/5-najwazniejszych-roznic-pomiedzy-pla-a-abs/. Accessed 16 Nov 2021

# Development Process of Customised Products, Supported by Technologies, a Case of Tailor-Made Furniture

Wiesław Urban, Krzysztof Łukaszewicz(✉), and Elżbieta Krawczyk-Dembicka

Bialystok University of Technology, Wiejska 45A Street, 15-351 Bialystok, Poland
{w.urban,k.lukaszewicz}@pb.edu.pl

**Abstract.** There is a noticeable general trend of reorientation of businesses towards so-called mass customisation of production. Adopting a strategy characterised by mass customisation and customisation of products requires a greater emphasis on customer interaction and support. In this regard, it is important to have efficient contact with the customer throughout all the activities related to product development, especially at the stage setting the production.

The aim of the study is to analyse the process of preparation and performing of a customised furniture products and to develop the concept of changes in this process with the purpose of improving its flow and ensuring higher satisfaction with the end results.

The proposed changes in technological and organisational solutions in the examined production of furniture units enable much higher customisation of products, also enabling more efficient exchange of information, both with customers and in the process flow between the particular stages of product and technology design, and manufacturing. The concept of process changes includes, among other things, the use of general purpose IT solutions to support the preparation of production. Research indicates that the use of a proposed preparation and realisation concept enables furniture manufacturers to achieve a high level of cooperation with their customers.

**Keywords:** Product development · Customised manufacturing · Communication with customers · Furniture production

## 1 Introduction

Today, companies are facing a strong and increasing pressure to reduce costs, shorten time-to-market and add value to their products by investing in product development. These pressures have led to an expansion of activities related to the development and improvement of production processes [1]. However, highly customised production requires high flexibility and responding to customer needs in a timely fashion [2]. Close attention must be paid to this particular stage, taking into account that 80% of a product's costs are generated at the phase of development [3].


J. Trojanowska et al. (Eds.): MANUFACTURING 2022, LNME, pp. 90–104, 2022.
https://doi.org/10.1007/978-3-030-99310-8_8

A number of interconnected digital technologies are emerging within the concept of Industry 4.0, such as: Artificial Intelligence, Internet of Things, Big Data, Cloud Computing and Blockchain, altering the supply chain and remodelling industrial competition [4]. Manufacturing companies are increasingly often using digital technologies to create a digitisation strategy [5]. Digitisation is becoming one of the most important strategies for companies to survive and grow in an increasingly competitive environment [6].

To increase customer satisfaction while reducing development time under these conditions, the product development process (PDP) must become smarter. One of the main challenges in implementing smart PDPs is to adapt quickly to changing conditions and requirements [7].

In the case of today's mechanical engineering companies, however, the increasing interdisciplinarity of processes and development teams and distributed company locations lead to longer production times and more complex PDPs, although the required development times are shortened [8].

In such circumstances, it can be observed that manufacturers looking for competitive advantage often move from offering products to offering advanced product-service-software systems, a transition referred to as digital servitisation [9–11]. Evolving customer needs, financial challenges, increased competition, product commoditisation and advances in cyber technology and digital systems mean that manufacturers cannot rely solely on innovative, breakthrough products. Therefore, they need to combine products, services and software into smart solutions to increase customer value and internal efficiency [10].

Undoubtedly, an extremely important aspect is the implementation of all actions towards sustainable development. Kuhlman and Farrington [12] emphasise that sustainable development concerns well-being of future generations and, in particular, irreplaceable natural resources - as opposed to the gratification of present needs which they call well-being. At the same time, they point to the need to find a balance between the two states. The development of digitisation, by increasing the range and accuracy of data and generally improving analysis, should favour the achievement of this balance.

Due to the importance of the production preparation stage, as noted earlier, there is a need to search for and develop methods to improve all activities, including the development of efficient design methods and successful exchange of information.

The aim of the article is to analyse the customised furniture preparation and realisation process and to develop the concept of changes that improve its execution and help better satisfy the needs of customers.

The analysis of the customised furniture production preparation and execution process was carried out with particular emphasis on the flow of information with customers/users and the support of IT tools. Afterwards, a concept of changes in the examined process was developed towards improving communication with customers, leading to increased customisation of products and efficiency of the entire process, which requires general purpose IT solutions supporting the preparation of production (among other things). This is because new information technologies are opening up completely new opportunities for cooperation with customers at the stage of product design and enable further customisation.

For the in-depth study using the case study method, the intention was to analyse the production of custom-made furniture, since by definition it involves close and repeated contacts with customers during the execution of each customised order, satisfying the individual preferences of each customer and installation at customers' premises. In this case, the process of product design and manufacture are closely linked due to the sequence and take a relatively short time. The adopted study object allows cross-sectional observation of the manufacturer's interaction with the customer during product development; it also allows the creation of the relevant concepts and theoretical generalisations within the adopted scope.

## 2 Views on the PDP in Literature

A large amount of complex information is processed during product development activities. This is not only due to requirements from external and internal sources of the company. This is also because PDP activities affect and are affected by all areas of the organisation [13].

The growth of digitisation and simulation processes at every stage of production creates opportunities for organisations to achieve higher productivity [14]. Zweber et al. [15] report that intensive use of information technology to design, tweak and then manufacture products (i.e. Digital Manufacturing) potentially enables reuse of older work, more informed decision making, and more reliable planning and estimation. In other words, the quality of information is improved, resulting in shorter deadlines and lower costs, and higher customer satisfaction. Development of computer technology that leads to an increase in efficiency and the development of information systems enable an integration of functional tasks.

Improvement of PDPs has been the subject of manufacturing studies for a long time. Integration, interoperability and sustainability are emphasised throughout the lifecycle to enhance business agility [16]. As a practical strategic approach, product lifecycle management (PLM) promises to holistically incorporate product data and information into design activities and core business processes [17]. In the context of Industry 4.0, intellectual capital processing is valued increasingly as the key to increasing business agility and reaping the benefits of smart manufacturing. It is also the basis for product improvement and innovation. Of course, properly selected IT solutions that play a key role in the company's operations are the crucial prerequisite [18, 19].

In practice, successful adaptive design means finding the most reliable and the least time-consuming ways to design a product that fully meets the customer's demands and preferences. The implementation of information-assisted design is based on the integration of 3D modelling software with parametric design, expert system, computer-aided analysis and knowledge management [20].

Iaksch and Borsato [21] note that current industry practices still show an isolation of knowledge domains, as evidenced by a simple transfer of information to the people responsible for the subsequent activities within the PDP, which is contrary to the behaviour advocated by Concurrent Engineering and Design for Manufacturing.

## 3 The Furniture Industry and Customised Products

Furniture production is an area of business that generally requires mass customisation. However, there are significant differences between businesses producing standard, off-the-shelf furniture and businesses producing bespoke, custom-made furniture. For the latter, the production process is much longer and more complicated. It starts with a research and development phase, followed by sales (business contracting), participatory design, production, and ends with the delivery and installation at the customer's premises [22]. This process faces multiple challenges, from the processing of non-standard orders to very precise production and management of very large amounts of data, which means that implementation of appropriate IT systems is indispensable. In order to maintain information connectivity throughout the furniture manufacturing process, businesses involved in the production of highly customised products should use Industrial Internet of Things technologies that connect industrial assets to business processes and information systems, providing businesses with real-time data monitoring, collection, exchange and analysis capabilities [23].

Customised furniture production requires, first and foremost, an individual approach to design and the use of modern and unique designs to create innovative, ground-breaking and original products. This approach is a characteristic feature of Design-Driven Innovation (DDI), which plays a special role in the furniture industry. This is because industrial design is considered a major factor of competitiveness and design is a driving force of innovation [24, 25]. Thanks to DDI, a business can design more valuable products [26] and expand its capabilities beyond participatory design and broaden its knowledge of technological and product development, which plays a significant role in both new and developing markets. However, as noted by De Goey et al. [27], design-driven innovation requires businesses to develop the communication skills and other capabilities necessary to create and promote new products.

An important factor for the furniture industry is the current global trend indicating the responsibility of furniture manufacturers for environmentally friendly production, which is why environmental criteria are more and more often taken into consideration with respect to production processes the finished products [28]. They also appear in companies' production strategies, with the indication of production technology. Among the advanced methods of furniture production, 3D printing is gaining popularity as an additive manufacturing technique. Subtractive Manufacturing is another technique. Each of these techniques has its own advantages and disadvantages [29]. In the context of Industry 4.0, the application of new techniques supporting product realisation and effective information exchange becomes an extremely important issue due to a great emphasis on product customisation.

## 4 Methods

Given the stated study objective, which includes an in-depth exploration of a selected type of product development process, the case-study method was adopted, which allows for a detailed exploration of the selected area [30, 31]. One of the assumptions of the study was to focus on the process intended to provide products with a high degree of customisation in unit production, with intense cooperation with the customer, including information exchange. Therefore, production of custom-made furniture on individual order was chosen for the in-depth study, since due to its nature, it involves intensive cooperation between the manufacturer and the customers at each stage of the process. The subject of the study allows for an in-depth observation of the manufacturer's interaction with the customer at all stages of the product development process. The case study involved tracing all the activities in the process, describing the internal process flows and the interaction with the customer/recipient. A flow chart was also used for process modelling.

The second stage of this study is the formulation of concepts based on the identified process state. Formulation of a conceptual framework is an important research task that involves identifying structures and relationships between concepts, objects and ideas that are of interest to the researcher [32, 33]. A precisely identified process of customised furniture production was the starting point for defining concepts for its operation using the available information technologies and/or newly emerging technologies that are being gradually adopted in selected sectors of the economy.

## 5 A Typical PDP in Customised Furniture Production

The process of designing and manufacturing a custom-made furniture product was adopted for the research. It is a personalised manufacturing process intended to ensure a perfect fit into the premises (interior) where the furniture is to be used by the customer. This type of furniture production is a large and important segment of the industry and is performed by micro and small enterprises, usually on local markets. This process was identified at one business of that type. The definition of the product realisation process included careful tracing of customer contacts and IT system support. The product realisation process in the company selected for the study was presented in the form of a diagram shown in Fig. 1.

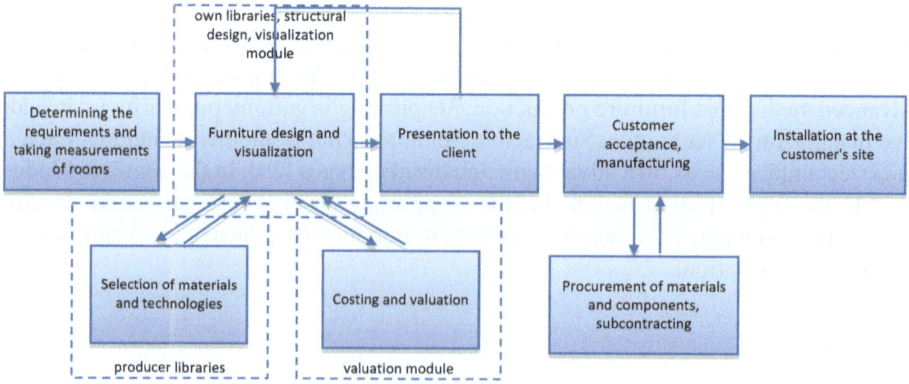


**Fig. 1.** A custom-made furniture production process.

The initial contact between the customer and the company representative can be treated as the beginning of the process. It can be a visit to the company's premises, a phone call or an e-mail. The customer is asked about the vision, preferences and expectations concerning the designed furniture and preliminary arrangements are made as to the order completion date. More often than not, the specific characteristics of the future require taking appropriate measurements at the customer's premises. In such cases, a mutually convenient measurement date is agreed. Additional guidelines are determined in relation to the designed furniture, a preliminary hand-drawn sketch of the design is made, and an approximate quotation for the product is provided.

Once the design concept is approved by the customer, the actual product design stage is can begin. The basic product modelling is carried out by the designer, who creates a model of the product on the basis of the provided guidelines and draws up its design documentation. Dedicated software for the furniture industry (KD Max) is used, based primarily on libraries of typical/normalised parts. Work at this stage requires taking many aspects into consideration in relation to construction, manufacturing techniques and cost estimation. At the analysed company, the designer communicates with the production engineer and the pricing specialist for this purpose. If there is a need for changes due to the selected of materials and technology, the production engineer provides verbal and/or written comments to the designer, who modifies the CAD geometric model.

When the presented form of the product meets the customer's expectations, the furniture design is sent to the production department. At this company, the production engineer determines the type and quantity of semi-finished products and decides which processing methods to choose, or alternatively uses the services of a subcontractor. The paper-based documentation is also used at this stage – the technical drawings of the parts to be machined. The company also uses CNC machine tools. In the case of machining simple parts, the NC program is generally created directly by the machine operator at the machine control panel. When processing of more complicated parts is required, such as complex ornamental milling patterns, the operator can use a separate computer workstation, where they generate a program and transfer it to the appropriate CNC machine.

During order fulfilment, the company often halts the production process if the customers change their preferences with regard to the functionality or look of the furniture they ordered. Therefore, such changes cause delays that extend the waiting period for the finished products. Dimensional changes in the designs also often occur in the case of furniture ordered during ongoing room renovations.

In the product realisation process at the company, standardised/typical elements are used to a large extent. The company's software provides broad support for the selection of such objects, including its own libraries. Clear difficulties arise in the case of non-standard elements, especially decorative ones, such as non-standard designs of handles, brackets, worktop milling, cabinet fronts, etc.

Customers are often unsatisfied with the proposed look of the furniture, finding particular surface structures, decorative patterns or colours unappealing. A known problem for the company is the proper visual representation of the designed products at the time when they are shown to the customer. The appearance of the finished product often differs from the look expected by the customer, indicated at the model presentation stage. This causes customer dissatisfaction and, in some cases, the need modify the design accordingly.

# 6 Improvement Concept

## 6.1 Specification of Proposed Modifications

Analysing the current product realisation process at the company (Fig. 1), it seems beneficial to increase the scope of interaction/collaboration between the manufacturer and the consumer at individual stages of production. From the point of view of improvement of information exchange, it is proposed to establish an efficient contact channel for two-way exchange of design data, including quick updates and visualisation of the product.

Figure 2 presents a model of the preparation process of furniture products, including the use of integrated product design and technology software supporting also selected manufacturing elements, as well as a proposed module for visual interaction with the customer and building a virtual model of the object in the premises where the customised furniture is to be installed.

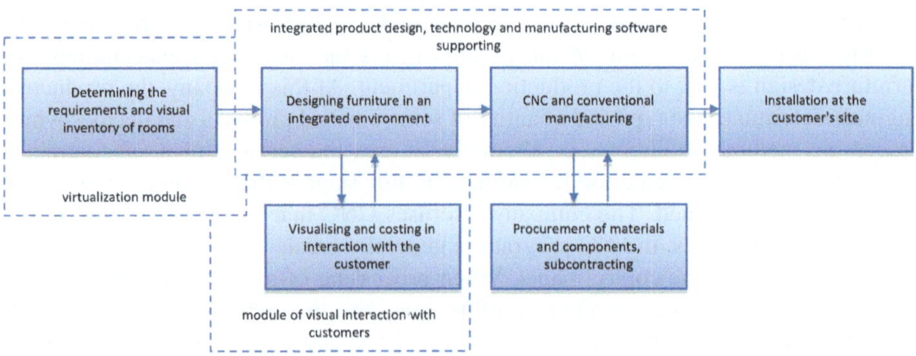


**Fig. 2.** A furniture production process enriched by technologies.

The concept of the process presented in Fig. 2 is based on supplementing the software environment that is currently used at the company by general purpose software with additional modules that enable an appropriate level of product virtualisation and visualisation. It is important to strengthen the connections between the particular stages of the process using digital-only information. This also applies to the stage at which the requirements are determined and the premises are measured. The use of appropriate technological and software solutions, such as special cameras, allows for an effective visual inventory of the premises and successful virtualisation of the project. The room virtualisation module provides not only a dimensioned geometric model of interior areas with their contents, in which the designed product will be installed, but also a full 3D graphic visualisation of the current appearance of rooms, including the predominant lighting conditions. Such a model is the only way to imitate the future user experience after the product is installed, and also allows the user to be involved in the production process. Currently, there are IT solutions for the construction of virtual room models and support in selected areas. For example, one of the simplest tools supporting the creation of virtual models is an inventory module with laser rangefinder support, which aids the designer in creating digital models.

By eliminating data transfer on paper, we can ensure uniformity of the digital model in the entire process of production setup and enable an attractive presentation of the product with efficient modifications at each stage of production. To some extent, the presented concept (Fig. 2) allows us to reduce the number of stages of product realisation in relation to those realised using traditional methods (Fig. 1). Ongoing presentation of progress for the customer and increasing the level of interaction during the design and quotation process enables the acceptance (approval) stage to be moved towards the beginning of the process, which offers more flexibility for introducing changes.

It is estimated that the solution has a large potential in relation to the production of non-standard articles of furniture, such as unusual decorative milling on furniture fronts or other custom decorative elements made to order.

Developing an interactive wizard and linking it to the design software can be an example of streamlining the data exchange process. Such a wizard would be provided to the customer, for example via the company website. This solution makes it possible to partially automate the uniform digital model creation process, which serves as the basis for quotations, construction documentation and planning the processing operations.

A PDP environment and a centralised database can be used to efficiently manage technical documentation. Such a database can be used for making product variations. This significantly reduces the time needed to search for previously created designs and supports making changes according to customer guidelines. These functions enable real-time collaboration between designers and production engineers. Operations in the CAM environment connected with generating CNC machining codes cane be performed along with designing the structural form of the product and verification of compliance with the technological process. Thus, generated program can be sent directly to the machine tools using DNC capabilities. The ability to simulate processing on the computer screen is an important aspect in this regard.

## 6.2 Software Selection

Today, the computer software market offers many solutions for the furniture industry. Specific products support production at the stages of development, preparation, supervision, and distribution of finished products. In each area, there are several alternative products with varying degrees of functionality. The software most often supports such functionalities as interior design with product libraries, quotation, optimisation of panel cutting and to preparation of CNC machine tool element processing programs.

The furniture industry seeks to satisfy individual needs of the customers, which means that custom-made products account for a large portion of production. The capabilities of dedicated furniture design software are constantly expanding, but general-purpose software is successfully used for manufacturing non-standard products. According to the consulting agency Apps Run The World, the worldwide PLM and engineering applications market is practically dominated by 10 companies, with an 80.5% share of the entire global product lifecycle management and engineering applications market [34]. These companies are: Siemens PLM Software, Dassault Systèmes (DPS), Synopsys,

Autodesk, Cadence, PTC, SAP-PLM, Ansys, Mathworks, and Oracle-PLM. PLM software was developed with particular emphasis on total product definition. The system's capabilities include management of design, engineering and simulation data, EBOM, and quality and document management. PLM is also designed to communicate engineering changes across the manufacturing and supply chain in real time.

The implementation of a high-end integrated PLM system requires the significant investments, which is why SMEs are usually not interested in solutions of this kind. On the other hand, general purpose mid-level modelling programs are popular among SMEs, such as: Solidworks, Inventor and Solid Edge. Given the adopted study area, these three most popular systems were taken into account for the purpose of the concept.

Such software offers basic functions ranging from 3D modelling, creation of 2D drawing documentation, support for material selection, handling of sheet metal and welded structures, carrying out simple simulations, integration of machining processes and enabling data import and export.

Due to the similar capabilities of these environments, it is difficult to clearly identify the right one. The choice of a particular solution is determined by the specific circumstances of the company, current and future needs, as well as the characteristics of the product itself. All things considered, the Inventor Professional software is worth taking into account in the context of PDPs in the furniture industry, which is analysed here. It offers a range of possibilities and maintains a high degree of flexibility. Although it is a general-purpose program, it is suitable for modelling most furniture design solutions. The software enables the parameterisation of models and also allows easy exchange of components, which makes it useful for designing non-standard variants of typical furniture products. In order to increase the functionality, it is suggested to use the additional module Woodwork for Inventor. After installation, this module becomes an integral part of Inventor software. It ensures preservation of all functionalities of the system and provides new tools and design methods. According to the software manufacturer, it allows to automate work and reduce the time needed to prepare technical documentation by up to 50%. Additionally, there is a possibility of cooperation of the programme with systems dedicated to optimisation of cutting panels. The program also provides extensive functionality for creating material structures and lists, and for interfacing with the database environment. The software is relatively user-friendly, with customisable interface and recurring processes to streamline tasks. Its large number of modules allows for integration of CAD, CAM and CAE processes, which significantly increases the productivity of production setup. It also enables real-time collaboration between different departments by integrating project data in one place.

## 7 Discussion

It can be generalised that many of the tasks performed during the furniture design process can be considered routine. Software specifically dedicated for the furniture industry is perfect for this purpose. However, there are a number of components which are designed from scratch with specific assumptions and design requirements in mind. In such cases, it is more efficient to use general purpose software. This also supports the realisation of more customised products. The paradigm of mass customisation currently prevails in the Industry 4.0 environment, Ding, et al. [22] conclude that mass customisation is difficult to achieve on a large scale, since fulfilling individual customer orders can affect production costs and overall complexity. At the same time, they show a need for new information technologies. These are the challenges of manufacturers moving towards mass customisation.

In industrial production environments, where manufacturing requires detailed product development, the time taken to prepare production greatly affects the delivery dates. Grijota et al. [35] present a case study that suggests a very high potential in reducing product delivery time by approximately 10–20% as a result of reduced PDP lead times. The traditional custom order processing approach currently used at the company is very time-consuming. Furniture design and the associated quoting prolong the lead time. Making changes requires repeating the customer – designer – production engineer – pricing specialist process loop, which is why in practice creating a new variant additionally extends the time of completion of finished products. In the case of using paper-based documentation, all the information provided to the customer and the information required for the production processes must be minimised to the simplest possible form.

Table 1 compares the characteristics of the PDP currently used at the company and expected characteristics with references to proposed concept.

Greater customer interaction in the product creation process allow the furniture design to be corrected quickly, accelerating the process and reducing the number of major revisions and delays resulting therefrom. It should be noted, however, that no typical CAx software is recommended for this task on the customer's side, since it requires special skills and high-end equipment, which makes it usable primarily for professionals.

Constraints on the design of non-standard products may be due to the machines owned by the company, such as different machine tool software systems. Using CAM modules eliminate this problem to a certain extent, since they have multiple integrated drivers for many manufacturers of equipment. The introduction of general purpose software can also reduce the number of subcontractors, which enables making some of the products in-house.

Scientific research provides proposals of methods that faithfully represent objects in virtual space. For example, Xu [36] and Fu, et al. [37] state that the problem of effectively simulating different interior designs and thereby establishing a coherent vision for the designer and customer can be solved using VR technology.

**Table 1.** Comparison of PDP characteristics.

| Characteristic variables | Current process | Proposed concept of the process |
|---|---|---|
| Overall process productivity | High time investment due to the information exchange used and designing custom furniture components | Reduced time spent on customer contacts and designing custom furniture components |

(*continued*)

**Table 1.** (*continued*)

| Characteristic variables | Current process | Proposed concept of the process |
|---|---|---|
| Degree of product customisation | Difficult implementation of non-standard products, e.g. in terms of ornaments, the need for subcontracting | Capability of manufacturing products that satisfy all the customer requirements |
| Process efficiency, making changes and corrections and the number thereof | A large number of dimensional and visual changes resulting from preliminary measurements of the room and unmet customer expectations with regard to the look must be taken into consideration | The number of changes and corrections in the final version of the product is reduced because of their ongoing implementation and better visualisation of the product |
| User involvement in the product design | Limited involvement, restricted to defining the general assumptions and amendments to the preliminary design | High level of involvement, with the ability to suggest changes at any stage of the product |
| Meeting the expectations of users/ordering parties | Fully satisfying customer expectations is difficult, especially in terms of the visual aspect of the finished product and on-time delivery | Fully satisfying customer expectations is easier, including on-time delivery. High customer involvement |
| Current process implementation costs (excluding materials and overheads, etc.) | Coordination of activities in the process is costly, with downtime between activities, making corrections and changes, also after the installation of the product in the intended interior | Much more efficient coordination ensures lower ongoing costs of the process, reducing rework |
| Capital expenditure | Moderate inputs, known and widely used technologies | Advanced software requires more resources; selected module functions are not yet available, high acquisition costs are to be expected |

Swaminathan et al. [38] show methods to compute textured 3D models and measure real world objects based on single images that clearly reflect the impression of objects for the purposes of furniture design. Other authors [39] evaluate the results of interior 3D scanning obtained with Matterport, relatively popular software. Although the results are very promising, the study a number of mapping quality shortcomings. It should be noted that Matterport is a leader in building virtual twins of architectural objects, but is not yet used on a larger scale in the design of custom-made furniture.

Freitag et al. [40] present a use case that enables employees to select new office furniture using VR/AR technology, stating that the use of VR technology is now becoming increasingly important for manufacturing companies. They help increase customer interaction in the product development process and reduce market risk.

Canvas provided by tech start-up Occipital is another interesting software for creating 3D models using a camera connected to a tablet, or even just with the smartphone. Canvas allows users to scan rooms with the camera and send the recorded model in CAD format for further use in environments using this standard. There is plenty evidence that this solution is not yet fully stable, but it can nonetheless be an important experience and a step towards easy, intuitive and highly reproducible virtual room models for the furniture industry. There is reason to believe that a solution consistent with the assumptions set out in this article that is ready for a wide-scale deployment will appear in the near future.

## 8 Conclusions

Small furniture manufacturers support the coverage of the existing demand for custom-made products, which fits well with the current trend towards mass customisation and one of the paradigms of Industry 4.0.

The introduction of additional general-purpose software with additional modules that can be adapted to the needs of the furniture industry into the company's processes can greatly simplify and better synchronise product realisation.

The presented concept allows for efficient production of primarily non-standard products, also using standard components. The key is to increase customer interaction and streamline the data exchange during product development, which enables the project requirements to be quickly defined and makes it easy to make corrections, which in turn accelerates the process, reduces delays and increases customer satisfaction. However, increasing customer interaction requires the development of IT solutions that are more accessible to the user without requiring software-specific training and engineering skills in the area of design.

It seems that the direction of development is to ensure full integration of general purpose software with the currently used dedicated software for the furniture industry, therefore maximising the use of unified components and typical materials and technological processes.

In the future, more emphasis should be placed on developing methods that more faithfully represent the model in the existing interior, for example using AR/VR technologies.

**Acknowledgments.** This research was funded by a grant from the Ministry of Education and Science received by the Bialystok University of Technology, grant number WZ/WIZ-INZ/3/2019.

## References

1. Letens, G., Farris, J.A., Van Aken, E.M.: A multilevel framework for lean product development system design. Eng. Manag. J. **23**(1), 69–85 (2011)

2. Zheng, P., Chen, C.H., Shang, S.: Towards an automatic engineering change management in smart product-service systems–a DSM-based learning approach. Adv. Eng. Inform. **39**, 203–213 (2019)
3. Liu, W., Zeng, Y.: Conceptual modeling of design chain management towards product lifecycle management. In: Chou, S.Y., Trappey, A., Pokojski, J., Smith, S. (eds.) Global Perspective for Competitive Enterprise. Economy and Ecology, pp. 137–148. Springer, London (2009). https://doi.org/10.1007/978-1-84882-762-2_13
4. Ardolino, M., Rapaccini, M., Saccani, N., Gaiardelli, P., Crespi, G., Ruggeri, C.: The role of digital technologies for the service transformation of industrial companies. Int. J. Prod. Res. **56**(6), 2116–2132 (2018). https://doi.org/10.1080/00207543.2017.1324224
5. Sklyar, A., Kowalkowski, C., Tronvoll, B., Sörhammar, D.: Organizing for digital servitization: a service ecosystem perspective. J. Bus. Res. **104**, 450–460 (2019). https://doi.org/10.1016/j.jbusres.2019.02.012
6. Ferreira, J.J., Fernandes, C.I., Ferreira, F.A.: To be or not to be digital, that is the question: firm innovation and performance. J. Bus. Res. **101**, 583–590 (2019). https://doi.org/10.1016/j.jbusres.2018.11.013
7. Meißner, M., Jacobs, G., Jagla, P., Sprehe, J.: Model based systems engineering as enabler for rapid engineering change management. Procedia CIRP **100**, 61–66 (2021). https://doi.org/10.1016/j.procir.2021.05.010
8. Riedel, R., Jacobs, G., Wyrwich, F., Siebrecht, J.: Identification of dependencies between product parameters and process stakeholders. Procedia CIRP **100**, 247–252 (2021). https://doi.org/10.1016/j.procir.2021.05.063
9. Huikkola, T., Kohtamäki, M., Rabetino, R., Makkonen, H., Holtkamp, P.: Overcoming the challenges of smart solution development: co-alignment of processes, routines, and practices to manage product, service, and software integration. Technovation, 102382 (2021). https://doi.org/10.1016/j.technovation.2021.102382
10. Hsuan, J., Jovanovic, M., Clemente, D.H.: Exploring digital servitization trajectories within product–service–software space. Int. J. Oper. Prod. Manag. **41**(5), 598–621 (2021). https://doi.org/10.1108/IJOPM-08-2020-0525
11. Kohtamäki, M., Parida, V., Oghazi, P., Gebauer, H., Baines, T.: Digital servitization business models in ecosystems: a theory of the firm. J. Bus. Res. **104**, 380–392 (2019). https://doi.org/10.1016/j.jbusres.2019.06.027
12. Kuhlman, T., Farrington, J.: What is sustainability? Sustainability **2**(11), 436–3448 (2010). https://doi.org/10.3390/su2113436
13. Persson, J.-G.: Current trends in product development. Procedia CIRP **50**, 378–383 (2016). https://doi.org/10.1016/j.procir.2016.05.088
14. Choi, S., Jun, C., Zhao, W.B., Do Noh, S.: Digital manufacturing in smart manufacturing systems: contribution, barriers, and future directions. In: Umeda, S., Nakano, M., Mizuyama, H., Hibino, H., Kiritsis, D., von Cieminski, G. (eds.) APMS 2015. IAICT, vol. 460, pp. 21–29. Springer, Cham (2015). https://doi.org/10.1007/978-3-319-22759-7_3
15. Zweber, J.V., Kolonay, R.M., Kobryn, P., Tuegel, E.J.: Digital thread and twin for systems engineering: requirements to design. In: 55th AIAA Aerospace Sciences Meeting (2017). https://doi.org/10.2514/6.2017-0875
16. Wynn, D.C., Clarkson, P.J.: Process models in design and development. Res. Eng. Design **29**(2), 161–202 (2017). https://doi.org/10.1007/s00163-017-0262-7
17. Wang, R., et al.: Ontology-based representation of meta-design in designing decision workflows. ASME J. Comput. Inf. Sci. Eng. **19**(1), 011003 (2019). https://doi.org/10.1115/1.4041474
18. Riezebos, J., Klingenberg, W.: Advancing lean manufacturing, the role of IT. Comput. Ind. **60**(4), 235–236 (2009). https://doi.org/10.1016/j.compind.2009.01.005

19. Hamrol, A., Gawlik, J., Sładek, J.: Mechanical engineering in Industry 4.0. Manag. Prod. Eng. Rev. **10**(3), 14–28 (2019). https://doi.org/10.24425/mper.2019.129595
20. Varl, M., Duhovnik, J., Tavčar, J.: Customized product development supported by integrated information. J. Ind. Inf. Integr., 100248 (2021). https://doi.org/10.1016/j.jii.2021.100248
21. Iaksch, J.S., Borsato, M.: Method for digital evaluation of existing production systems adequacy to changes in product engineering in the context of the automotive industry. Adv. Eng. Inform. **42**, 100942 (2019). https://doi.org/10.1016/j.aei.2019.100942
22. Ding, J., Wang, M., Zeng, X., Qu, W., Vassiliadis, V.S.: Mass personalization strategy under Industrial Internet of Things: a case study on furniture production. Adv. Eng. Inform. **50**, 101439 (2021). https://doi.org/10.1016/j.aei.2021.101439
23. Boyes, H., Hallaq, B., Cunningham, J., Watson, T.: The industrial internet of things (IIoT): an analysis framework. Comput. Ind. **101**, 1–12 (2018). https://doi.org/10.1016/j.compind.2018.04.015
24. Conti, E., Chiarini, A.: Design-driven innovation: exploring new product development in the home appliances and furniture industry. TQM J. **33**(7), 148–175 (2021). https://doi.org/10.1108/TQM-12-2020-0313
25. Qiu, X., Cano-Kollmann, M., Mudambi, R.: Competitiveness and connectivity in design innovation: a study of Norwegian furniture industry. Compet. Rev. **27**(5), 533–548 (2017). https://doi.org/10.1108/CR-03-2017-0025
26. Verganti, R.: Design-Driven INNOVATION: Changing the Rules of Competition by Radically Innovating What Things Mean. Harvard Business Press, Boston (2009)
27. De Goey, H., Hilletofth, P., Eriksson, L.: Design-driven innovation: a systematic literature review. Eur. Bus. Rev. **31**(1), 92–114 (2019). https://doi.org/10.1108/EBR-09-2017-0160
28. Papadopoulos, I., et al.: Market potential and determinants for eco-smart furniture attending consumers of the third age. Compet. Rev. **26**(5), 559–574 (2016). https://doi.org/10.1108/CR-06-2015-0058
29. Ma, J., Li, Z., Zhao, Z.-L., Xie, Y.M.: Creating novel furniture through topology optimization and advanced manufacturing. Rapid Prototyping J **27**(9), 1749–1758 (2021). https://doi.org/10.1108/RPJ-03-2021-0047
30. Creswell, J.W.: Research Design: Qualitative, Quantitative, and Mixed Methods Approaches. Sage, Thousand Oaks (2013)
31. Thomas, G.: Doing case study: abduction not induction, phronesis not theory. Qual. Inq. **16**, 575–582 (2010)
32. Grant, C., Osanloo, A.: Understanding, selecting, and integrating a theoretical framework in dissertation research: creating the blueprint for your 'House.' Adm. Issues J. Connect. Educ. Pract. Res. **4**(2), 12–22 (2014). https://doi.org/10.5929/2014.4.2.9
33. Peshkin, A.: The goodness of qualitative research. Educ. Res. **22**(2), 23–29 (1993)
34. APPSRUN, Pang, A., Markovski, M., Markovska, A.: Top 10 PLM and Engineering Software Vendors. Market Size and Market Forecast 2019–2024 (2021). https://www.appsruntheworld.com/top-10-product-lifecycle-management-engineering-software-vendors-and-market-forecast/. Accessed 21 Nov 2021
35. Grijota, C.G., Acero, R., Yagüe-Fabra, J.A.: Product development methodology 'scalability.' Procedia CIRP **100**, 571–576 (2021). https://doi.org/10.1016/j.procir.2021.05.125
36. Xu, X.: Interior design decoration simulation based on wireless sensors. Microprocess. Microsyst. **82**, 103813 (2021). https://doi.org/10.1016/j.micpro.2020.103813
37. Fu, Q., Chen, X., Wang, X., Wen, S., Zhou, B., Fu, H.: Adaptive synthesis of indoor scenes via activity-associated object relation graphs. ACM Trans. Graph. **36**(6), 1–13 (2017). https://doi.org/10.1145/3130800.3130805
38. Swaminathan, R., Schleicher, R., Burkard, S., Agurto, R., Koleczko, S.: Happy measure: augmented reality for mobile virtual furnishing. Int. J. Hum. Comput. Interact. **5**(1), 16–44 (2013). https://doi.org/10.4018/jmhci.2013010102

39. Virtanen, J-P., et al.: Depth camera indoor mapping for 3D virtual radio play. Photogram. Rec. **33**(162), 171–195 (2018). https://doi.org/10.1111/phor.12239
40. Freitag, M., Westner, P., Schiller, C., Nunez, M.J., Gigante, F., Berbegal, S.: Agile product-service design with VR-technology: a use case in the furniture industry. Procedia CIRP **73**, 114–119 (2018). https://doi.org/10.1016/j.procir.2018.03.305

# Pick Performance System as an IT Support for Order Completing – A Case Study

Norbert Kawęcki[1](✉) and Arkadiusz Gola[2]

[1] University of Maria-Curie Sklodowska, Plac Marii Curie-Skłodowskiej 5, 20-031 Lublin, Poland
n.kawecki@googlemail.com
[2] Department of Production Computerisation and Robotisation, Lublin University of Technology, ul. Nadbystrzycka 36, 20-618 Lublin, Poland

**Abstract.** The main goal of the article is to present a Pick Performance System (PPS) as an IT support used for order complementing in an example enterprise. The system helps to keep productivity of employees on a required level and in worst case scenario it leads to dismissal of employee. It is accurate and provides data necessary to manage big distribution center. The first part of the work is overview of the literature written in this matter. Second part will present characteristics of the order complementation process in the analyzed enterprise. Third part will show characteristic of the IT support system for the order complementation process. The last part of the article will introduce summary and conclusion.

**Keywords:** Logistics · Pick performance · Order complementing · IT system · Support

## 1 Introduction

Logistic centers, distribution centers and warehouses are important parts of logistic systems, and the success of the their operations affects the entire supply chain performance. In warehouse operations, order picking is a very costly process and should be carried out efficiently [1]. In terms of operating costs, the process of picking customer orders is the most important activity that takes place in distribution warehouses [2]. The picking is very laborious, forming up to 55% of the costs associated with the warehouse operation, and up to 50% of the picking time is consumed by the movement of operators to reach goods localized at picking positions [3, 4]. To make the pick operations more efficient companies introduce more and more often both automated and robotized equipment [5, 6] and more and more sophisticated IT systems [7, 8].

It must be emphasized that designing orders picking systems, including IT support systems is one of the most difficult roles of logistics planning because its considerable complexity [9]. The system performance is depended on availability of resources such warehouse racking, transport and handling equipment and also of availability of labor force [10–12]. The effectiveness of orders picking is ensured by application of the specific information systems such as the warehouse management system (WMS) [3]. Such class


J. Trojanowska et al. (Eds.): MANUFACTURING 2022, LNME, pp. 105–115, 2022.
https://doi.org/10.1007/978-3-030-99310-8_9

of systems provides data and information enabling control of the goods flow from receipt to dispatch. As a consequence such class of systems become the basis for the other departments, such as purchase, distribution and sales [13]. To make the proper level of efficiency all information systems in the company (also WMS system) must be connected and able to communicate with other information systems. [14]. Unfortunately classical WMS systems usually do not provide support for orders picking processes [15, 16].

The goal of this paper is to present an original Pick Performance System (PPS) developed directly for the purposes of a specific logistic center located in England. The provided study presents both the functionality and also advantages and disadvantages of the IT system dedicated for picking processes support.

## 2 Literature Analysis

The term "peak performance" was firstly used for describing mentoring approach an individually-based development and that allows to raise the performance of individuals. This phrase was widely used in in sales, that is typical individual team activity. However, relatively quickly this concept has been also extended to team-based working, and organizational [17].

In the 1980-s, C. Garfield defined six aptitudes and capacities of peak performers [18]:

- mission that motivate to act,
- results obtained in real time,
- self-management and self-mastery,
- team building, team operation and team playing,
- correction of the course,
- continuous change management.

Pan, Shih, Wu and Lin presented in their work a storage assignment problem (SAP) [19]. This work emphasizes that it is not easy to find an effective way for proper localization of products in a warehouse to improve the operational efficiency of order picking process. Because the storage assignment is belongs to the group of NP-hard problems, many different methods and algorithms have been proposed. At the beginning researches usually focused on picker-to-parts warehousing systems. Many of them also dealt with the problems of automated storage and retrieval systems for such purposes. However, it must be noted that pick-and-pass systems play an important role and allow faster delivery of small and frequent orders. In this aspect, especially two factors lead to idle time of pickers: imbalance of picking line and products' shortage replenishment. In particular the Authors developed a heuristic method based on genetic algorithms that allows to solve SAP for a pick-and-pass system with multiple pickers. This method gives opportunity to balance the workload of each picking zone and to determine the appropriate storage space for each product. In consequence the system's performance can be improved. The proposed heuristic algorithm was implemented as a simulation model using FlexSim software and was used to compare the throughput for different storage assignment methods. The obtained results proved the efficiency of the proposed heuristic policy in a pick-and-pass system.

R. de Koster presented in his work the analysis of pick-to-belt order picking system [20]. In particular he developed an approximation method that allows to obtain rapid insight in the performance of pick-to-belt order picking systems. This method can be used for evaluation the effects of changing the layout of the system. In particular the property of the number of pickers, the number of picking stations, the conveyor speed, the number of order lines per bin, the number of bins to be processed per day, etc. can be evaluated. This method can be preferably used when goods are prepacked in cartons: in such case barcodes are easily attached to cartons and therefore can be easily read by automated stations (Fig. 1).

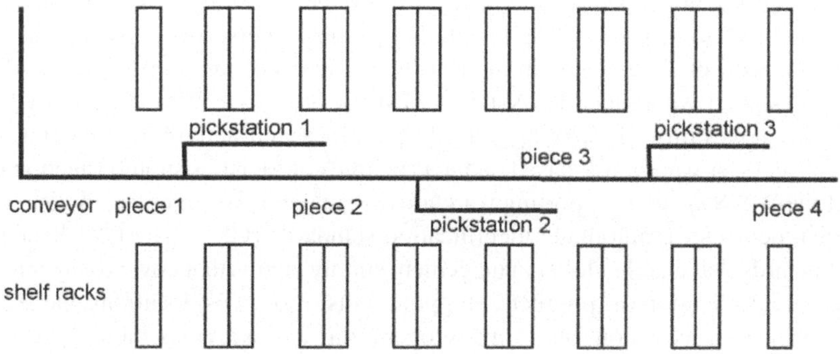


**Fig. 1.** Process of picking by PPS [20].

Du et al. described the process of order complementing in an example enterprise [21]. They present, that the order is the foundation of the information flow in the logistics system. The order fulfillment cycle consists of four basic activities: placing an order, processing orders, preparing the ordered goods and shipping ordered products. The first phase is to place the order. The time in this case depends on the choice of the folding method orders, which means that this stage may last from several days when placing an order by mail, to several minutes when processing by telephone, or even up to a few seconds when using the method electronic data interchange. Nowadays, thanks technological progress, more and more enterprises are taking advantage from applications that allow you to place orders online, which results shortening the order fulfillment cycle and increasing quality customer service. Order processing includes pre-treatment and processing, i.e. all activities related to checking terms of delivery (price, method of delivery), the customer's creditworthiness and credibility, providing information to sales register, checking the availability of goods and sending orders to the inventory storage area of the warehouse. On this at the level, the carrier, means and route of transport are selected, the route of transport, costs and delivery date are determined, and the necessary documents related to the shipment of goods are prepared. Thanks EDI systems can perform several activities at the same time. Preparation of ordered goods can be a process simple or complex, depending on the type of goods prepared for shipment. In this phase, the warehouse documents are issued and the goods are summarized in the warehouse. Completion The load can be performed manually or in an automated manner, using auxiliary means and devices, for example forklift trucks, which significantly affects the time of completion the entire process. The last element in the order fulfillment cycle the customer is shipping the ordered goods. This stage covers the period from placing the load on the means of transport until the order is delivered to the recipient and the goods are unloaded on site destination. Success in this episode largely depends on the qualifications of shippers and carriers. Companies today, striving to maintain their position on the market place great emphasis on the course of the contract that consists of many important factors. In order to achieve results companies aim to create and execute the perfect order. Execution of a perfect order, i.e. delivered in full, on time, along with a complete, correctly completed documentation, is possible thanks to the optimization of logistics and transport processes, i.e. the introduction of solutions that will improve the course of the process in the best possible way. Optimization is made due to certain factors, among which they belong others: time, cost, quality.

Papoutskdaki et al. describe in his article standard IT warehouse systems [22]. A properly designed warehouse IT system - MSI (or otherwise WMS) should take into account the possibility of carrying out all operations and activities, occurring in the storage process. A warehouse based on the principles of logistics should function on the basis of the WMS system supporting warehouse management, which is efficient uses IT and global devices for automatic identification standards (GS1). The GS1 System is a set of standards that enable global management supply chains that cover many industries by unique identification of: products, shipping units, resources, locations and services. The warehouse process consists of many operations and activities taking place in the warehouse, which are related to the four main phases of this process:

- receiving,
- storage,
- picking,
- issuing.

In the IT-managed warehouse, each activity is registered in the system information technology and the course of work is recorded and updated on the present day document used [23]. Completing the activity is also there confirmed in the IT system. The IT warehouse system of the WMS type primarily tells us and reminds the warehouse keeper of the appropriate operating procedures during implementation warehouse activities. WMS system, taking into account any possible use logistic criteria, he supervises and improves the work of a warehouse keeper by indicating, for example:

- symbols of storage places of individual logistic units in the warehouse during receptions and releases,
- ways of issuing goods during picking,
- the order in which the assortments are released from the warehouse.

Technically, the installation model of the WMS system, assisted by techniques ADC covers all possible types of methods and devices for remote communication with the use of: scanners, batch or radio terminals, type computers PalmTop using various communication techniques (radio, mobile telephony, Internet). In order for it to function well, it must be provided with appropriate technical means and facilities; in the analyzed company such devices are [24]:

- computer terminals installed on forklifts,
- barcode readers, placed on the handles for safety self-returnees,
- four access points, placed in different parts of the warehouse for the purpose ensuring a network signal across the entire warehouse area,
- receivers on forklifts, placed to ensure communication a terminal on a forklift truck with an access point transmitter,
- two printers for printing customer address labels and pallet labels, informing about the articles and the numbers of these articles placed on a given pallet,
- printer for printing warehouse documents,
- computer network ensuring connectivity with the server.

Wasilewski and Wasilewski described in their article integration of IT systems in an e-commerce farm which includes [25]:

- sending orders from the online store to the warehouse system,
- sending information about the payment status from the payment operator to the online store,
- sending information about the availability of products from the warehouse system to online store,
- sending order fulfillment statuses from the warehouse system to the online store,
- sending a request to generate a consignment note and collecting a consignment note label from the carrier's system,
- sending information about sales to the financial and accounting system,
- sending information about changes in stock levels,
- sending information about product price changes.

Hybrid e-commerce farm integration solutions using ESB and iBPMS are already used in practice, e.g. by Fast White Cat SA, e-commerce farm operator, operating several online stores, incl. Tchibo Cafissimo, Primamoda, Kari. Currently Fast White Cat, thanks to the combination of muleESB and IBM BPM 8.0.1, is able to ensure high quality customer service and reliability of integration, even with several thousand shipments per day.

According to author the hybrid architecture of integration, using the ESB bus and the iBPMS class system, is a response to the specific requirements set for integration of many different IT systems for the purposes of running e-commerce farm. This solution has the advantages of both combined approaches, i.e. it is fast when data transfer speed matters, it is trustworthy and reliable, and it gives you full control over multi-instance support various business processes of end customer service. However, it should be remembered that such a solution is not - and probably will not - be available to small companies, because the cost of implementing an iBPMS class system is large and it pays off investment in such a system requires time and a large scale of operation. Hence, it should be assumed that the described hybrid integration architecture will only be used by the largest and most dynamic e-commerce farms [25].

## 3 Characteristics of the Order Complementation Process in the Anlayzed Enterprise

The presented study was provided in a distribution center located in England. The study was divided into three stages (Fig. 2).

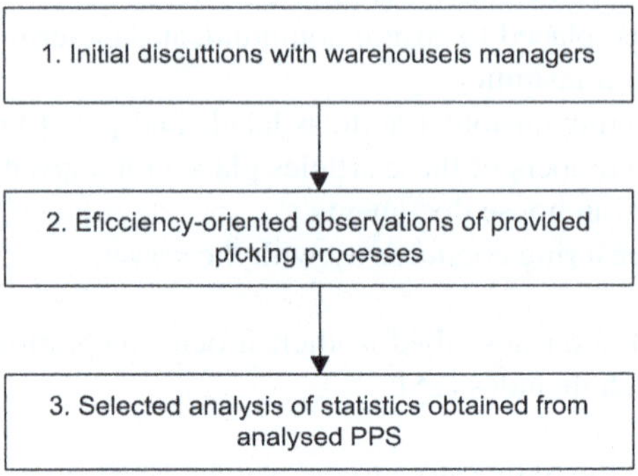


**Fig. 2.** A logic diagram of provided study.

The analyzed process of orders picking is closely related to Pick Performance System. It contains the following five stages (Fig. 3):

1. label scanning by scanner whilst picking,
2. after scanning label scanner shows the location of goods,
3. the whole order is being picked on the pallet,
4. the ready pallet is ready to be put away,
5. the process of order is closed in the scanner and all data are saved for pick performance

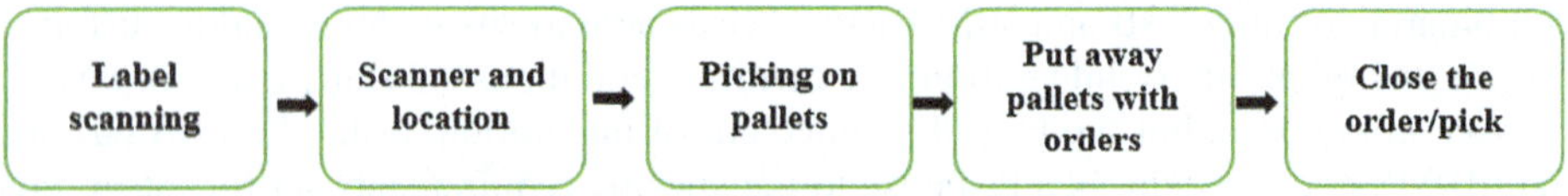


**Fig. 3.** Process of orders picking by PPS.

Every single employee, before starting work on its shift, receives a scanner signed on issue sheet. Once the scanner is on, an employee goes to pick desk where he/she gets label with order. The label is being scanned and in the scanner location for single item to be picked is being shown. The employee starts picking directed by info on a scanner screen. Picking aisles and the route (Fig. 4) is designed to improve picking service.

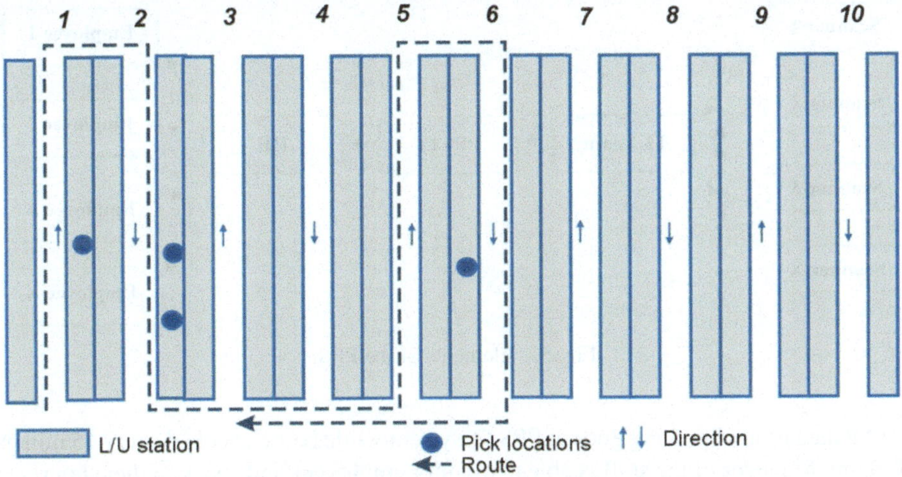


**Fig. 4.** Process of picking by PPS.

All picks are being picked on the pallets. Once the order is ready it is being palletized and put away to the proper location for each store. Completed order is being closed and the label with barcode is being issued on each pallet. After all pallets are being put away and loaded by forklift driver to proper truck ready to go to store.

## 4 Characteristics of the IT Support System for the Order Completing Process

Pick Performance System (PPS) is an IT system which was especially designed for management employees in DHL in Great Britain. The system helps manager to monitor and manage employees based on empirical data coming from scanners whilst employees doing picking.

Pick Performance System PPS was especially designed to manage people in big warehouse distribution in United Kingdom. The place employed in total around 700 staff, which around 85% was managed by PPS. PPS is an IT system which with connection of scanners and IT team helps to monitor employees whilst they preparing picking for orders. It is being used in put away, colleting orders in/out, stock taking and tagging.

PPS consists of following elements (Fig. 5):

- scanners,
- IT team,
- PPS IT,
- HR Team,
- employees.

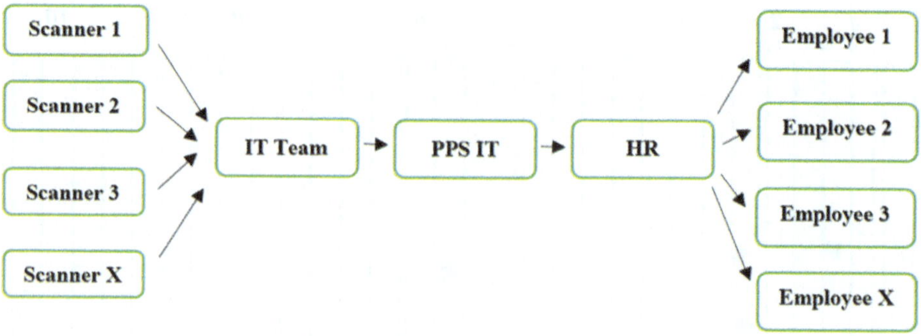


**Fig. 5.** Elements of the PPS.

All data base of picking goes to PPS IT system while being checked every 15 min by IT Team. Manager of the staff is able to monitor employees and speak to them hourly to increase productivity on every shift. If the issue is much more serious then it goes to HR Department to take an example employee on Pick Performance Management Procedure. For this process PPS IT is the most important tool. It helps to generate data base about employee or employees on every shift (morning AM, afternoon PM, night NI). By input initials of employee it comes up empirical info about pick performance measures hourly, daily, weekly and monthly. The system helps to generate data base immediately so the manager can monitor staff on each shift properly (Fig. 6).

| Employee Name | Initials | | | Pick | Performance | Sheet | | | |
|---|---|---|---|---|---|---|---|---|---|
| | | Mon | Tue | Wed | Thur | Fri | Sat | Sun | Total |
| Adam Smith | AS | 97% | 101% | 98% | 99% | 102% | 103% | 97% | 99.6% |
| Piotr Kowalski | PK | 101% | 100% | 99% | 102% | 103% | 105% | 104% | 102% |
| Eve Kulciene | EK | 98% | 99% | 102% | 103% | 104% | 101% | 96% | 100.4% |
| Rene Isu | RI | 96% | 104% | 103% | 100% | 99% | 98% | 97% | 99.6% |
| Susie Tarry | ST | 102% | 104% | 96% | 94% | 87% | 88% | 93% | 94.9% |
| ... | | | | | | | | | |

**Fig. 6.** Pick performance sheet.

Once received individual data the PPS IT system affords to develop empirical information about employees on each shift. This data helps to create reports, assess employees and provide accurate information's both for supplier and customer (Fig. 7). It helps to manage employees and in recruitment and selection process to do improvement.

Daily total Pick Performance (Fig. 8) helps to create weekly and monthly reports. It is used for improvement the whole service for both client and supplier. Manager can make decision in improvements in picking as whole piece in the warehouse.

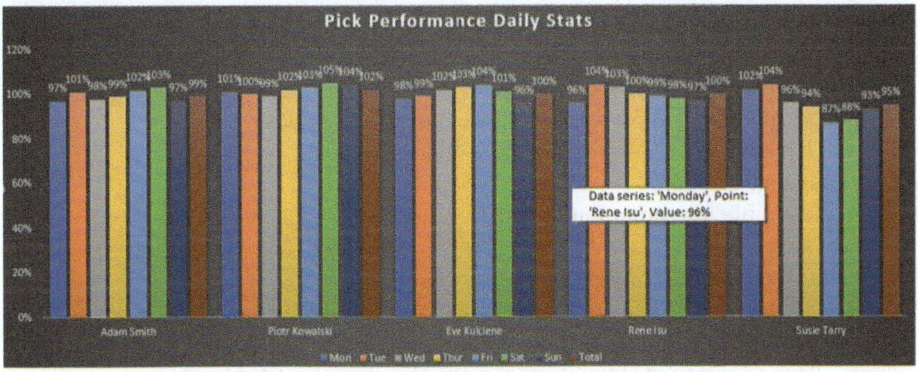


**Fig. 7.** Pick performance daily stats.

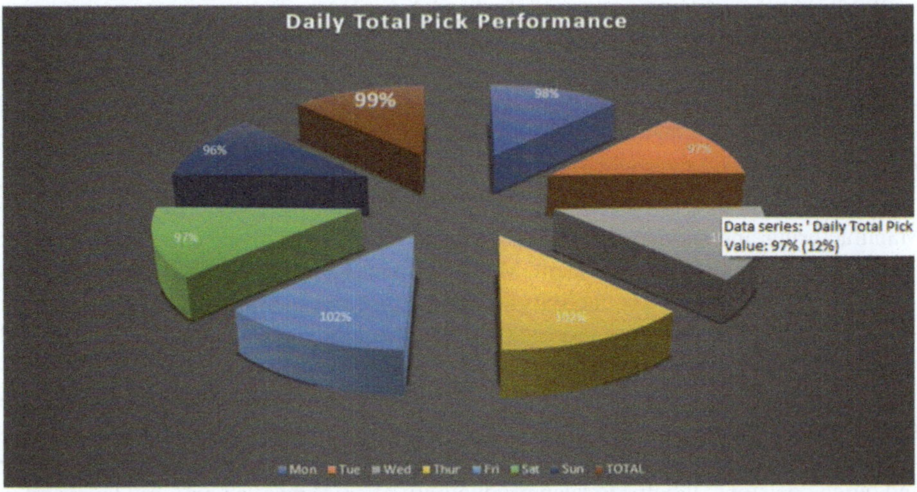


**Fig. 8.** Daily total pick performance.

All statistics generated by the PPS are very useful for managers and allows to optimize the completion process. They supports the process of defining individual norms and to make the motivation procedures for employed people.

## 5 Advantages and Disadvantages of PPS System

The Pick Performance System PPS has advantages and disadvantages.

As advantages it can be pointed out as follows:

- hourly, daily, weekly and monthly employees monitoring,
- empirical data for disciplinary procedure,
- based on PPS creating weekly reports for Top Management,
- easy and fast service by login,

- update of empirical data every 15 min,
- failure rate close to zero,
- large selection of empirical data,
- data necessary to solve pick performance employees issues,
- pie and bar charts,
- very easy data reading,
- clear and simple interpretation of data.

Identified disadvantages for PPS are as below:

- lack of employees photos which in similarity of names can cause a mistake,
- skips – lack of goods in stock decreases employee pick performance,
- special tasks are excluded in PPS which decreases employee pick performance and manager has to monitor and access employee individually,
- PPS serves only for measuring employee pick performance,
- lack of other options – necessity of using other additional systems,
- PPS do not shows employee location,
- PPS do not show scanner number used by employee.

## 6 Summary and Conclusions

Manufacturing and service companies confront different challenges to fulfil customer orders and distinguish between them. Many companies consider reducing and eliminating costs and improving productivity in their logistic system important improvement activities. Moreover more and more sophisticated equipment and software is developed and implemented to increase the level of effectiveness of provided processes. The combination of human capital management in outsourcing in cooperation with an IT system specially designed for this type of activity is an element of supporting effective management for the manager and the speed of access to information as well as making accurate and quick decisions. It helps to keep productivity of employees on required level and in worst case scenario it leads to dismissal of employee. The system is accurate and provides data necessary to manage big distribution center. Continues improvement through IT systems should be developed to manage big volume of people whilst achieving targets.

In future research we are going to provide more detailed analysis of the presented PPS system. In particular we will try to focus on its weaknesses and identify problems and bottlenecks that could be eliminated. Moreover, the future works that are going to improve the system are planned.

## References

1. Cergibozan, C., Tasan, A.S.: Genetic algorithm based approaches to solve the order batching problem and a case study in a distribution center. J. Intell. Manuf. **33**, 137–149 (2022)
2. Tompikins, J.A., White, J.A., Bozer, Y.A., Tanchoco, J.M.A.: Facilities Planning. Wiley, New York (2010)
3. Kasparova, P., Dyntar, J.: Effective designing of order picking systems using dynamic simulation. Acta Informatica Pragensia **10**(1), 108–120 (2021)

4. De Koster, R., Le-Duc, T., Roodbergen, K.J.: Design and control of warehouse order picking: a literature review. Eur. J. Oper. Res. **182**(2), 481–501 (2007)
5. Stączek, P., Pizoń, J., Danilczuk, W., Gola, A.: A digital twin approach for the improvement of and autonomous mobile robots (AMR's) operating enviroment – a case study. Sensors **23**(21), 7830 (2021)
6. Musiał, K., Kotowska, J., Górnicka, D., Burduk, A.: Tabu search and greedy algorithm adaptation to logistic task. In: Saeed, K., Homenda, W., Chaki, R. (eds.) CISIM 2017. LNCS, vol. 10244, pp. 39–49. Springer, Cham (2017). https://doi.org/10.1007/978-3-319-59105-6_4
7. Xuan, L.: Computer aided optimization of picking system in logistics distribution center. Comput. Aided Des. Appl. **19**(S4), 46–55 (2022)
8. Brzozowska, J., Gola, A.: Computer aided assembly planning using MS Excel software – a case study. Appl. Comput. Sci. **17**(2), 70–89 (2021)
9. Altarazi, S.A., Ammouri, M.M.: Concurrent manual-order-picking warehouse design: a simulation-based design of experiments approach. Int. J. Prod. Res. **56**(23), 7103–7121 (2018)
10. Gu, J., Goetschalckx, M., McGinnis, L.F.: Research on warehouse operation: a comprehensive review. Eur. J. Oper. Res. **177**(1), 1–21 (2007)
11. Kłosowski, G., Gola, A., Thibbotuwawa, A.: Computational Intelligence in control of AGV multimodal systems. IFAC PapersOnline **51**(11), 1421–1427 (2018)
12. Borucka, A.: Empirical analysis of transportaiton systems availability using the semi-Markov process. In: Proceedings of the 29th European Safety and Reliability Conference ESREL 2019, Hannover, pp. 834–838 (2020)
13. Relich, M., Nielsen, I.: Estimating production and warranty cost at the early stage of a new development project. IFAC PapersOnLine **54**(1), 1092–1097 (2021)
14. Bocewicz, G., Nielsen, I., Gola, A., Banaszak, Z.: Reference model of milk-run traffic systems prototyping. Int. J. Prod. Res. **59**(15), 4495–4512 (2021)
15. Guo, X., Chen, R., Du, S., Yu, Y.: Storage assignment for newly arrived items in forward picking areas with limited open locations. Transp. Res. Part E Logistics Transp. Rev. **151**, 102359 (2021)
16. Keshavarz, A.R.A., Jaafari, D., Khalaj, M., Dokouhaki, P.: A survey of the literature on order-picking systems by combining planning problems. Appl. Sci. **11**(22), 10641 (2021)
17. https://www.ims-productivity.com/page.cfm/content/Peak-Performance. Accessed 30 Apr 2021
18. Garfield, C.A.: Peak Performance: The New Heroes of American Business. William Marrow Paperbacks, California (1986)
19. Pan, J.C.H., Shih, P.H., Wu, M.H., Lin, J.H.: A strorage assignment heuristic method based on genetic algorithm for a pick-and-pass warehousing system. Comput. Ind. Eng. **81**, 1–13 (2015)
20. De Koster, R.: Performance approximation of pick-to-belt order picking systems. Eur. J. Oper. Res. **72**(3), 558–573 (1994)
21. Du, M., Wang, X., Geng, J., Ye, L.: A review of the optimization of the manual picking system. ACM Int. Conf. Proc. Ser. **175046**, 661–6666 (2021)
22. Papoutskikakis, M., Sigala, M., Simeonaki, E., Tseles, D.: Innovative IT systems for material management in warehouses. In: AIP Confrence Proceedings, vol. 1872, p. 020010 (2017)
23. Gajsek, B., Dukic, G., Kovacic, M., Brezocnik, M.: A multi-objective genetic algorithms approach for modelling of order picking. Int. J. Simul. Model. **20**(4), 719–729 (2021)
24. Klumpp, M., Gesenius, M., Meyer, O., Ruiner, C., Gruhn, V.: Production logistics and human-computer interaction – state of the art, challenges and requirements for the future. Int. J. Adv. Manuf. Technol. **105**(9), 3691–3709 (2019)
25. Wasilewski, A., Wasilewski, M.: SOA approach in e-commerce integration. In: Soliman, K.S. (ed.) Vision 2020: Sustainable Economic Development and Application of Innovation Management, pp. 8506–8515 (2018)

# Declarative Models of Periodic Distribution Processes

Grzegorz Bocewicz[1](✉), Izabela Nielsen[2], Robert Wójcik[3], and Zbigniew Banaszak[1]

[1] Faculty of Electronics and Computer Science, Koszalin University of Technology, Koszalin, Poland
grzegorz.bocewicz@tu.koszalin.pl
[2] Department of Materials and Production, Aalborg University, Aalborg, Denmark
[3] Department of Computer Engineering, Faculty of Information and Communication Technology, Wrocław University of Science and Technology, Wrocław, Poland

**Abstract.** The presented approach employs a declarative modeling-based methodology to assess possible mesh-like distribution networks carrying out multimodal processes flow. In order to accomplish this study, the grid network topology concept is used to model, analyze, and design supply distribution networks incorporating the multimodal processes paradigm. To guarantee the congestion-free flow of multimodal processes, encompassing the movement of items following delivery routes, a grid network free of deadlocks is required to support them. The functionality of a network consisting of different modes of local transportation processes has to be guaranteed. A declarative model of a mesh-like network, describing the network's local use and multimodal processes, provides a framework enabling us to determine the conditions sufficient to guarantee its cyclicity (i.e., blockage-free course. The main problem lies in determining the conditions guaranteeing assumed processes' performance while matching the topological constraints encountered by the network structure and examining the constraints resulting from different speeds at which local processes are supported. The results are illustrated in case studies concerning AGVS and milk-run systems operation in different mesh-like layouts of distribution networks, including grid and fractal topologies.

**Keywords:** Multimodal process · Mesh-like structure · Concurrent cyclic processes · Declarative modeling

## 1 Introduction

The main objective of this chapter is to propose a declarative modeling-based methodology aimed at assessing possible mesh-like distribution networks, carrying out multimodal processes flow. The mesh network topology concept is used to model, analyze, and design supply distribution networks incorporating a multimodal processes paradigm. A feature of multimodal processes [4] is that their flows (e.g., following delivery routes) comprise fragments of local flows supported by different means of modalities. Examples


J. Trojanowska et al. (Eds.): MANUFACTURING 2022, LNME, pp. 116–129, 2022.
https://doi.org/10.1007/978-3-030-99310-8_10

of such processes can be found in different domains covering Automated Guided Vehicle Systems (AGVS); train networks, ship and airline, and also data and supply media flows [1, 4, 9, 10, 12, 14, 20].

The smooth (undisturbed) flow of multimodal processes is conditioned by the deadlock-free implementation [2] of the local unimodal processes supporting them. Deadlocks in processes cause conflicts in other situations due to their simultaneous access to shared resources, e.g., bus and train stations, transportation hubs, workstations, and warehouses. Consequently, to avoid blockages, it is necessary to introduce appropriate mechanisms, e.g., employing dispatching rules that synchronize the processes while guaranteeing a cyclic steady-state behavior of concurrently executed local processes. In the context of cyclic flow being deadlock-free, the NP-hard problem of deadlock handling may be treated as equivalent to the problem of cyclically-executed local processes synchronization. The delivery period in a network formed by a set of local carriers depends on the cycle of this network.

The distribution network structure under consideration adopts topological assumptions based on numerous reports concerning commonly found regular networks, widely observed in different application domains (intercity and urban passenger transport). These kind of networks have been considered for over twenty years and to be gained from their layout being vital to improving manufacturing (public transport) [3, 5, 6, 15, 19].

The above mentioned process classes and synchronization problems are represented in a declarative modeling framework. Assuming that a given distribution network's mesh-like structure limits the local cyclical processes carried out in it, the main problem is defining the conditions that guarantee that the local processes are deadlock-free and support their flow of multimodal processes. The task of searching for conditions matching the topological constraints of the network structure, and constraints resulting from different speeds of the local processes carried out in them, is aimed at uncovering assumed performance measures.

To summarize, the new contributions provided to the currently available literature are (i) presenting a declarative modeling-based methodology aimed at assessment of mesh-like distribution networks carrying out multimodal processes flow, (ii) elaborating a declarative reference model for mesh-like distribution networks and conditions sufficient to guarantee a congestion-free flow of delivery processes, and (iii) developing methods that allow planning multimodal deliveries in the distribution networks with grid or fractal topological structures.

The paper is organized as follows. Section 2 elaborates on modeling and controlling concurrently-flowing discrete processes, focusing on both cyclic and multimodal processes. Section 3 presents the declarative approach to performance modeling of the mesh-like distribution networks, focusing on periodicity conditions. Section 4 provides the conclusions followed by the description of future research.

## 2 Concurrent Discrete Processes

Processes occurring in many manufacturing areas, such as product flows in factories, city traffic flow, data flow, etc., have one common feature—they need to be executed concurrently. This property means that concurrency is a common feature in these systems, where several processes can execute different actions simultaneously; however, their actions often overlap in time. In cases when the processes' routes share the common resources, the blockade and (or) congestion-free flow of interacting processes depends on assumed dispatching rules controlling (synchronizing) their access to such resources.

### 2.1 Systems of Concurrently Flowing Cyclic Processes

The railway system is an example of the class of Systems of Concurrently Executed Cyclic Processes (SCCP) [1, 15, 17, 18]. Local cyclic processes correspond to particular vehicles circulating in the connection network, stopping according to a specified timetable, at the stations deployed along the route traced in the network.

The timetable determining the arrival/departure time of the vehicles stopping at a given station is a solution to a cyclic scheduling problem and has been a subject of many studies [7, 8, 13].

Consider the system shown in Fig. 1a and its graph model Fig. 1b. A solution representing a workstations' delivery services order while minimizing the total production cycle is sought.

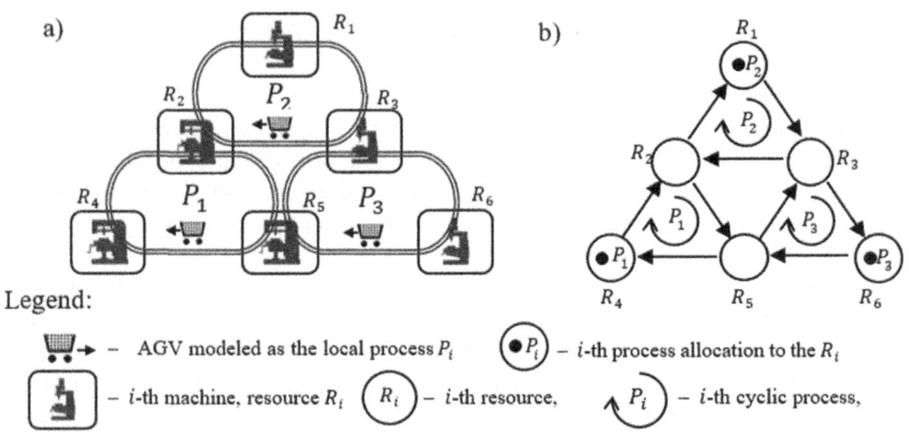


**Fig. 1.** a) Exemplary layout of flexible manufacturing system and b) its graph model representation.

**Deadlock-Freeness.** For deriving the cyclic conditions of SCCP class systems, let's consider the structure of an example shown in Fig. 1. Assumed allocation of processes carried out in a given instance of SCCP is represented by the sequence $(PA_i(R_k), \ldots, PA_j(R_l), \ldots, PA_w(R_d))$ describing the current resource allocation, where:

$$PA_i(R_k) = \begin{cases} P_i \text{ if the i-th process is allocated to the k-th resource} \\ \Delta \text{ if no process is allocated tothe k-th resource} \end{cases}$$

Processes allocation (see Fig. 1b) illustrates the following sequence $PA_0$, where the symbol Δ indicates no resource assignment.

$$R_1,\ R_2,\ R_3,\ R_4,\ R_5,\ R_6$$

$$PA_0 = (P_2,\ \Delta,\ \Delta,\ P_1,\ \Delta,\ P_3)$$

Assuming that $PA_0 = (P_2, \Delta, \Delta, P_1, \Delta, P_3)$ sequence is the initial allocation of processes in the system under consideration, it can be shown that it leads to a deadlock allocation [2] in which the process $P_1$ requests access to the resource $R_3$, that the process $P_2$ requests access to the resource $R_1$ etc.. The closed-loop of mutual requests is illustrated in Table 1. In the considered case, processes block results from the adopted method of resource conflicts resolution defining the order of processes allocation to each of the i-th shared resources. The priority dispatching rule controls the scheme of resource allocations $\sigma_i = (P_j, P_a, P_b, \ldots, P_k, \ldots, P_n)$ and thus determine the order in which, after execution of an operation performed by process $P_j$ on the resource $R_i$, then the subsequent access to $R_i$ is allocated to $P_a$, and then to $P_b$ and so on. Note that such synchronization rules can be seen as processes coordinating processes competing for access to shared resources. Therefore, the set of Priority Dispatching Rules (PDRs) $\sigma_2 = (P_1, P_2)$, $\sigma_3 = (P_2, P_3)$, $\sigma_5 = (P_3, P_1)$ results in the behavior described by the sequence of processes allocation collected in Table 1a.

**Table 1.** The sequence of processes allocation reached from $PA_0$ following priority dispatching rules: $\sigma_2 = (P_1, P_2), \sigma_3 = (P_2, P_3), \sigma_5 = (P_3, P_1)$ a) and $\sigma_2 = (P_1, P_2), \sigma_3 = (P_2, P_3), \sigma_5 = (P_1, P_3)$.

a)

| | $R_1, R_2, R_3, R_4, R_5, R_6$ |
|---|---|
| $PA_0$ | $(P_2, \Delta, \Delta, P_1, \Delta, P_3)$ |
| $PA_1$ | $(\Delta, P_1, P_2, \Delta, P_3, \Delta)$ |
| $PA_2$ | $(\Delta, P_1^*, P_2^*, \Delta, P_3^*, \Delta)$ |

b)

| | $R_1, R_2, R_3, R_4, R_5, R_6$ |
|---|---|
| $PA_0$ | $(P_2, \Delta, \Delta, P_1, \Delta, P_3)$ |
| $PA_1$ | $(\Delta, P_1, P_2, \Delta, \Delta, P_3)$ |
| $PA_2$ | $(\Delta, \Delta, P_2, \Delta, P_1, P_3)$ ← |
| $PA_3$ | $(\Delta, P_2, \Delta, P_1, \Delta, P_3)$ |
| $PA_4$ | $(P_2, \Delta, \Delta, P_1, P_3, \Delta)$ |
| $PA_5$ | $(P_2, P_1, P_3, \Delta, \Delta, \Delta)$ |
| $PA_6$ | $(P_2, \Delta, \Delta, \Delta, P_1, P_3)$ |
| $PA_7$ | $(\Delta, \Delta, P_2, P_1, \Delta, P_3)$ |
| $PA_8$ | $(\Delta, P_2, \Delta, P_1, P_3, \Delta)$ |
| $PA_9$ | $(P_2, \Delta, P_3, P_1, \Delta, \Delta)$ |
| $PA_{10}$ | $(P_2, P_1, \Delta, \Delta, \Delta, P_3)$ |

$P_i^*$ means that the i-th process occupying the relevant resource is awaiting to release another, $P_i$ means that the i-th process is executed on the resource, Δ means that no process is executed nor awaiting on the resource.

In general, however, in a given structure of cyclic processes, PDRs may result in different processes flow, leading or not to its blockade. Therefore, the system cycle (its cyclicality) is determined by choice of priority rules used in it, and in general, the adoption of different dispatching rules may lead to periodic behaviors characterized by different repeatability periods T. The described course illustrates the sequence of processes allocations collected in Table 1b. In the considered case, see Fig. 3b for the initial allocation $PA_0 = (P_2, \Delta, \Delta, P_1, \Delta, P_3)$ and the following set of PDRs $\{\sigma_2 = (P_1, P_2), \sigma_3 = (P_2, P_3), \sigma_5 = (P_1, P_3)\}$ the cyclic steady state of processes allocations containing repeating sequence of allocations $PA_2$–$PA_{10}$ is followed by a transient period consisting of two allocations $PA_0$ and $PA_1$, respectively. Because processes allocations $PA_2$–$PA_{10}$ repeat every nine allocations, the system cycle thus calculated is equal to $T = 9$.

To sum up, the study of the relationships linking the behavior of the SCCP need to set conditions enabling the synthesis of distributed control procedures for SCCP.

**Mesh-Like Network Structures.** As an introduction to mesh-like class structures, let's consider structures as shown in Fig. 2, composed of clusters, which are identical repeating substructures, called Elementary Marquing Structures [5]. Each EMCS consists of a set of cyclical processes representing the means of transport, thus enabling the passenger flow consisting of the infrastructure of the multi-modal processes system. The protocols managing processes to access shared resources while guaranteeing its cyclic steady-state behavior (i.e., resulting in the repetitive character of the material flow) have to follow the assumptions:

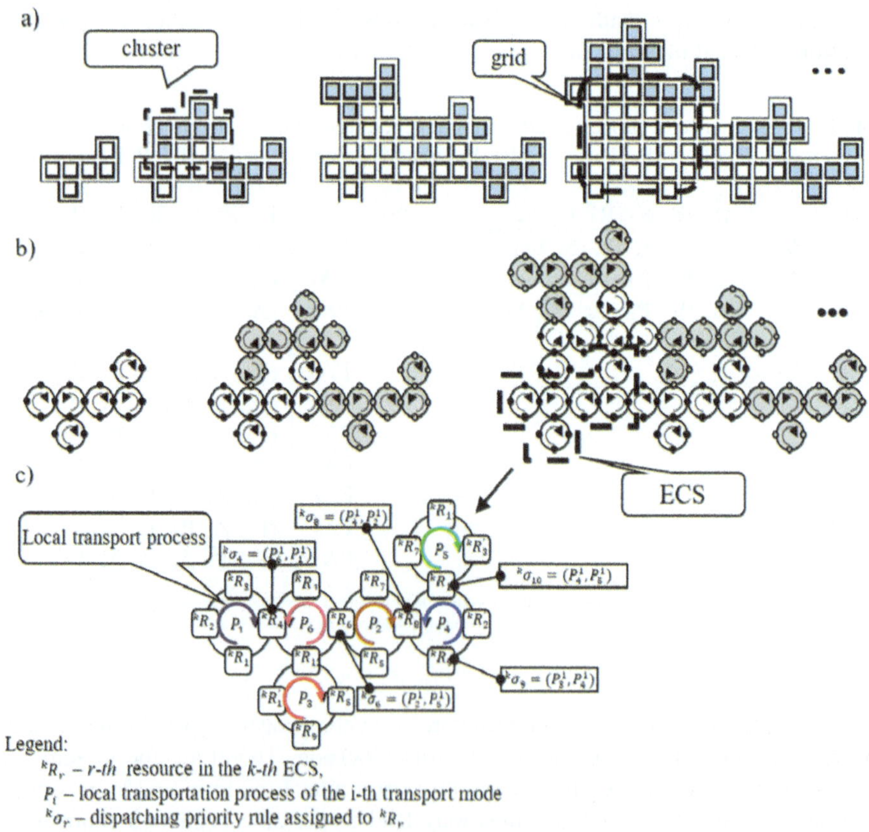


**Fig. 2.** Grid-like structure a) composed of EMCSs b) created from local processes.

- Each EMCS represents a repetitive substructure performing a non-empty sequence of pick-up and place operations,
- Only one operation can be executed at a time on each system resource,
- Coupled processes that compete for access to the system‘s shared (non-preemptive) resources can be either executed or suspended on one of the associated resources.

A particular role in this class of systems is played by the structure of the transport systems included in them [16, 19]. The advantages of periodic structures are flexibility and robustness. Moreover, urban agglomerations' development is subject to the laws of recursion, which are best modeled by fractal structures. Typical problems aim to minimize travel costs or the total travel distance required to visit all customers while following the assumed schedule of frequency visits within a given time horizon. These kind of problem are an NP-hard [11–14].

According to the current taxonomy, the methods for solving such problems can be partitioned into two main categories: approximation and exact methods [22]. The approximation methods try to find the best possible solution while providing no guarantees on its quality, i.e., they prefer quick solutions over optimal ones. In turn, exact methods are guaranteed to find the best solution to the problem given enough time, i.e., they are oriented towards the search for the optimal solution at the expense of the time incurred to obtain it. In the approximation approach, some methods implement heuristic algorithms (e.g., metaheuristics driven like Variable Neighborhood Search (VNS), Simulated Annealing (SA), and Tabu Search (TS)) and population algorithms (such as Particle Swarm Optimisation (PSO), Ant Colony Optimization (ACO), Artificial Bee Colony (ABC) and evolutionary algorithms, e.g., Memetic Algorithms (MMA), Genetic Algorithm (GA)) [21, 24]. Exact approach leverage intelligent forms of enumerative search, such as Dynamic Programming (DP), Mixed Integer Linear Programming (MILP), Branch-and-Bound (BB), and Constraint Programming (CP) [4–6, 23].

Transportation processes executed by particular lines are usually cyclic, hence the passenger flow supported by them also have a periodic character. In consequence, the period of multimodal passenger flow depends on a period of the relevant transportation network, i.e., on EMSCs. It should be noted that the passenger travel schedules can be estimated easily while considering the cyclic behavior.

## 2.2 Multimodal Processes

As already mentioned, the course of multimodal processes—their cycle and the delivery takt - depend on the structure and the cycle time of the SCCP supporting them. The topological structure of a considered SCCP system can be a mosaic of different grid structures synchronized by appropriately chosen dispatching rules [17, 18].

**Distribution Networks.** Consider SCCP being of the fractal network, shown in Fig. 3, which can be seen as composed of "+" shaped EMSC. Two multimodal processes are considered: $mP_1$, $mP_2$.

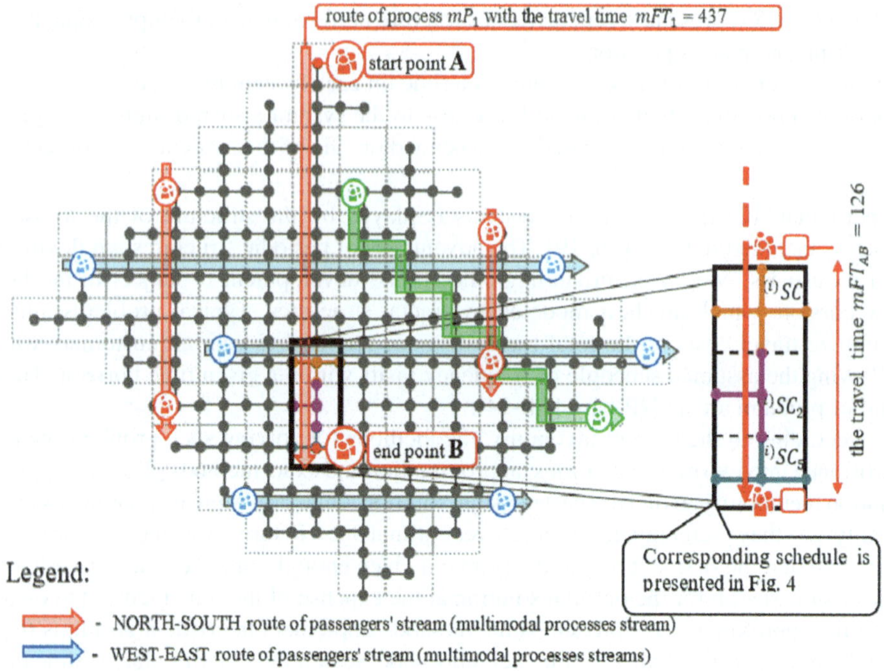


**Fig. 3.** Fractal SCCP.

The transportation route depicted by the red line corresponds to the multimodal process $mP_1$ and the blue line corresponds to the multimodal process $mP_2$ are supported by different transportation modes (busses, trams, metro, etc.). This way of operation means that the transportation routes can be considered composed of parts of the routes of local cyclic processes. In order to ensure the periodicity of multimodal processes implemented in the systems considered (EMSC), it is enough to demonstrate a method to organize them that guarantees the cyclic behavior of SCCP. In a system with the structure shown in Fig. 3, we are faced with a question: do there exist routes of transportation means and the associated transport schedules guaranteeing timely and deadlock-free transport of passengers?

**Flow Schedules.** Let's focus on fractal SCCP from the previous section, which is related to the multimodal process and distinguished by the red line; see Fig. 3. The EMSC cycle time should be calculated to determine the time needed to travel on the road connecting points A and B. In the case under consideration, it is 48 u.t. resulting from the value of the cycle time of the elementary process (equal to 12 u.t.); see Fig. 5. Consequently, the takt time of the multimodal process $mP_1$ is equal to 48 u.t. The resultant schedule illustrating the operation of transportation modes and how they provide passengers' transportation routes between points A and B is shown in Fig. 5. The travel time between points A and B (route of the multimodal process $mP_1$), is equal to 126 u.t. ($3 \times 48 - 18$), where: $c \cdot {}^{(i)}m\alpha_1 - \Delta F_1$, ${}^{(i)}m\alpha_1$ – the period of multimodal processes executed by networks of regular structures. $\Delta F_k$ is the difference between the completion time of a trip and the start time of the subsequent one, and $c$ is the number of EMSCs included in a transportation route. Therefore, the travel time along the route of $mP_1$ is equal to 437 u.t.

## 3 Performance Modeling

The quality of the functioning evaluation of a given system depends decisively on which model is used to assess it. Due to the nature of the research carried out, covering both the analysis and synthesis of SCCP issues. Available model classes are divided into those using an imperative method and those based on a declarative approach. Differences between models rest because imperative approaches focus on how (to solve) while declarative focus on what (to describe/specify).

### 3.1 Declarative Modeling

A constraint programming approach (declarative modeling) offers the methods unavailable in mathematical programming but efficient for scheduling and other combinatorial problems. Users declaratively state the constraints on the feasible solutions for a set of decision variables. In other words, the constraint programming being a representative of the exact approach can be seen as a paradigm for solving combinatorial problems (Fig. 4).

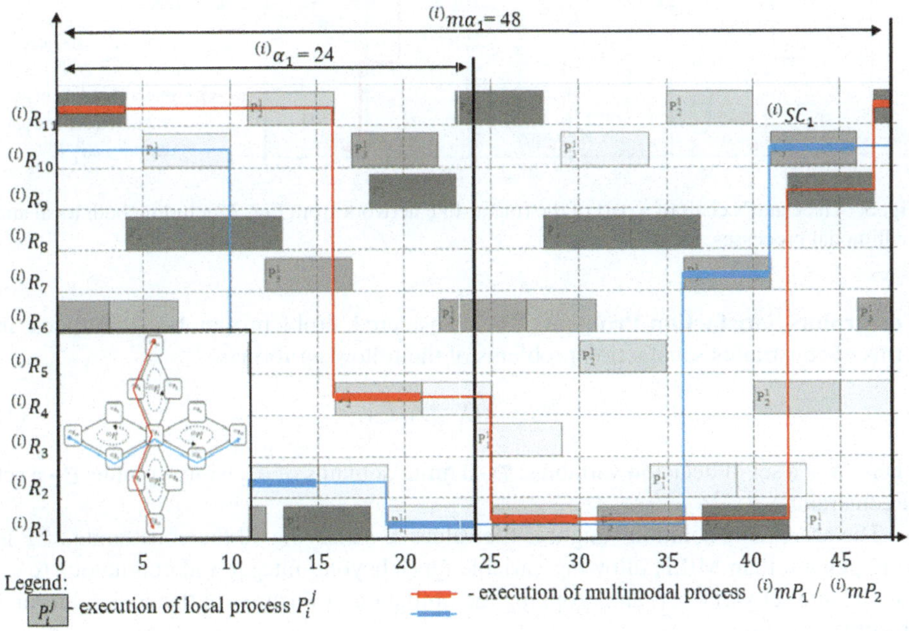


**Fig. 4.** The schedule ${}^{(i)}X_1'$ of SCCP following ${}^{(i)}SC_1$ from Fig. 3.

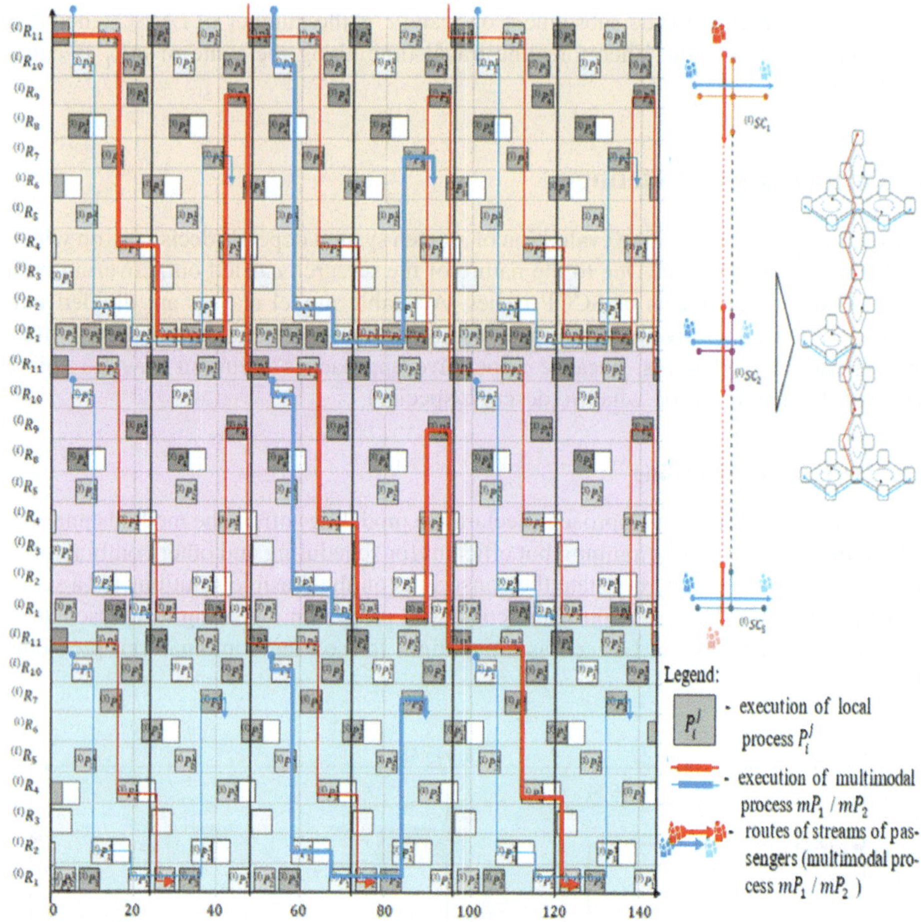


**Fig. 5.** The Gantt's chart of a part of the fractal-like network from Fig. 3 including both local and multimodal processes.

**Constraints Satisfaction Problem.** The considered problem can be formulated in terms of constraints satisfaction problems of the following form:

$$PS = ((X, D), C), \tag{1}$$

where: $X$ – a set of decision variables; $\mathcal{D}$– a finite domains of decision variable, $\mathcal{C}$– a set of constraints.

The acceptable solution sought is the values $X$ satisfying all the constraints . CP is more general than MILP, allowing variable types beyond integer and continuous (e.g., interval and set variables) and dropping the limitation of linearity in the constraints and objective function.

The reference problem (1) can be formulated as a analysis or synthesis. For further discussion, it is assumed the analysis problem for which the following question is considered: Is it possible to make supplies satisfying the customer demands? [8] (2):

$$PS_{A} = ((X, \mathcal{D}), \mathcal{C}ST), \tag{2}$$

where:

$X = (X', Xs', \alpha')$ - a cyclic schedule, where: $X' = (x_\lambda | \lambda = 1, \ldots, \omega)$, $x_\lambda$ – starting time of $o_\lambda$ transport operation at resource, $Xs' = (xs_\lambda \; | \lambda = 1,\ldots,\omega)$, $xs_\lambda$ – time of the release of a resource, $\alpha'$– takt time.

$\mathcal{D}$– a finite set of decision variable domains $\{X', Xs', \alpha'\}$, where: $\alpha', x_\lambda \in \{0,\ldots,T\}$, $xs_\lambda \in \{0,\ldots, 2T\}$, and $T$ stands for a planning horizon.

$\mathcal{C}$ST – the relationships between processes operations.

The resulting schedule $X$ determines the admissible timetable of transportation means, guaranteeing timely and deadlock-free travel of passengers.

In turn, the synthesis problem is related with the question: Is there a transport network structure that ensures deliveries satisfying customer demands? [8] (3):

$$PS_{RE} = ((\{X, B, F\}, \mathcal{D}), \mathcal{C}R), \tag{3}$$

where:

$X$ – a cyclic schedule, $X = (X', Xs', \alpha')$, $B$, $F$ – routes of processes.

$\mathcal{D}$– domains of variables: $X, B, F$: $x_\lambda \in \{0, \ldots, T\}, xs_\lambda \in \{0, \ldots, 2T\}, b_\lambda \in \{0,\ldots,\omega\}$, $f_\lambda \in \{1, \ldots, \omega\}$.

$\mathcal{C}R$ – the relationships between the processes operations and transport network structure (procesess routes).

Sought are $B$, $F$ determining the routes of local processes guaranteeing the schedule $X$. In that context the solution of $PS_{RE}$ (3), it is enough to find values of $B$, $F$ and $X$, for which all constraints $R$ are satisfied.

Implementing the above presented CSP-driven approach in an interactive DSS environment allows for online prototyping of concurrently executed flows.

### 3.2 Periodicity Conditions

Periodicity or cyclicality shows a system feature whose behavior is described by repetitive changes occurring at regular intervals. An example of this type of system is tram and bus lines running cyclically along closed loop routes and stopping at shared depots. Similar networks form metro lines, suburban transport, and long-distance rail. Another system in which intralogistics solutions play a dominant role in manufacturing systems is inter-operational transport through AGVS and (or) milk-run systems. All these cases have the same feature: the occurrence of cyclically recurring repetitive processes, sharing common system resources. The consequence of this is the need to guarantee their admissible, i.e., smooth (congestion and deadlock-free) flow.

**Deadlock Prevention Conditions.** Deadlock-freeness is a unique property, stating that the system can never be in a situation where no further progress is possible. In other words, such a system's property supposes its capability to run indefinitely. However, the considered problem belongs to NP-hard class. In this study, assuming that the guarantee of deadlock-free system behavior provides a guarantee of its cyclicality, we focus mainly on the problem of cyclic systems synthesis instead of searching for deadlock-avoidance control methods. We focus on systems with regular topology structures, searching for appropriate polynomial computational complexity methods. The acceptable solutions obtained in this way do not exhaust all possible ones. This result means that the conditions under which they were obtained are merely sufficient. They, therefore, correspond to the deadlock prevention conditions and not to deadlock avoidance conditions, thereby guaranteeing analysis of all possible cases (i.e., corresponding to necessary and sufficient conditions).

Let's consider supply networks of the grid-like structure obtained by clusters aggregation - see Fig. 6. This network EMSC (Fig. 6b) is a connected digraph comprised of elementary substructures occurring in the regular structure.

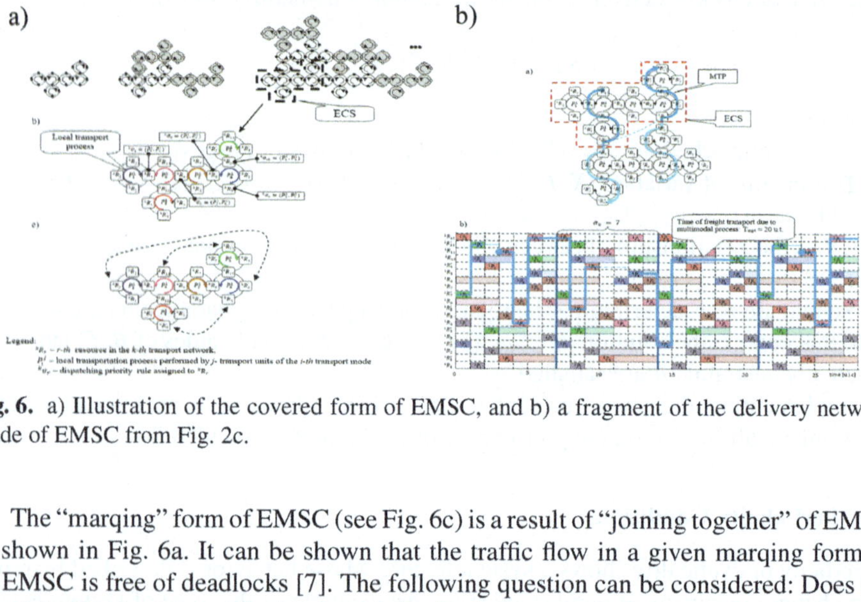


**Fig. 6.** a) Illustration of the covered form of EMSC, and b) a fragment of the delivery network made of EMSC from Fig. 2c.

The "marqing" form of EMSC (see Fig. 6c) is a result of "joining together" of EMSC as shown in Fig. 6a. It can be shown that the traffic flow in a given marqing form of an EMSC is free of deadlocks [7]. The following question can be considered: Does the schedule $X$ exist that enable timely and deadlock-free travel of passengers? The answer to the above question is provided by feasible solution shown in Fig. 7 (solution of $PS_{\mathrm{A}}$ (2)). The presented schedule can be determined based on the schedule of its EMSCs.

Thus the cyclic behavior of the EMSC implies cyclic behavior of the regular-structure distribution network; see Fig. 7.

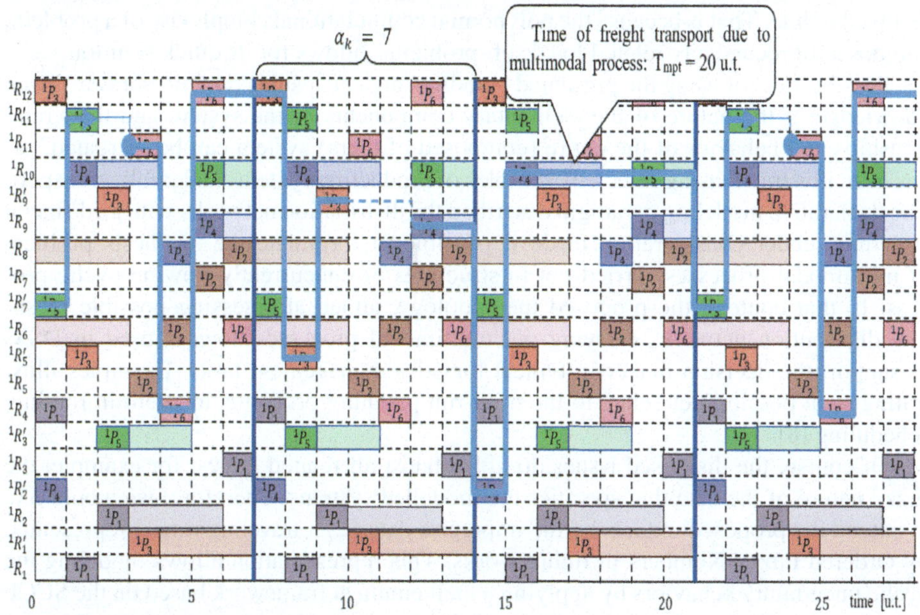


**Fig. 7.** A representation of an MTP performed in an SN given in Fig. 3(a), Gantt's chart of local transport processes, and the MTP (b).

## 4 Concluding Remarks

Chapter shows an alternative way to synthesizing systems that guarantee the expected behavior. The class of systems under consideration includes cases that are often found in practice and characterized by concurrent cyclical processes implemented in them. Occurring in this area problems: analysis (searching for conditions that guarantee assumed system behavior with given constraints imposed by its structure) and synthesis, belong to NP-hard class of problems. These problems require the need for a laborious and individualized search for an approximate solution by the specificity of the instance.

The presented approach assumes that the behavior of its elementary structures fully characterizes the behavior of systems with "regular" structures. In other words, the behavior of a system established from standard, homogeneous, recursively-combined subsystems can be described using parameters specifying its constituent elements. This assumption is confirmed by the examples observed in practice: systems with grid and (or) fractal structures (typical for the development of urban areas, modular solutions for electronic equipment structures based on the plug-and-go principle, and others). In solutions of this type, the abovementioned analysis and synthesis problems boil down to the following questions: 1) Does a given system structure guarantee its assumed behavior? 2) Can a given system's functioning be reachable in its structure (composed of assumed components)? The computational complexity of problems formulated in this way allows one to solve them online in situations encountered in practice. This approach is possible since the small scale of the elementary component problems can be resolved online. That is because the polynomial computational complexity of a problem, composed of recursively related local sub-problems, allows for its quick solution.

Possibilities for using the presented paradigm are encapsulated in the phrase: "From knowledge of the nature of the elementary components of the system and their relationships, the behavior of the entire (composed of them) system can be recreated," as has been illustrated using selected examples of production systems and public transport. The declarative modeling paradigm described the distribution networks with grid/fractal structures under consideration made it possible to formulate the synthesis problem of multimodal processes carried out in structures of concurrently-flowing cyclic process. In that context, the proposed methodology, aimed at assessing possible mesh-like distribution networks carrying out multimodal processes flow, is used in DSSs to support the decision-maker in typical logistics management tasks. Examples illustrating such possibilities relate to the milk-run planning problems inherent in routing, scheduling [8].

Of course, the discussed issues do not exhaust all related issues, for example, the fuzzy nature of the data that specifies the modelled systems. Therefore, we would like to extend the proposed models by the imprecise data (e.g., traveling times) represented by Ordered Fuzzy Numbers in future works. This representation allows modeling the cyclic uncertainty behaviors by applying a mathematical framework based on the SCCP systems concept. It makes it possible to formulate a new problem that boils down to the mutual reachability of cyclic steady states in the space of states achieved in a given system structure [6]. Also, determining the transition states between two given cyclic steady states remains a challenge. Due to the integers domain, these kinds of problems can be classified as Diophantine problems.

**Acknowledgments.** This research was co-founded by the National Science Centre, Poland, grant number 2019/03/X/ST6/00429.

## References

1. Banaszak, Z.: Distributed bottleneck control for repetitive production systems. J. Intell. Manuf. **8**, 415–424 (1997)
2. Banaszak, A.Z., Krogh, H.B.: Deadlock avoidance in flexible manufacturing systems with concurrently competing process flows. IEEE Trans. Robot. Automat. **6**(6) (1990). https://doi.org/10.1023/A:1018510300148
3. Barakou, F., Koukoula, D., Hatziargyriou, N., Dimeas, A.: Fractal geometry for distribution grid topologies. IEEE Eindhoven PowerTech, Eindhoven, 1–6 (2015). https://doi.org/10.1109/PTC.2015.7232496
4. Bocewicz, G., Banaszak, Z.: Multimodal processes scheduling in mesh-like network environment. Arch. Control Sci. **25**(2), 237–261 (2015). https://doi.org/10.1515/acsc-2015-0016
5. Bocewicz, G., Jardzioch, A., Banaszak, Z.: Modelling and performance evaluation of fractal topology streets network. In: Omatu, S., et al. (eds.) 13th International Conference. Advances in Intelligent Systems and Computing, vol. 474. Springer, Cham (2016). https://doi.org/10.1007/978-3-319-40162-1_52
6. Bocewicz, G., Banaszak, Z., Nielsen, I., Muszyński, W.: Re-scheduling of AGVs steady state flow. IFAC-PapersOnLine **50**, 3493–3498 (2017). 20th IFAC World Congress. https://doi.org/10.1016/j.ifacol.2017.08.935

7. Bocewicz, G., Banaszak, Z.: Designing mass-customized network of passenger services subject to grid topology constraints. In: Szewczyk, R., Zieliński, C., Kaliczyńska, M. (eds.) ICA 2017. AISC, vol. 550, pp. 120–131. Springer, Cham (2017). https://doi.org/10.1007/978-3-319-54042-9_11
8. Bocewicz, G., Nielsen, P., Banaszak, Z.: Declarative modeling of a milk-run vehicle routing problem for split and merge supply streams scheduling. In: ISAT 2018. Advances in Intelligent Systems and Computing, vol. 853. Springer, Cham (2019). https://doi.org/10.1007/978-3-319-99996-8_15
9. Crama, Y., Kats, V., Klundert, J.: Cyclic scheduling in robotic flowshops. Ann. Oper. Res. **96**, 97–124 (2000). https://doi.org/10.1023/A:1018995317468
10. Dawande, M.W., Geismar, H.N., Sethi, S.P.: Dominance of cyclic solutions and challenges in the scheduling of robotic cells. SIAM Rev. **47**(4), 709–721 (2005)
11. Kamoun, H., Sriskandarajah, C.: The complexity of scheduling jobs in repetitive manufacturing systems. Eur. J. Oper. Res. **70**(3), 350–364 (1993)
12. Kats, V., Levner, E.: Cyclic scheduling of operations for a part type in an FMS handled by a single robot: a parametric critical-path approach. Int. J. Flex. Manuf. Syst. **10**, 129–138 (1998). https://doi.org/10.1023/A:1008045414118
13. Levner, E., Kats, V., Pablo, D., Cheng, T.C.E.: Complexity of cyclic scheduling problems: a state-of-the-art survey. Comput. Ind. Eng. **59**(2), 352–361 (2010). https://doi.org/10.1016/j.cie.2010.03.013
14. Li, Q., Pogromskiy, A.Y., Adriaansen, A.C., Udding, J.T.: A control of collision and deadlock avoidance for automated guided vehicles with a fault-tolerance capability. Int. J. Adv. Rob. Syst. **13**(2) (2016). https://doi.org/10.5772/62685
15. Majdzik, P., Obuchowicz, A.: Prototyping of concurrent cyclic processes. In: 3rd IFAC Workshop on Discrete-Event System Design, IFAC Proceedings, vol. 39, pp. 257–262 (2006). https://doi.org/10.3182/20060926-3-PL-4904.00043
16. Montreuil, B.: Fractal layout organization for job shop environments. Int. J. Prod. Res. **37**(3), 501–521. https://doi.org/10.1080/002075499191643
17. Polak, M., Wojcik, R., Majdzik, P., Banaszak, Z.: Prototyping of distributed control procedures in concurrent cyclic processes systems. In: 2003 IEEE Conference on Emerging Technologies and Factory Automation, vol. 2, pp. 261–268 (2003)
18. Polak, M., Majdzik, P., Banaszak, Z.A., Wójcik, R.: The performance evaluation tool for automated prototyping of concurrent cyclic processes. Fund. Inform. **60**(1–4), 269–289 (2004)
19. Venkatadri, U., Rardin, R.L., Montreuil, B.: A design methodology for fractal layout organization. IIE Trans. **29**, 911–924 (2001)
20. Wallace, M., Yorke-Smith, N.: A new constraint programming model and solving for the cyclic hoist scheduling problem. Constraints **25**(3–4), 319–337 (2021). https://doi.org/10.1007/s10601-020-09316-z
21. Slowik, A., Kwasnicka, H.: Evolutionary algorithms and their applications to engineering problems. Neural Comput. Appl. **32**(16), 12363–12379 (2020). https://doi.org/10.1007/s00521-020-04832-8
22. Stork, J., Eiben, A.E., Bartz-Beielstein, T.: A new taxonomy of global optimization algorithms. Nat. Comput. (2020). https://doi.org/10.1007/s11047-020-09820-4
23. Xue, F., Anqi, P., Zhengyun, R., Zhiping, F.: Hybrid driven strategy for constrained evolutionary multi-objective optimization. Inf. Sci. **585**, 344–365 (2022). https://doi.org/10.1016/j.ins.2021.11.062
24. Zhangjie, F., Jingnan, Y., Guowu, X., Yiming, C., Yuanhang, M.: A heuristic evolutionary algorithm of UAV path planning. Wireless Commun. Mob. Comput. **2018**, 11 (2018). Article ID 2851964. https://doi.org/10.1155/2018/2851964

# Analysis and Improvement of the Foreman's Work Methods

Paulina Rewers(✉) and Przemysław Paluszyński

Poznan University of Technology, 60-965 Poznan, Poland
paulina.rewers@put.poznan.pl

**Abstract.** The article presents the analysis and actions taken to improve work as a foreman. The literature part introduces the concepts of introducing improvements, methods and tools used in production management (work standardization) and quality management (5Why, Ishikawa diagram). The research methodology based on the analysis of activities performed by foremen, in particular, focus was on additional activities that do not bring added value. The analysis used the timing method. An assessment of the reliability in collecting information required in formal duties was also carried out. The 5Why analysis and the Ishikawa diagram were performed in order to indicate the root causes of the need for additional activities by foremen. Proposed actions were aimed at eliminating the sources of the problems and after it the foremen's work was analyzed again in order to check the effects of the improvements.

**Keywords:** Standardization of work · Ishikawa diagram · 5Why analysis

## 1 Introduction

In today's economy, manufacturing companies, regardless of the industry, are struggling with constantly growing competition and increasing customer requirements. Therefore, they are looking for ways to increase production efficiency without the need to significantly invest capital, such as in the case of hiring additional employees or expanding the machine park [1, 2]. Currently, modern management concepts are gaining more and more popularity, the purpose of which is to improve the quality of manufactured products, reduce production costs, increase the availability of resources and timely deliveries. In order to achieve the above goals, quality management tools and methods [3–6] and Lean Manufacturing (LM) [7–10] are most often used.

In the case of the current market situation, a large turnover of production workers is observed. This is, of course, associated with many difficulties, ranging from the time devoted to training a new employee, through the costs of acquiring and learning him, in many cases with a decrease in the quality of manufactured products and reduced production efficiency, as well as with the increasing responsibilities of foremen, who often at the expense of their own duties, spend a lot of time checking and training new employees. This situation is particularly troublesome in enterprises where there are no work standards that would allow new employees to learn to work in a given position


J. Trojanowska et al. (Eds.): MANUFACTURING 2022, LNME, pp. 130–143, 2022.
https://doi.org/10.1007/978-3-030-99310-8_11

on their own. The work of a foreman most often consists in controlling employees and the flow of products from production, reporting the state of production to superiors and helping production workers in emergency situations. It often happens that foremen are the only people who are able to perform complex machining or retooling. The foreman is also a representative of production employees, who often has a great impact on the employee's assessment, bonuses, awards and dismissals. In many companies, the work of a foreman is based on the employee's list of duties. Unfortunately, due to the lack of work standards and high rotation of production workers, the foreman, instead of e.g. retooling or reporting, is forced to "put out fires" at a workstation [11–13].

The purpose of this article is to analyze and improve the work of foremen in a certain production company. The formal scope of responsibilities of the foreman was discussed. An analysis of all performed activities was carried out, including additional activities that are not included in the duties of this employee. The analysis used the timing method. An assessment of the reliability in collecting information required in formal duties was also carried out. The 5Why analysis and the Ishikawa diagram were performed in order to indicate potential reasons for the foreman to perform additional activities. Successively, improvement actions were proposed and the foreman's work was analyzed again in order to check the effects of the improvements.

## 2 Analysis of the Literature

Lean Manufacturing is recognized as a modern production management concept. The goal of LM is to eliminate any waste occurring in production, i.e. to reduce losses in human effort, in inventories, during the introduction of products to the market and the proper organization of the production space. Actions aimed at improving production processes are undertaken in order to be able to quickly respond to changing customer needs, while producing high-quality products in the most efficient and economical way [14, 15].

Waste elimination is understood as all the events that involve an employee or material and do not bring added value. In the context of work improvement, it is associated with the indication of an additional, extra-compulsory activity, which disturbs the work flow. One of the tools that is used to analyze and measure the time of activities is timing method. In its basic, it is a method consisting in measuring the times of an operation or its elements (procedures, activities, working movements) in order to determine, on the basis of a certain sample, the proper, standardized duration, rational for the normal pace of work [16]. Nevertheless, thanks to the possibility of adapting this tool to individual needs, it has found a much wider application than just standardizing working time. It is suitable for determining the time-consuming nature of processes or studying process variability. In the enterprise where the processes are tested and improved, the aim is to determine the conditions for the stability of the activities performed, as they bring better results [17].

The timing is also a source of a lot of valuable information about the activities performed. However, it does not show the root causes and the main focus of the problems. Therefore, an in-depth analysis leading to their determination is necessary. For this purpose, you can use quality improvement tools - the Ishikawa diagram [18, 19] and the

5Why technique (five questions why) [20, 21]. They allow you to find the real causes of problems and effectively eliminate them.

The LM concept also includes a number of methods and tools that support enterprises in the process of eliminating waste. These include, among others, work standardization, Poka Yoke, 5S, SMED or the workplace control process [22].

Standardization of work is one of the commonly used Lean Manufacturing tools in production plants and defines the best practices in performing a given activity [23–25]. In many cases, a well-defined standard implies the correct course of the process. Correct application of the standard supports safe work performance, reduces costs and the number of errors, allows to maintain high quality of products and allows new employees to adapt faster to the performed process [26]. By documenting the best practices, information is created about the course of the process, which is the basis for further continuous improvement. The standard is not set once and for all, it is modified with changes in product technology and changes proposed by employees.

Introducing any improvements on the basis of the conducted analyzes, specific root causes, or the applied tools for improving processes and positions, is associated with the need to apply the mainstream, which is a kind of rhythm of actions for problem solving - the PDCA cycle, developed by A. Deming [14]. In practice, these are the next stages of the procedure, starting from identifying the problem (P - Plan) through the use of available resources to solve it (D - do) and assigning process elements to the decision made (C - Check), to increasing the level and quality of work thanks to the applied solution (A - Act) [27, 28].

## 3 Characteristics of the Foreman's Work

The analyzed company manufactures ventilation ducts and fittings made of galvanized sheet. Modern machines equipped with PLC controllers and numerical control are used for the production of products. The company's production is based on the principles of production to order (MTO). This forces a high pace of work and significantly hinders the symmetrical distribution of the load on individual machines.

The role of people supervising the work of production line employees is played by foremen. They assign duties to employees and control the way tasks are performed, taking into account the even division of work, health and safety and the current load on the line. The foremen work in shifts according to the production shifts. The company has formally defined key responsibilities for foremen, which form the basis for further research.

1. Pay particular attention to bottlenecks (on each line) and their monitoring and reporting of production runoff. In particular, identifying activities ensuring the continuity of their work, including to appoint operators who are to handle the bottleneck during scheduled breaks, or to move the operator from another position to achieve a specific production goal.
2. Control of overproduction and preventing the generation of excessive inter-operational stock. To prevent overproduction, it is important to gather relevant information when walking around the hall, anticipate events that could have a negative

impact on production, and report data collected during the walk to leaders and management. Three critical questions are distinguished here, to which the foreman must be able to answer each time: Does the welding machine operator have material for the current and next order? Is there a stock of gaskets on the ROE machine? When will retooling occur?

3. Production runoff control. When the production runoff is too low, and there are no problems in production, the foreman is obliged to react by issuing a red report. This is the first level of the employee warning, the next red report results in the withdrawal of the employee's bonus. However, if the process generates more than 10% of waste, the foreman is obliged to stop the production.

In the further part of the article, an analysis of the current state will be presented, a proposal for improvements and an analysis of the foreman's work after the improvements have been implemented. The reason why the topic of analysis and improvement of foremen's work is important for the analyzed company is the foreman's frequent lack of knowledge about the current operation of the line, and also problems with the flow of products from production caused by stoppages on the line. This entails delays in deliveries to customers, as well as the accumulation of materials in inter-operational warehouses and finished goods in the warehouse.

## 4 Current State Analysis

### 4.1 Research Methodology

Before starting the research, the scope of responsibilities of foremen was carefully analyzed. Then it was assumed that all measurements would be performed for four different foremen. They were given numbers from 1 to 4. Foreman 1 and 2 control area 1, while foreman 3 and 4 supervise area 2. Due to the fact that they work in shifts, the measurements were carried out so that the production load was as similar as possible and that each of them work with the same leader - that is, during the first shift from 6:00 to 14:00. The image of 3 full working shifts for each foreman was adopted as the appropriate sample for the research. The research scheme is shown in Fig. 1.

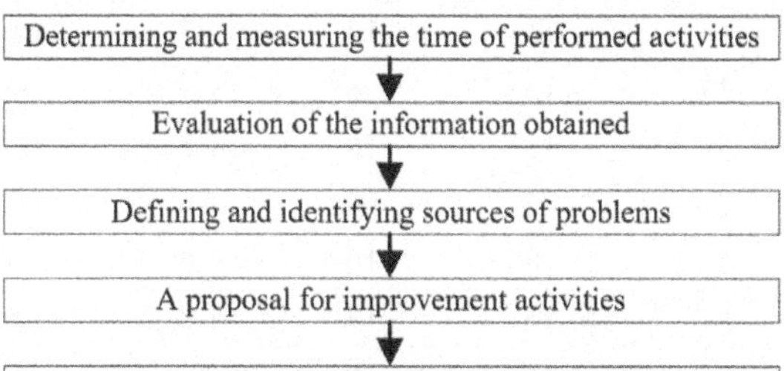


**Fig. 1.** Scheme of performing the test.

The research began with recording and measuring the time of activities undertaken by foremen during the analyzed period. In the next stage, based on the observations and available materials of the company, a list of key information that the foreman should collect during the line control was developed and it was checked whether it was obtained. The regularity of the hourly monitoring of the runoff was checked in a similar way. If the workstation was inspected within a specified period, it was referred to as correct inspection, if not - incorrect inspection. In the third stage, analyzes were made regarding the identification of the sources of the problems. For this purpose, the 5Why analysis and the Ishikawa diagram were used. Then, solutions were proposed. Their aim were eliminating the causes of the problems. At the last stage, the activities performed were analyzed again to check whether the proposed solutions brought the expected results.

### 4.2 Record of Performed Activities

In the first stage of the research, a detailed record of activities performed by foremen 1–4 was made. The research was carried out until a total of twelve entries of a full eight working hours were achieved (3 entries for each foreman). The timing method was used for this purpose, and 1 min was taken as the unit of time. Additionally, during the measurements, the popular Endomondo application was used, which counts the total distance traveled during the day. During the data collection, no steps were taken that interfered with the natural behavior of the foremen. The results obtained in this way were analyzed in terms of their compliance with the official duties of the employee. Any tasks not included in the schedule of responsibilities were considered as having no added value, i.e. additional activities. In addition, the route of the foreman during his work in a given hour is presented. Table 1 shows the obtained results for foreman 2.

**Table 1.** Record of the foreman's working time 2.

| Foreman 2 | | | | | | |
|---|---|---|---|---|---|---|
| Hour of work | Time for additional activities [minutes] | | | Distance traveled [kilometers] | | |
| | Day 1 | Day 2 | Day 3 | Day 1 | Day 2 | Day 3 |
| 6:00–7:00 | 17 | 11 | 11 | 2,2 | 2 | 1,8 |
| 7:00–8:00 | 16 | 13 | 14 | 1,7 | 1,6 | 1,7 |
| 8:00–9:00 | 23 | 13 | 8 | 1,9 | 1,4 | 1,5 |
| 9:00–10:00 | 19 | 13 | 16 | 1,5 | 1,9 | 1,9 |
| 10:15–11:00 | 13 | 12 | 10 | 1,5 | 1,1 | 1,4 |
| 11:00–11:55 | 9 | 18 | 16 | 1,6 | 1,7 | 1,6 |
| 12:00–13:00 | 13 | 15 | 13 | 1,8 | 1,6 | 1,6 |
| 13:00–14:00 | 11 | 12 | 15 | 1,9 | 1,8 | 1,7 |
| Average | 15 | 13 | 13 | 1,76 | 1,64 | 1,65 |
| Daily [minutes] | 121 | 106 | 103 | 14,1 | 13,1 | 13,2 |

The data collected for all foremen show that the waste of time amounts to up to 30% of the total working time. The measurements showed that it is not dependent on the person and the environment, because in all the measurements carried out, the average value did not drop below 17.7% - Fig. 2. Information about the time of additional activities may be helpful when implementing improvements, as focusing on the biggest problems generates the greatest potential for improvement.

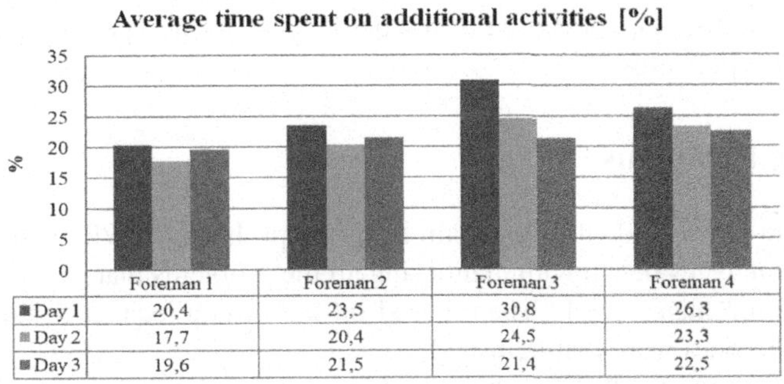


**Fig. 2.** Average daily time spent on additional activities.

### 4.3 Assessment of Information Gathering

The next step was to assess the reliability in collecting information from the subordinate production area. For this purpose, an auxiliary table with questions related to the duties of the foreman was made. The evaluation criterion was zero-one. If the foreman collected the information, 1 was recorded, if not - 0. For each foreman, 50 single assessments were made. The values in Table 2 were then calculated as the ratio of the number of ratings "1" to "0". Consequently, the higher the percentage in the cell, the more often the information has been correctly collected. Columns marked as B1, B2, B3 and B4 refer to individual foremen, as previously numbered.

**Table 2.** Key information collected by the foreman.

| No | Question | Result [%] | | | | |
|---|---|---|---|---|---|---|
| | | B1 | B2 | B3 | B4 | Average |
| 1 | Is the production plan for the workstation up-to-date? | 64 | 60 | 67 | 65 | 64 |
| 2 | Does the operator know the production sequence and the number of products that should be made? | 54 | 49 | 53 | 41 | 48 |
| 3 | Is material available for the next production order? | 40 | 38 | 67 | 61 | 55 |
| 4 | Has information been collected on a potential changeover? | 72 | 80 | 68 | 78 | 75 |
| 5 | Is there a stock of gaskets on the ROE machine? | 100 | 91 | 96 | 94 | 94 |
| 6 | Does the bottleneck employee know the rules of work of his workplace? | 63 | 69 | 75 | 72 | 72 |
| 7 | Has the order of the workstation been noted? | 81 | 90 | 74 | 85 | 83 |
| 8 | Is the amount of waste monitored? | 89 | 78 | 81 | 86 | 82 |
| 9 | Has the table for hourly monitoring of product run-off been completed? | 58 | 56 | 54 | 57 | 56 |
| 10 | Have the data for hourly monitoring of product run-off been read from the meter? | 36 | 27 | 41 | 35 | 35 |

The data in the table shows that, regardless of the monitored area, the biggest problem is the clear completion of the cards with the production plan, i.e. informing employees about the work sequence, finding materials and planning production changeovers. The fields that require only minimal correction are the care for the order of the machine's working area and the re-explanation of the bottlenecks to the operators of the rules of working at these positions.

## 5 Defining Problems

Two tools were used to identify the sources of the problems: 5Why and the Ishikawa diagram. The analyzes began with identifying all the problems that occurred during the monitoring of foremen's work. Then, based on the time allocated to them, the most time-consuming ones were distinguished.

Table 3 summarizes all additional activities that were noted during the recording of activities performed by foremen and completing the checklist. However, in order to select the most important activities that constitute an obstacle to the correct and efficient fulfillment of duties, the time spent on each of them should be analyzed, therefore each of the listed activities was assigned a sum of minutes allocated to them by all foremen.

**Table 3.** Additional activities of the foremen.

| No | Additional activity, which is an obstacle to the efficient fulfillment of duties | Total time spent on a task by all foremen [min] |
|---|---|---|
| 1 | Ordering missing components from a warehouse | 88 |
| 2 | Long waiting time for the machine set up | 140 |
| 3 | Organizing components needed for production | 83 |
| 4 | Re-filling an unreadable production plan | 129 |
| 5 | Updating the production plan at the operator's place | 87 |
| 6 | Repeated need to instruct employees on the bottleneck | 171 |
| 7 | Cleaning the inter-operational warehouse | 98 |
| 8 | Ordering missing sheets for production | 140 |
| 9 | Multiple checking of material availability | 122 |
| 10 | Contact with the Leader in making basic decisions | 82 |
| 11 | Chaos in movement, unnecessary walking | 100 |

Each of the above-mentioned activities was analyzed using the 5W methods (Fig. 3) and the Ishikawa diagram (Fig. 4). To illustrate the problem analysis process, an example analysis will be presented for additional activity No. 6: Multiple necessity to instruct employees in the area of work of the bottleneck, marking cartons and labels for products. This activity was chosen because of the greatest labor-consumption.

The main conclusion from the analysis of all activities/problems is the lack of work standards, both for foremen and operators. This determines the need to prepare training materials covering all the most important elements of the operator's work, so that the operator can perform the indicated activities himself, without the interference of the foreman and the introduction of clear and accurate standards for the work of foremen. Important elements will also be the correct placement of the instructions and informing the management about the current knowledge of the operators, as a consequence of carefully paying attention to the training of new employees.

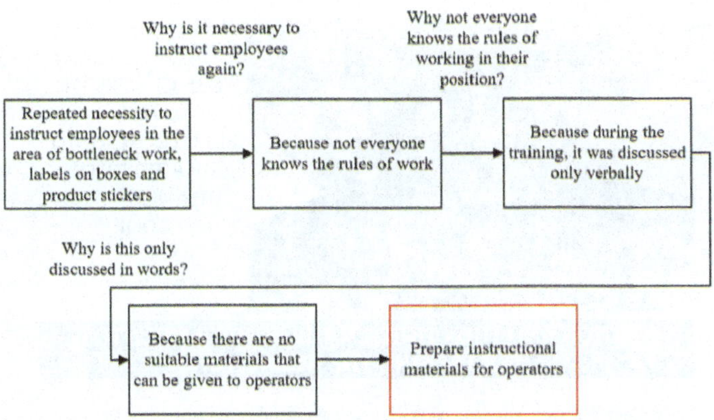


**Fig. 3.** 5W analysis of additional activity no. 6.

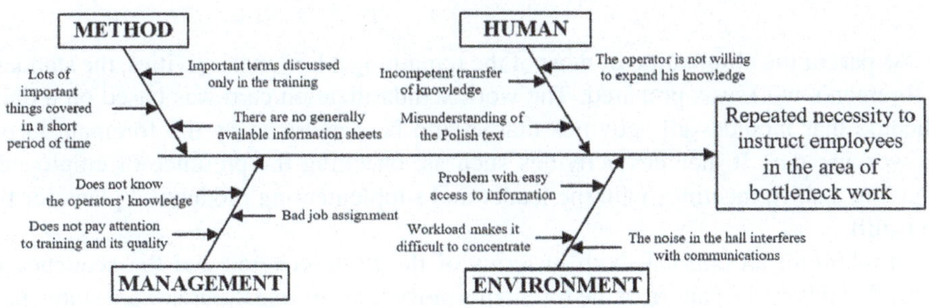


**Fig. 4.** Ishikawa diagram for activity no. 6.

## 6 Solutions of Defined Problems

The key factor determining the occurrence of additional activities in the work of foremen is the lack of work standards and information/training materials for operators, and the lack of clear and legible standards of foremen's work.

In order to eliminate the longest additional activity performed by the foreman (additional activity no. 6, Table 3), information sheets for positions were proposed, containing key information for the job, such as work rules or information on packaging materials. Pictures of labeled items that operators know from their daily work and surroundings have also been added.

As part of the elimination of additional activity no. 2 (see Table 3), an instructional video on retooling was prepared for machine operators. The instructions are based on descriptions of the activities performed and a video showing the step of changing the tools. The substantive part was consulted with the most experienced toolmaker and the company's technical department and prepared in accordance with the SMED method. An excerpt from the video is shown in Fig. 5.

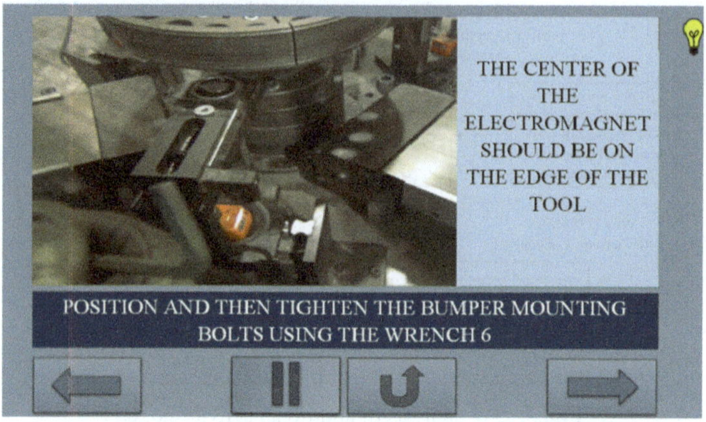


**Fig. 5.** An excerpt from instructional video on retooling.

As part of the elimination of most of the remaining additional activities, the standard of foremen's work was prepared. The work standardization card was based on a daily schedule that includes all activities that should be performed by the foreman during the working day. It includes activities such as: checking the presence of employees, receiving a shift, meetings with the leader and supplementing production plans for the next shift.

In addition, an optimal, both in terms of the route covered and the sequence of controlled positions, pattern of the foreman's movement around the area was established. According to the assumptions, the foreman will walk this route every full hour, filling in the tables for hourly kayaking monitoring. During such a round, he will also pay particular attention to the parameters of key importance for the operation of the line. The collected information will be recorded by the foreman, thanks to which, after the inspection is completed, he will receive a clear picture of the hall and all the most important problems. Figure 6 shows a fragment of the foreman's work standardization card, showing the scheme of moving around the hall.

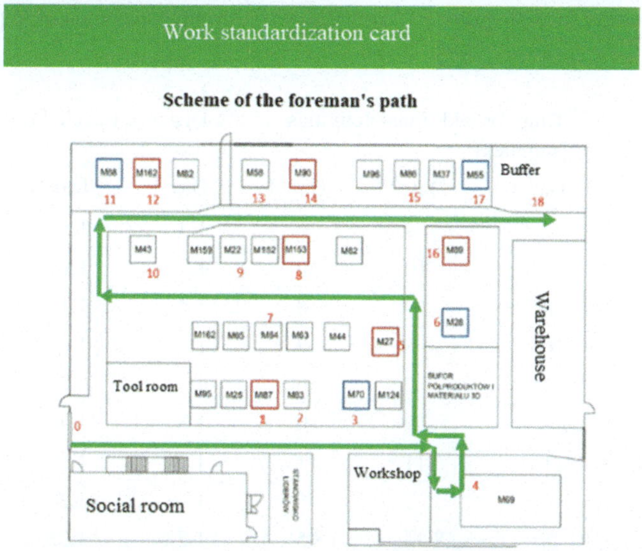


**Fig. 6.** A fragment of the foreman's work standardization card.

In addition to the guidelines contained in the work standardization cards, a basic decision-making scheme was prepared in consultation with the production leader, so that the actions performed by the foremen were carried out in the same way in each case. The diagram shows the steps to be taken when inspecting the position. By influencing the process in this way, the difference in problem-solving skills of individual people was also limited.

In addition to the above-mentioned improvements, the additional activity no. 4 was also leveled by implementing the standard of completing the production plan. The standard also took the form of an instruction that contains the individual steps of completing the production plan and guidelines for the most common problems.

## 7 Analysis of the Impact of Improvement Actions

After the improvements were implemented, the activities performed by the foremen were re-analyzed. These studies are intended to determine how much influence the applied solutions had on the work of foremen. The observation pattern remained unchanged.

The analysis shows that the foreman 2 supervising area 1, for which the results of the current state of analysis were presented in the previous chapters, after introducing all the improvements and standards, changed his work so much that the results of the analyzes were similar to those of Foreman 1, who achieved the best results in initial analyzes. On average, during the day, he spends 8 min an hour on additional activities and travels about 10 km. Compared to the original condition, this represents an improvement of approximately 40 min and 13 km per day. Table 4 shows the obtained results for foreman 2.

**Table 4.** Record of working time of foreman 2 after the implementation of improvements.

| Foreman 2 | | | | | | |
|---|---|---|---|---|---|---|
| Hour of work | Time for additional activities [minutes] | | | Distance traveled [kilometers] | | |
| | Day 1 | Day 2 | Day 3 | Day 1 | Day 2 | Day 3 |
| 6:00–7:00 | 4 | 4 | 6 | 1,3 | 1,5 | 1,5 |
| 7:00–8:00 | 9 | 10 | 10 | 1,2 | 1,3 | 1,4 |
| 8:00–9:00 | 8 | 9 | 8 | 1,4 | 1,1 | 1,2 |
| 9:00–10:00 | 10 | 8 | 9 | 1,2 | 1,4 | 1,3 |
| 10:15–11:00 | 7 | 7 | 6 | 1 | 0,8 | 1,1 |
| 11:00–11:55 | 8 | 9 | 10 | 1,3 | 1,4 | 1,3 |
| 12:00–13:00 | 11 | 11 | 11 | 1,4 | 1,3 | 1,4 |
| 13:00–14:00 | 7 | 7 | 8 | 1 | 1,2 | 1 |
| Average | 8 | 8,13 | 8,5 | 1,23 | 1,25 | 1,28 |
| Daily [minutes] | 64 | 65 | 68 | 9,8 | 10 | 10,2 |

Summarizing the data collected, time is wasted up to 15.9% of the total working time. Compared to the previous result, this is an improvement by almost half a percentage point (from 30% to 15.9%). The actions taken affected each of the foremen, which proves that the problems were correctly defined and that appropriate remedial steps were applied. In general, for all records of working time, the results are shown in Fig. 7. The percentage of time spent on additional activities fluctuates around 14%. Moreover, this level for each of the foremen is very similar, despite the fact that previously the disproportions were significant.

As discussed earlier, the foreman during the control of individual positions is obliged to collect information on the production. The most important of them were included in the work standardization card, no problems were found with obtaining the remaining ones. The percentage results on how many cases the foremen collected the relevant information are presented in Table 5.

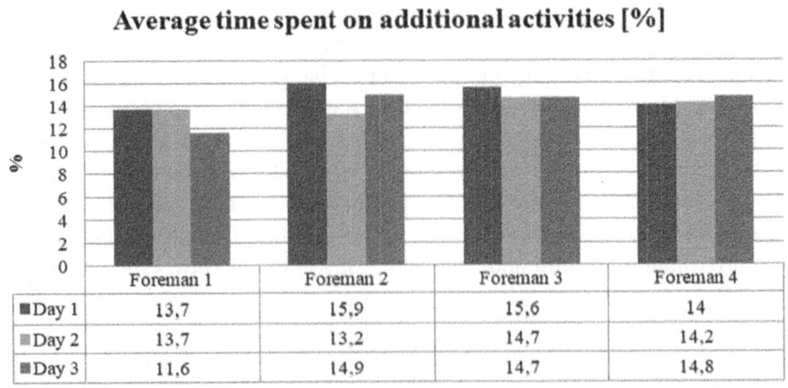


**Fig. 7.** Average daily time spent on additional activities.

**Table 5.** Key information collected by the foreman.

| No | Question | Result [%] | | | | |
|---|---|---|---|---|---|---|
| | | B1 | B2 | B3 | B4 | Average |
| 1 | Is the production plan for the workstation up-to-date? | 90 | 87 | 86 | 87 | 87 |
| 2 | Does the operator know the production sequence and the number of products that should be made? | 86 | 83 | 82 | 83 | 83 |
| 3 | Is material available for the next production order? | 92 | 86 | 80 | 75 | 80 |
| 4 | Has information been collected on a potential changeover? | 95 | 92 | 90 | 81 | 88 |
| 5 | Is there a stock of gaskets on the ROE machine? | 93 | 90 | 88 | 71 | 83 |
| 6 | Does the bottleneck employee know the rules of work of his workplace? | 90 | 87 | 85 | 89 | 87 |
| 7 | Has the order of the workstation been noted? | 90 | 87 | 88 | 89 | 88 |
| 8 | Is the amount of waste monitored? | 94 | 91 | 92 | 93 | 92 |
| 9 | Has the table for hourly monitoring of product run-off been completed? | 89 | 89 | 94 | 85 | 89 |
| 10 | Have the data for hourly monitoring of product run-off been read from the meter? | 68 | 69 | 71 | 83 | 72 |

Compared to the status before the improvements, the greatest improvement can be seen in points 2 and 3, which concern, in turn, the knowledge of the production sequence and checking the availability of material for the next order. This improvement is 35% for question number 2 and 25% for question number 3. The measures taken also had a positive impact on providing operators with an up-to-date production plan - point 1, an increase of 23%.

The influence of the introduced standards and, above all, the milkman's way can also be seen in the systematic control of the hourly monitoring of the runoff. The regularity of hourly monitoring of the runoff fluctuates around 90%. Compared to the situation before the improvements, this is an improvement of approximately 30%. Due to the characteristics of the work and the narrow time window adopted for the measurements, this value is satisfactory. In addition, the frequency of reading data from the meter increased significantly, i.e. by about 40%, and thus also the quality of this data.

## 8 Conclusions

The article concerns the improvement of the foreman's work, which ultimately also brought positive results in the efficient operation of production and the work of operators. The work of production foremen was analyzed, then the areas of work requiring improvement were identified and improvement actions were introduced. The results of the work carried out are presented in Table 6. For each foreman, there was an improvement, the average values are 40 min less time per day spent on additional activities and on average 3 km less per day, doing the same work.

**Table 6.** The results of the improvements.

| Foreman | Average time spent on additional activities during the day [min]: | | Average distance traveled during the day [km]: | |
|---|---|---|---|---|
| | Before the improvements | After the improvements | Before the improvements | After the improvements |
| Foreman 1 | 88 | 59 | 12,3 | 9,7 |
| Foreman 2 | 110 | 65 | 13,3 | 10 |
| Foreman 3 | 118 | 68 | 13,3 | 9,8 |
| Foreman 4 | 102 | 66 | 11,7 | 9,7 |

Thanks to the introduced improvements, foremen can focus on performing the activities described in their scopes of duties, and this in turn translates into greater production control, in particular the work of the bottleneck, and faster ability to react appropriately to emerging problems.

The conducted research and the introduced improvements also clearly show how important it is to implement and comply with the standards. This has a positive effect not only on the company in terms of reduced waste, but also on employees and foremen. They gain a calmer working environment, know the guidelines of their work precisely and can fulfill it satisfactorily. Nevertheless, individual improvement actions are only a temporary solution, as management control over compliance with the rules and employees' self-control are necessary. What's more, after some time, when the company changes or the standard is well adopted, there is room for further improvement, in line with the concept of continuous improvement.

**Acknowledgments.** The research work presented here was made possible by the Ministry of Education and Science of Poland.

## References

1. Henao, R., Sarache, W.: Lean manufacturing and sustainable performance: trends and future challenges. J. Clean. Prod. **208**, 99–116 (2019)
2. Ikumapayi, O.M., Akinlabi, E.T., Mwema, F.M., Ogbonna, O.S.: Six sigma versus lean manufacturing – An overview. Mater. Today Proc. **26**(2), 3275–3281 (2020)
3. Kujawińska, A., et al.: Methodology supporting the planning of machining allowances in the wood industry. In: Pérez García, H., Alfonso-Cendón, J., Sánchez González, L., Quintián, H., Corchado, E. (eds.) SOCO/CISIS/ICEUTE -2017. AISC, vol. 649, pp. 338–347. Springer, Cham (2018). https://doi.org/10.1007/978-3-319-67180-2_33
4. Anh, N.N., Phan, A.C., Thoa, P.T.X., Matsui, Y.: Quality information and quality performance: roles of process control as a mediator and shop floor leadership as a moderator. Manag. Prod. Eng. Rev. **12**(4), 13–26 (2021)
5. Kujawińska, A., Diering, M.: The impact of the organization of the visual inspection process on its effectiveness. Int. J. Adv. Manuf. Technol. JAMT-D-20–02941R1 **112**, 1295–1306 (2021). https://doi.org/10.1007/s00170-020-06543-9

6. Sabry, I.: Six Sigma methodology used to improve the mechanical properties for friction stir welding of aluminum pipes. Manag. Prod. Eng. Rev. **11**(1), 73–78 (2020)
7. Kotliar, A., et al.: Ensuring the economic efficiency of enterprises by multi-criteria selection of the optimal manufacturing process. Manag. Prod. Eng. Rev. **11**(1), 52–61 (2020). https://doi.org/10.24425/mper.2020.132943
8. Trojanowska, J., Żywicki, K., Varela, M.L.R., Machado, J.: Shortening changeover time—an industrial study. In: 10th Iberian Conference on Information Systems and Technologies (CISTI), pp. 1–6 (2015)
9. Rewers, P., Trojanowska, J., Diakun, J., Rocha, A., Reis, L.P.: A study of priority rules for a levelled production plan. In: Hamrol, A., Ciszak, O., Legutko, S., Jurczyk, M. (eds.) Advances in Manufacturing. LNME, pp. 111–120. Springer, Cham (2018). https://doi.org/10.1007/978-3-319-68619-6_11
10. Turconi, G., Ventola, G., González-Prida, V., Parra, C., Crespo, A.: A literature review on lean manufacturing in the Industry 4.0: from integrated systems to IoT and smart factories. In: Verma, J.K., Saxena, D., González-Prida, V. (eds.) IoT and Cloud Computing for Societal Good. EAI/Springer Innovations in Communication and Computing. Springer, Cham (2022)
11. Steckler, N., Fondas, N.: Building team leader effectiveness: a diagnostic tool. Organ. Dyn. **23**(3), 20–35 (1995)
12. Marksberry, P.W.: A new approach in analysing social-technical roles at Toyota: the team leader. Int. J. Hum. Resour. Dev. Manag. **10**(4), 395–412 (2010)
13. Åhlström, P.: Sequences in the implementation of lean production. Eur. Manag. J. **16**(3), 327–334 (1998)
14. Pavnaskar, S.J., Gershenson, J.K., Jambekar, A.B.: Classification scheme for lean manufacturing tools **41**(13), 3075–3090 (2003)
15. Palange, A., Dhatrak, P.: Lean manufacturing a vital tool to enhance productivity in manufacturing. Mater. Today Proc. **46**(1), 729–736 (2021)
16. Kiyatkina, E.P., Kornilova, A.D., Kirsanova, M.V.: Production Management. Science Book Publishing House LLC, Yelm (2014)
17. Krajewski, L.J., Ritzman, L.P.: Operations Management: Strategy and Analysis. Prentice Hall, New York (2002)
18. Luca, L.: A new model of Ishikawa diagram for quality assessment. Mater. Sci. Eng. **161**(1), 012099 (2016)
19. Ishikawa, K., Loftus, J.H.: Introduction to Quality Control. 3A Corporation, Tokyo (1990)
20. Ershadi, M.J., Aiasi, R., Kazemi, S.: Root cause analysis in quality problem solving of research information systems: a case study. Int. J. Prod. Qual. Manag. **24**(2), 284–299 (2018)
21. Prashant, G.: A systematic approach to root cause analysis using 3 × 5 why's technique. Int. J. Lean Six Sigma (2019)
22. Rewers, P., Trojanowska, J.: Production management by using tools of lean manufacturing. Logist. Manag. Dev. trends **2008**, 43–56 (2016)
23. Sundar, R., Balaji, A.N., Satheesh Kumar, R.M.: A review on lean manufacturing implementation techniques. Procedia Eng. **97**, 1875–1885 (2014)
24. Gomathi Prabha, V.M.: Work standardization and line balancing in a windmill gearbox manufacturing cell: a case study. Mater. Today Proc. **46**(19), 9721–9729 (2021)
25. Feng, P., Ballard, G.: Standard work from a lean theory perspective. In: Annual Conference of the International Group for Lean Construction (2008)
26. Productivity Press Development Team: Standard Work for the shopfloor. Productivity Press (2016)
27. Realyvásquez-Vargas, A.: Applying the Plan-Do-Check-Act (PDCA) cycle to reduce the defects in the manufacturing industry. A case study. Appl. Sci. **8**(11), 2181 (2018)
28. Ruano, J.: Lean school: a learning factory for training lean manufacturing in a physical simulation environment. Manag. Prod. Eng. Rev. **10** (2019)

# The Overall Labour Effectiveness to Improve Competitiveness and Productivity in Human-Centered Manufacturing

Andrea Bonci[1(✉)], Dorota Stadnicka[2], and Sauro Longhi[1]

[1] Department of Information Engineering (DII), Polytechnic University of Marche, via Brecce Bianche, 12, 60131 Ancona, Italy
{a.bonci,sauro.longhi}@univpm.it
[2] Faculty of Mechanical Engineering and Aeronautics, Rzeszów University of Technology, al. Powstańców Warszawy 12, 35-959 Rzeszow, Poland
dorota.stadnicka@prz.edu.pl

**Abstract.** The assessment of labour efficiency is important for manufacturing companies. Even though employees are a very important resource of the organization, at the same time they constitute a cost for the company. So, it is vital to make the most of their operational readiness. In this study, the possibilities of using the overall labour effectiveness indicator at the level of the entire enterprise are analysed. The revised OLE (ROLE) indicator is also discussed along with its use in the evaluation of overall production effectiveness (OTE) to address the monitoring and diagnosis of factory-level performance also considering workforce issues. It was also proposed to introduce a new LEAN-ROLE indicator, which not only assesses the effectiveness of human resources, but also identifies the percentage of work done by employees that creates value for the customer. By introducing digitization in enterprises and registering the work performed by employees in databases as well as the duration of these works, it is possible to measure the effectiveness of employees' work on an ongoing basis. This will allow to identify weaknesses in a system. Then, after identifying the causes of decreased efficiency, appropriate actions can be introduced to improve employee involvement in value creation.

**Keywords:** Overall Labour Effectiveness · Key Performance Indicators (KPI) · Lean Manufacturing

## 1 Introduction

Digitization is becoming a must and ubiquitous in all phases of industrial, manufacturing and craft production. The so-called cyber-physical systems (CPSs) address the integration of the physical parts of manufacturing systems with the logical and computational parts (the cyber world) using digital technologies to improve productivity, quality, accuracy, flexibility, and efficiency [1]. For example, a Kanban system can be introduced to facilitate the flow of products in the production system while minimizing the work in


J. Trojanowska et al. (Eds.): MANUFACTURING 2022, LNME, pp. 144–155, 2022.
https://doi.org/10.1007/978-3-030-99310-8_12

progress (WIP) [2]. At the same time, there is an emerging need to ensure that the human operator plays a key role in the digitized and integrated production phases of the new factories of the future [3]. In this context, human-centered manufacturing focuses on the role of humans in new production systems and their increasingly close integration and collaboration with the pervasive intelligent and autonomous entities introduced by new digital technologies [4, 5]. In the human-centered manufacturing paradigm collaborative work efficiency, human safety, ergonomics, or advanced applications concerning brain signals are of utmost importance, e.g., studies on intention detection [6] or applications of brain computer interfaces [7]. Although at first glance these aspects may seem contradictory, they are only different levels of a future socially responsible automation (SRA) [8] which is widely recommended to help technology developers and business leaders drive the evolution of new automation for the good of society. In this vision, principles to be satisfied includes a) the preservation of the central and fundamental role of humans in the workplaces of the future, through their ability to control, complement and support technological solutions; and b) the role of automation, artificial intelligence, machine learning and related technologies only as tools to improve and enrich human livelihoods and lives. The SRA identifies four factory automation levels according to their business goals, which are: (1) cost-focused automation, (2) performance-driven automation, (3) human-centered automation and (4) SRA. This paper will focus on the second level of automation, performance-centric automation, with the main goal of incorporating the human factor into performance evaluation by considering humans as an integral part of human-machine collaboration systems. This will ensure that overall process performance is not lost sight of in the next level of automation, human-centric automation, where the focus is on developing the critical and valuable role of people in human-machine collaboration. In the authors' opinion, the overall evaluation of processes performance still misses the role of humans in the loop. Common saying states that "you get what you measure", so the first step in getting a workforce performance assessment is to introduce a measurement metric. Surprisingly, even though manufacturers track attendance and time, they rarely have a method for measuring, or understanding, how workforce actions directly affect productivity. Identifying ways to help the workforce become more productive presents both a major challenge and an opportunity to effectively pursue the goals of human-centered automation and allow manufacturers to find other ways to further increase competitiveness. To date, most measurement systems focus on machine effectiveness or production output. Powerful tools that can be used to measure performance and also perform diagnostics at the equipment and factory level are the OEE (Overall Equipment Effectiveness) [9] and OTE (Overall Throughput Effectiveness) [10] metrics, respectively. OEE has been recognized as a relevant method for measuring equipment performance since the late 1980s [11], today OEE is undoubtedly the most widely known and accepted key performance indicator (KPI) [12–14] for measuring the productivity of individual production equipment in a factory. Although relevant, OEE is not sufficient to assess the efficiency of an integrated equipment system and monitor performance at the factory level. To this end, OTE metrics were initially proposed in [10, 15] and further developed and applied in the CPS context in [16]. OTEs are applied to subsystems by which the layout of the entire manufacturing system is classified. Such unique subsystems include "series", "parallel", "assembly" and "expansion" configurations [10].

Algorithms have been developed to automate the factory level performance monitoring and diagnostic process using OTE. Although OTE has its own formalization, it is not as widely used as OEE.

Despite the undoubted usefulness of these indicators, they properly address the equipment/machinery factor, neglecting any further investigation with respect to the workforce, energy, or materials factors. Beyond equipment-related operational failures, problems can often be traced to a variety of workforce issues that accumulated while the plants were operating at less-than-optimal capacity. Probable causes include difficulty in scheduling the right resources when and where specific skills were required. Absenteeism of some workers, or ineffectiveness or lack of training that limited quality and slowed production ramp-up, as well as product changes or new product launches. These factors highlight how labour is an equally critical element of production to be optimized in both the current and future demand-driven manufacturing systems.

In this regard a further KPI called Overall Labour Effectiveness (OLE) [17–19], and more recently a revised version of it the Revised OLE (ROLE) [20], have been introduced to consider labour effectiveness. They provide a holistic method for measuring work-related losses, while maintaining a certain formal similarity with Overall Equipment Effectiveness (OEE). Despite this kind of KPI is a useful tool to help the manager understand the labour effectiveness with respect to the overall productivity, it has never been adequately supported by IT (Information Technology) solutions thus it is not yet widely used to support factory level performance monitoring. Support in making decisions is one of the technological tools that are indispensable in the factories of the future [21]. Moreover, the indicator can be considered as one of the important indicators in the Intelligent Management system [22]. It can be taken into consideration in manufacturing production and supply planning as one of the important factors [23].

The main goal of the paper is to analyse the labour factor effects on the factory level indicators (OTEs) and present how the lean concept of taking only those actions that create value or are really necessary can be included in the indicator calculation.

The remainder of the paper is organized as follow. Section 2 presents a definition of OLE together with a structure of the loses which influence the OLE value. Then, the ROLE indicator is explained. Section 3 discusses an influence of ROLE at factory level and OTE evaluation. In Sect. 4 a LEAN-ROLE indicator as a new indicator is proposed.

## 2 The Overall Labour Effectiveness and Its Revised Version

The OLE original definition [17], in analogy with the OEE, measures the cumulative effect and the interdependency of availability, performance and quality, for individuals (it could be also enlarged to teams). Mainly the ROLE [20] is organised in the same way but it uses a different structure of losses and a new methodology for both collecting data and effectively measuring labour effectiveness.

Instead of the classical OEE subdivision, in ROLE the losses are related to four main categories (Fig. 1): structural losses (SL), abnormal losses (ANL), management-driven losses (MDL) and machine/workplace losses (MWL), which are in turn composed of further sub-categories [24, 25].

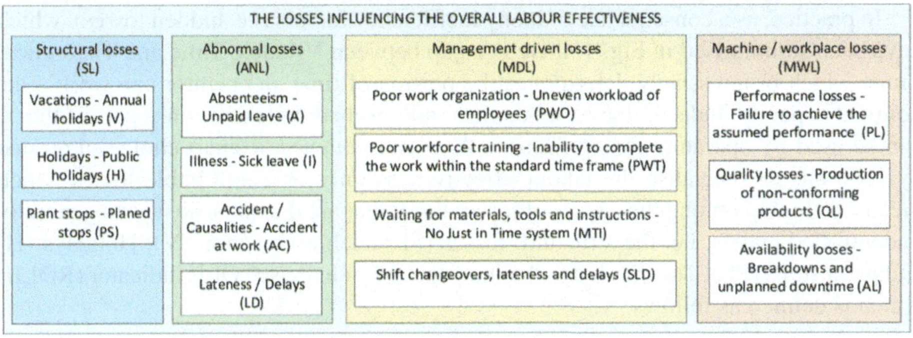


**Fig. 1.** The losses influencing the overall labour effectiveness.

Thus, on the one hand this allows to assess the sources of losses with their related impact and indicate improvement actions, while on the other hand the process of loss analysis become rather cumbersome and difficult to automate. For these reasons the ROLE is currently proposed as a tool to help the manager understand whether corrective actions have solved problems and improved overall productivity. Integrating worker effectiveness analysis into a cyber-physical system that integrates the cyber world (calculation, forecasting and analysis) with the physical world (processing, planning and production) requires synthesizing these indices efficiently and effectively.

By considering a generic production effectiveness indicator E, it can be theoretically defined as follows:

$$E = \frac{Cycle\ Time \cdot Valuable\ Time}{Cycle\ Time \cdot Theoretical\ Time} = \frac{Valuable\ Time}{Theoretical\ Time} = \frac{VT}{T_t} \quad (1)$$

where the Valuable Time (VT) represents the time in which a work activity is processed under optimal conditions, while the Theoretical Time ($T_t$) is the maximum interval of time that is ideally available for production.

Figure 2 presents temporal relationships between the causes of losses in the evaluation of the overall effectiveness.

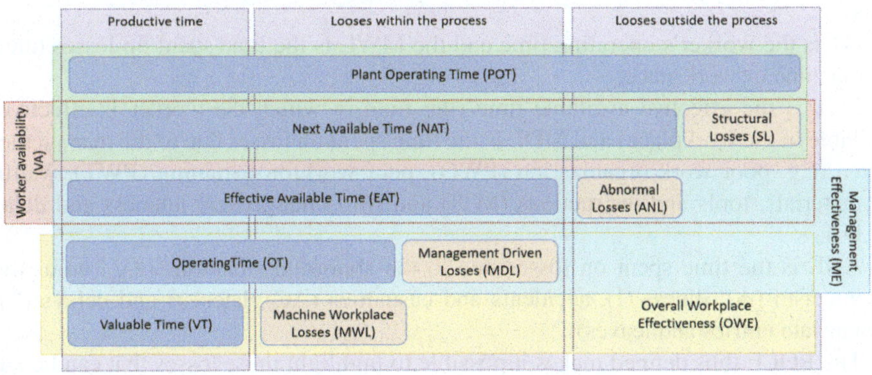


**Fig. 2.** Temporal relationships between the causes of losses in the evaluation of the overall effectiveness.

In practice, as a consequence of many inefficiency factors (i.e. hidden losses), which have been summarised in Fig. 1, there is a gap between Valuable Time and Theoretical Time, which progressively deteriorates the portion of time an operator can work at its nominal capacity. Indeed, due to scheduled and unscheduled stops, only part of the $T_t$ can be used by operators for production. As a consequence, a smart choice suggested by [17] was to specialise the labour effectiveness only on controllable losses which means excluding among the causes of losses those that are due to intrinsic and generally uncontrolled events, i.e. the structural losses (SL) such as vacation (V), Holidays (H) and plant stops (PS). This has led to the introduction of a revised OLE indicator (ROLE) which is defined as follows

$$ROLE = \frac{VT}{T_t} = \frac{VT}{NAT} \tag{2}$$

where the Net Available Time (NAT) is used as the $T_t$ under the assumption that only controllable losses can be tackled by means of improvement activities (i.e. NAT = POT − $Time_{(SL)}$, where $Time_{(SL)}$ = Time elapsed for structural losses). By analysing the time periods that make up the NAT and which are shown in Fig. 2, it is possible to express Eq. (2) as a function of three main causes of inefficiency, namely worker inefficiencies, management inefficiencies and worker availability

$$ROLE = \frac{VT}{NAT} = \frac{VT}{OT} \cdot \frac{OT}{EAT} \cdot \frac{EAT}{NAT} = OWE \cdot ME \cdot WA \tag{3}$$

where the three terms OWE, ME and WA respectively stand for Overall Workplace Effectiveness, Management Effectiveness and Worker Availability, and can be respectively computed as follows

$$OWE = \frac{OT - MWL}{OT} = \frac{VT}{OT} \tag{4}$$

$$ME = \frac{EAT - MDL}{EAT} = \frac{OT}{EAT} \tag{5}$$

$$WA = \frac{NAT - ANL}{NAT} = \frac{EAT}{NAT} \tag{6}$$

where:

OT is the worker's operating time and the MWL is the time spent on losses due to the machine or workspace;

EAT is the effective available time (the time in which the worker is effectively available in the workplace) and MDL is the time spent on losses due to the management driven (e.g. poor work organization (PWO), poor workforce training (PWT), waiting for materials, tools and instructions (MTI) and shift changeovers lateness and delays (SLD));

ANL is the time spent on losses due to the abnormal situations (e.g., unjustified absenteeism (A), illness (I), accidents and casualties (AC), lateness and delays (LD) such as late entries and leaves).

The ROLE thus defined makes it possible to highlight those losses that can be adequately observed and quantified. The formulation of the ROLE allows to evidence the losses in two different ways, in the first one the losses are due to the process (OWE and ME) and to those not dependent from it (WA). In the second one, WA is related to the worker's absence at the workplace, while both ME and OWE are related to the inefficiencies of workers within the workplace.

## 3 Influence of ROLE at Factory Level and OTE Evaluation

In the 4th industrial revolution, the role of human factor is a relevant topic and how to measure, on comparable metrics, the performance of humans and machines collaborating to allow an overall improvement is still an open issue. Even if the introduction of the ROLE can fill a gap at the tools level, it has been already pointed out [9] that the gains made at equipment level by introducing OEE, although important and still ongoing, are not enough to evaluate the overall production improvements because it is required to focus attention on the performance of the whole factory rather than limiting to the performance of individual tools.

Assuming ROLE as the indicator that provides a methodology to measure labour-related losses and highlighting its introduction to complement the limitations of the OEE metric that only measures machine performance, this section investigates the methods to take the ROLE indicator into account when evaluating aggregate factory-level indicators such as OTE.

Any factory layout can be thought of as consisting of some commonly occurring subsystems, four basic subsystems allow us to represent the most common factory layouts, these are the series, parallel, assembly and expansion. As a first attempt, the present paper will investigate the first two subsystems because among the possible combinations, series and parallel subsystems are widely used in real-world manufacturing systems. Moreover, the other types of subsystems can be deduced from these two basic types. Therefore, the formulas for calculating the OTE including ROLE for series and parallel type subsystems are presented.

Before going into the calculation of the OTE we will introduce some properties of the OEE that will be useful for introducing the ROLE in the OTE. Conventional or unit-based OEE is usually defined as:

$$OEE = A_{eff} \cdot P_{eff} \cdot Q_{eff} = \frac{T_u}{T_t} \cdot \left( \frac{T_p}{T_u} \cdot \frac{R_{avg}^{(a)}}{R_{avg}^{(th)}} \right) \cdot \frac{P_g}{P_a} \tag{7}$$

where, $A_{eff}$ = Availability efficiency (associated losses include non-scheduled down-time, breakdowns, set-up and adjustments, etc.), $A_{eff}$ = Performance efficiency (associated losses include idle, reduced speed, blockage, etc.), $Q_{eff}$ = Quality efficiency (associated losses include defects, rework, etc.), $T_u$ = equipment uptime, $T_t$ = total observation time of the equipment, $T_p$ = equipment production time, $R_{avg}^{(a)}$ = average actual processing rate for equipment in production for actual product output (for simplicity in the remainder of the paper it will be addressed as $R^{(a)}$), $R_{avg}^{(th)}$ = average theoretical processing rate for actual product output (for simplicity in the remainder of the paper it will be addressed as $R^{(th)}$), $P_g$ = good quality product output (units) from the equipment during $T_t$, $P_a$ = actual product units processed by equipment during $T_t$.

The Eq. (7) can be further simplified by considering that during $T_P$ the equipment might not be operating at its theoretical speed, so we can assume

$$R_{avg}^{(a)} \triangleq R^{(a)} = \frac{P_a}{T_p} \tag{8}$$

Thus, using Eqs. (7) and (8) leads the following alternative useful expression for OEE

$$OEE = \frac{P_g}{R^{(th)} \cdot T_t} \tag{9}$$

Assuming

$$P_a^{(th)} = R^{(th)} \cdot T_t \tag{10}$$

where, $P_a^{(th)}$ = actual attainable product output (units) that could have been produced according to the theoretical processing rate $R^{(th)}$ in total time $T_t$.

Based on Eq. (10) the OEE can be further simplified and re-defined as

$$OEE = \frac{P_g}{P_a^{(th)}} = \frac{good\ product\ output\ (units)\ in\ total\ time}{theoretical\ atteinable\ product\ output\ (units)\ in\ total\ time} \tag{11}$$

By means of this definition, OEE can be directly computed from measured $P_g$ and $P_a^{(th)}$ without using any other factors. Moreover, by extending the concept expressed in Eq. (11) to the factory level, it is possible to define the OTE as

$$OTE = \frac{P_g^F}{P_a^{(th)F}} = \frac{actual\ throughput\ (units)\ from\ factory\ in\ total\ time}{theoretical\ atteinable\ throughput\ (units)\ from\ factory\ in\ tot.time} \tag{12}$$

Now we have all the elements to introduce the ROLE in at the factory level, which means in the OTE metric.

Similarly, to the OEE described in Eq. (11), the ROLE can be further described in terms of products processed by labours rather than in terms of time as in Eq. (2), since we can assume.

$$ROLE = \frac{P_g}{P_a^{(th)}} = \frac{good\ product\ output\ (units)\ in\ net\ available\ time}{theoretical\ atteinable\ product\ output\ (units)\ in\ net\ available\ time} \tag{13}$$

Thus, by using (13), (10) and (2) and considering NAT as the total observation time, we have

$$P_g = ROLE \cdot R^{(th)} \cdot NAT \tag{14}$$

It is now possible to specialise the OTE for both the formulation of subsystems in series and in parallel, also considering the ROLE.

### 3.1 Series-Connected Subsystem

A series-connected subsystem consists of $n$ individual working stations connected as shown in Fig. 3.

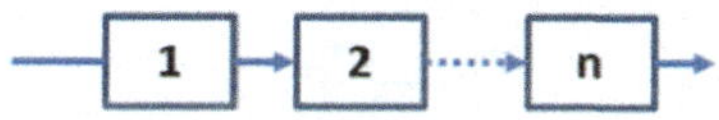


**Fig. 3.** Series-connected subsystem.

During the net available time NAT, by applying the theory of material flow conservation, the product output (units) with good quality of the working station $n$ must equal that of the series process, that is:

$$P_g^F = P_g^{(n)} \tag{15}$$

where $P_g^F$ *is the* good product output (units) of a factory during the period of NAT, while $P_g^{(n)}$ is the good product output (units) of the working station $n$.

Hence, by using (14) for the working station $n$, and substituting it in (15) we have

$$P_g^F = ROLE_{(n)} \cdot R_{(n)}^{(th)} \cdot NAT \tag{16}$$

Since in a series-connected subsystem the working station with the minimum processing rate is dominant in the production process, the theoretical processing rate of a series-connected subsystem in net available time NAT for actual product output (units) is the minimum between the n-subsystems, therefore Eq. (10) can be written as

$$P_a^{(th)F} = R_F^{(th)} \cdot NAT = min_{i=1,\ldots,n}\left\{R_{(i)}^{(th)}\right\} \cdot NAT \tag{17}$$

Analogously, in Eq. (16) also the term $ROLE_{(n)} \cdot R_{(n)}^{(th)}$ is limited by the minimum processing rate of the series, therefore by taking into account (17), Eq. (16) can be rewritten as

$$P_g^F = min\left\{min_{i=1,\ldots,n-1}\left\{ROLE_{(i)} \cdot R_{(i)}^{(th)}\right\}, ROLE_{(n)} \cdot R_{(n)}^{(th)}\right\} \tag{18}$$

Finally, by substituting (17) and (18) in (12) we have

$$OTE^{(s)} = \frac{min_{i=1,\ldots,n}\left\{ROLE_{(i)} \cdot R_{(i)}^{(th)}\right\}}{min_{i=1,\ldots,n}\left\{R_{(i)}^{(th)}\right\}} \tag{19}$$

It should be noted that since the quality of the work performed by the i-th operator is already included in the assessment of $ROLE_{(i)}$, there is no explicit dependence on the quality of work in the result.

### 3.2 Parallel-Connected Subsystem

A parallel-connected subsystem consists of $n$ individual working stations connected as shown in Fig. 4.

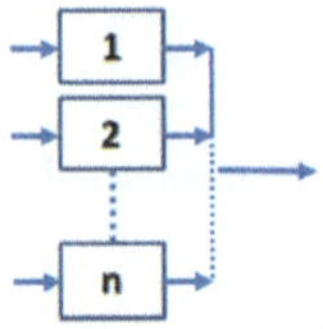


**Fig. 4.** Parallel-connected subsystem.

During the observation period of NAT, based on the theory of conservation of material flow, the product output (units) with good quality of all working stations must equal those of the parallel subsystem, while the actual product output (units) of all working stations must equal those of the parallel subsystem. Therefore

$$P_g^F = \sum\nolimits_{i=1}^{n} P_{g(i)} = \sum\nolimits_{i=1}^{n} ROLE_{(i)} \cdot R_{(i)}^{(th)} \cdot NAT \tag{20}$$

$$P_a^{(th)F} = \sum\nolimits_{i=1}^{n} P_{a(i)}^{(th)} = \sum\nolimits_{i=1}^{n} R_{(i)}^{(th)} \cdot NAT \tag{21}$$

by substituting (20) and (21) in (12) we have

$$OTE^{(p)} = \frac{\sum_{i=1}^{n} ROLE_{(i)} \cdot R_{(i)}^{(th)}}{\sum_{i=1}^{n} R_{(i)}^{(th)}} \tag{22}$$

## 4 A Proposal of LEAN-ROLE Indicator

In Lean Manufacturing all activities performed by employees can be divided into three groups: value added activities (VA), non-value added activities (NVA) and non-value added activities but necessary (NNVA) [26]. VA activities are directly connected with creating the value, i.e. the product or service, for a customer. NNVA activities are connected with other activities, such as e.g., changeover or trainings, which do not create value for the customers but are indispensable to realize production or deliver a service. NVA activities are pure waste. In Lean Manufacturing concept seven wastes are defined [27]: Overproduction (1), Transport (2), Inventory (3), Motion (4), Defects (5), Overprocessing (6), Waiting (7). Management Operative Effectiveness (MOE) (see Fig. 5) depends on what tasks will be set to be performed at the operational level and how the work at the operational level will be planned and organized.

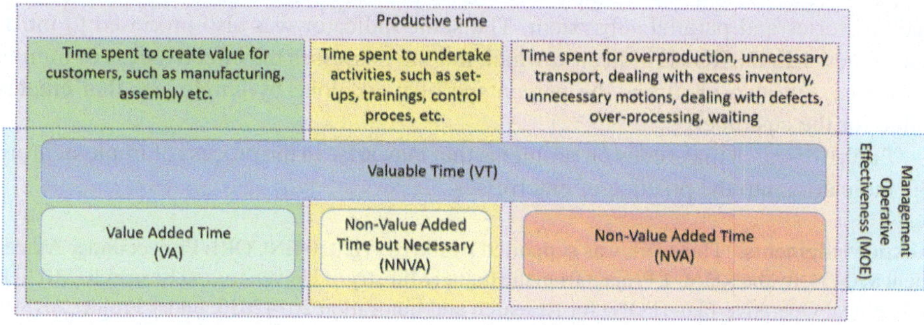


**Fig. 5.** Relationships between the Valuable Time and Value Added Time.

The ROLE calculates Overall Labour Effectiveness based on the tasks assigned to the employees without assessing whether they create value for the customers or not. The main conclusions that can be drawn from the ROLE value are whether or not the employee has worked and how effectively he has worked. In other words, it will be known whether the employee has completed the tasks entrusted to him and at what time. This time spent on a real work is called valuable time (VT), while this time does not have to be spent on performing value added activities. Therefore, we propose to improve the ROLE indicator by including into the equation value-adding activities (VA). The new equation will look as follow (see Eq. (7)).

$$LEAN - ROLE = \frac{VA}{NAT} = \frac{VA}{VT} \cdot \frac{VT}{OT} \cdot \frac{OT}{EAT} \cdot \frac{EAT}{NAT} \tag{23}$$

Additionally, for the purpose of data collecting it will be necessary to create a classification of the activities to indicate which are VA activities and which not. We suggest to treat as NVA/NNVA activities connected with transport (TR), defect production and elimination (DPE), changeover (CO).

The implementation of this indicator will not only help to understand how much the employees were working but also how much their work has contributed to creating value for the customer. We will also discover that only a percentage of employees create value.

## 5 Concluding Remarks

This paper deals with upcoming issues in implementing human-centered manufacturing solutions for the next factories of the future. It addresses the issue of performance-driven automation (the second level of automation) with the aim of incorporating the human factor in performance evaluation, considering people as an integral part of human-machine collaboration systems. It addresses how to adapt a well-known metric for monitoring and diagnosing the factory level performance (OTE), when considering problems associated with a range of workforce issues. For this purpose, the revised OLE (ROLE) indicator is discussed and a methodology is proposed to take it into account when modelling manufacturing systems for productivity improvement. The proposed methodology has been investigated with application to two typical subsystems widely used in manufacturing such as series and parallel subsystem. The same indicator was also proposed to introduce a new LEAN-ROLE indicator, which not only assesses the effectiveness of human resources, but also identifies the percentage of work done by employees that creates value for the customer.

Future research may focus on problems that may arise in the process of implementing the indicators into the practice of enterprises.

**Acknowledgments.** This work was supported by the EU H2020 ENCORE Project under Agreement 820434, by the REACT Project from the Italian Ministry of University and Research (MIUR)-in National Operative Plan (PON) for Research and Innovation 20142020, under Project ARS01 01031, and by the HD3Flab Project "Human Digital Flexible Factory of the Future Laboratory" EU ERDF (European Regional Development Fund), Regional Operative Plan (POR) MARCHE Region FESR (Fondo Europeo di Sviluppo Regionale) 2014/2020, AXIS 1, Specific Objective 2, ACTION 2.1.

## References

1. Colombo, A.W., Karnouskos, S., Kaynak, O., Shi, Y., Yin, S.: Industrial cyberphysical systems: a backbone of the fourth industrial revolution. IEEE Ind. Electron. Mag. **11**(1), 6–16 (2017)
2. Stadnicka, D., Bonci, A., Lorenzoni, E., Dec, G., Pirani, M.: Symbiotic cyber-physical Kanban 4.0: an approach for SMEs. In: 25th IEEE International Conference on Emerging Technologies and Factory Automation (ETFA2020), Vienna, Austria, 08–11 September, pp. 140–147 (2020)
3. Yildiz, Y.: Cyberphysical Human Systems. An introduction to the special issue. IEEE Control Syst. Mag. **40**(6), 26–28 (2020)
4. Dani, A.P., Salehi, I., Rotithor, G., Trombetta, D., Ravichandar, H.: Human-in-the-loop robot control for human-robot collaboration: human intention estimation and safe trajectory tracking control for collaborative tasks. IEEE Control Syst. Mag. **40**(6), 29–56 (2020)
5. Bonci, A., Cen Cheng, P.D., Indri, M., Nabissi, G., Sibona, G.: Human-robot perception in industrial environments: a survey. Sensors MDPI **21**(5), 1–29 (2021)
6. Tigrini, A., Pettinari, L.A., Verdini, F., Fioretti, S., Mengarelli, A.: Shoulder motion intention detection through myoelectric pattern recognition. IEEE Sens. Lett. **5**(8) (2021)
7. Bonci, A., Fiori, S., Higashi, H., Tanaka, T., Verdini, F.: An introductory tutorial on brain–computer interfaces and their applications. Electronics (Switzerland) **10**(5), 1–43 (2021). Art. no. 560
8. Sampath, M., Khargonekar, P.P.: Socially responsible automation: a framework for shaping the future. Nat. Acad. Eng. Bridge **48**(4), 45–52 (2018)
9. Stadnicka, D., Antosz, K.: Overall equipment effectiveness: analysis of different ways of calculations and improvements. In: Hamrol, A., Ciszak, O., Legutko, S., Jurczyk, M. (eds.) Advances in Manufacturing. LNME, pp. 45–55. Springer, Cham (2018). https://doi.org/10.1007/978-3-319-68619-6_5
10. Muthiah, K.M.N., Huang, S.H.: Overall throughput effectiveness (OTE) metric for fac-tory-level performance monitoring and bottleneck detection. Int. J. Prod. Res. **45**, 4753–4769 (2007)
11. Nakajima, S.: An Introduction to TPM. Productivity Press, Portland (1988)
12. Hansen, R.C.: Overall Equipment Effectiveness - A Powerful Production/Maintenance Tool for Increased Profits. Industrial Press, New York (2002)

13. ISO22400-1: Automation systems and integration - Key performance indicators (KPIs) for manufacturing operations management - Part 1: Overview, concepts and terminology (2014)
14. ISO22400-2: Automation systems and integration - Key performance indicators (KPIs) for manufacturing operations management - Part 2: Definitions and descriptions (2014)
15. Muthiah, K.M.N., Huang, S.H., Mahadevan, S.: Automating factory performance diagnostics using overall throughput effectiveness (OTE) metric. Int. J. Adv. Manuf. Technol. **36**(7–8), 811–824 (2008)
16. Bonci, A., Pirani, M., Longhi, S.: A database-centric framework for the modeling, simulation, and control of cyber-physical systems in the factory of the future. J. Intell. Syst. **27**(4), 659–679 (2018)
17. Braglia, M., Gabbrielli, R., Marrazzini, L.: Overall task effectiveness: a new Lean performance indicator in engineer-to-order environment. Int. J. Product. Perform. Manag. **68**(2), 407–422 (2019)
18. Deepak, V., Bhaskar, S., Balaji, M.: Enhancing overall labour effectiveness of CSD warehouse by adopting lean tools in construction equipment manufacturing process. Ind. Eng. J. **14**(1), 40–48 (2020)
19. Gordon, G.: Overall labour effectiveness: extending the principles of OEE to the workforce. Control Eng. Mag. Newslett. (2008). https://www.controleng.com/articles/overall-labor-effectiveness-extending-the-principlesof-oee-to-the-workforce/
20. Braglia, M., Castellano, D., Frosolini, M., Gallo, M., Marrazzini L.: Revised overall labour effectiveness. Int. J. Prod. Perform. Manag. **70**, 1317–1335 (2020)
21. Lupi, F., et al.: Toward a sustainable educational engineer archetype through Industry 4.0. Comput. Ind. **134**, 103543 (2022)
22. Łukasik, K., Stachowiak, T.: Intelligent management in the age of Industry 4.0–an example of a polymer processing company. Manag. Prod. Eng. Rev. **11**(2), 38–49 (2020)
23. Laciuga, M., Sęp, J.: Analytic optimization framework for resilient manufacturing production and supply planning in Industry 4.0 context-buffer stock allocation case study. Technologia i Automatyzacja Montażu **3**, 42–50 (2021)
24. Sakamoto, S.: Beyond World-Class Productivity: Industrial Engineering Theory and Practice. Springer, London (2010). https://doi.org/10.1007/978-1-84996-269-8
25. Naoum, S.: Factors influencing labor productivity on construction sites. Int. J. Product. Perform. Manag. **65**(3), 401–421 (2016)
26. Rother, M., Shook, J.: Learning to see: value stream mapping to add value and eliminate muda. Lean Enterprise Institute (2003)
27. Ohno, T.: Toyota Production System: Beyond Large-Scale Production. Productivity Press, Portland (1988)

# Relationships in the Human Lean Green Areas for the Benefit of Sustainable Production

Mariusz Bryke[1] and Beata Starzyńska[2](✉)

[1] Kaizen Institute Poland, 13 Koreańska Street, 52-121 Wrocław, Poland
[2] Faculty of Mechanical Engineering, Poznan University of Technology, 5 Maria Skłodowska-Curie Square, 60-965 Poznań, Poland
mbryke@kaizen.com

**Abstract.** In practical terms, the sustainable development of enterprises means a necessity to simultaneously achieve economic, environmental and social goals. These goals can be effectively attained using dedicated management concepts, e.g. Lean Manufacturing. However, the variety of goals faced by modern enterprises led to the creation of hybrid solutions supporting executive activity. One of such concepts is the Human Lean Green method. The article presents the results of research carried out using the Human Lean Green audit tool. The set of audit questions contained in the tool refers to the best of Human, Lean and Green (HLG) practices. In this way, the HLG method implements the operational concept of sustainable development involving the simultaneous implementation of all appropriately balanced activities of an economic, environmental and social nature. The new model of managing an enterprise must therefore take into account an appropriate and conscious shaping of relationships between the protection of employees' interests (Human), elimination of waste from processes (Lean) and environmental protection (Green). The aim of this article is to try to specify the character of these relationships on the basis of the results of previously conducted research.

**Keywords:** Sustainable development · Lean · Green · Human · Best practices · Manufacturing company · Services

## 1 Introduction

According to the Brundtland Commission Report, sustainable development meets the needs of modern societies without compromising the ability of future generations to meet their own needs [1]. Although sometimes it is still identified only with an environmental aspect, its other pillars are economic and social dimensions.

The concept of sustainable development can be seen from different perspectives and related to the functioning of various entities or objects. We can talk about the sustainable development of economy, an enterprise or sustainable production in a manufacturing enterprise.

A separate issue is the question whether a development strategy formulated in this way constitutes an indication of directions of activities or should it be understood as a certain postulated state sought to be achieved in a given system. In this case, it would be


J. Trojanowska et al. (Eds.): MANUFACTURING 2022, LNME, pp. 156–167, 2022.
https://doi.org/10.1007/978-3-030-99310-8_13

more appropriate to speak about obtaining a certain degree of equilibrium of an examined entity or object.

According to [2], sustainable production is a way of manufacturing products (or providing services) which are economically feasible, safe for employees and customers, and socially useful. Their production is carried out by means of economical and environmentally friendly processes and technical systems. Therefore, achieving the state of sustainability of an enterprise (production) requires the simultaneous (parallel) meeting of challenges of diverse nature: economic, environmental and social [3].

Dedicated management concepts such as Lean Manufacturing, pro-environmental management or the concept of humanization of work [4–6] are useful in the effective achievement of the above-mentioned goals. The variety of goals set for enterprises in the present day resulted in the creation of hybrid solutions supporting executive (manufacturing) activity. One of such concepts is the Human Lean Green method [7].

## 2 Research Problem

As previously emphasized, the sustainable development of an organization in practice means a necessity to "balance" activities in the achievement of economic, environmental and social goals. The research carried out so far by the authors in Polish enterprises and a review of the solutions functioning in them [8] indicate the lack of tools that would provide a comprehensive diagnosis of the condition of the organization, in terms of three key areas of sustainable development. Without the possibility of making such a diagnosis, it is impossible to consciously make decisions about actions aimed at strengthening effectiveness in terms of improving an enterprise's processes, its interaction with the natural environment or caring for a company's employees. All the more so, it is impossible to investigate how activities carried out in one of the areas affect the effects achieved in other areas.

On a macro scale, the new management model assumes appropriate and conscious shaping of relationships between economic growth, caring for the environment (not only natural) and quality of life (including human health). In a specific enterprise, this translates into shaping relationships between the protection of employees' interests (Human), elimination of waste from processes (Lean) and environmental protection (Green). The aim of this article is to try to specify the character of these relationships on the basis of the results of previously conducted research.

## 3 Literature Review

Nowadays, more and more enterprises are focusing in their development on obtaining high process efficiency while reducing the negative impact of an enterprise's operations on the natural environment. Process efficiency refers to basic processes directly related to production or provision of services. Increased efficiency is associated with the application of Lean Manufacturing principles oriented towards eliminating any form of waste from processes. An enterprise's efforts not to harm the environment are geared towards a goal that can be defined as "preserving the cleanest possible natural environment for humankind", which is obviously associated with the surrounding nature. Such an integrated approach is known in the literature as Lean Green [5, 9–13].

The literature related to the subject of lean management and pro-ecological management unanimously emphasizes the importance of integrating a human factor into systems. The issues related to protecting employees' interests are dealt with, among others, by Shah and Ward [14, 15]. Among the proposed practical activities, they recommend projects dedicated to increasing the level of work safety, developing multidisciplinary employee skills, building independent teams, investing in human resources, employee involvement, etc. A human factor is also considered here in the context of ergonomics - avoiding hazards in the workplace and losses for a company.

Over time, the basic assumptions of the philosophy of sustainable development have been implemented in the form of systems related to environmental protection, health and workplace safety, as well as corporate social responsibility management [16–18].

The analysis of the source literature shows that a human factor is taken into account to a limited extent as part of the policy related to the lean methodology and environmental protection. It is emphasized that managers should not perceive a "human factor" solely in the context of creating a comfortable and safe working environment for employees [19–22].

Taking into account the whole gamut of elements classified under the label of employees' interest protection, this area of a company's activity deserves attention on a par with lean and environmental protection.

The above suggestions became an inspiration to propose a new audit tool dedicated to management staff, the implementation of which allows to assess an enterprise declaring the introduction of the sustainable development strategy.

The analysis of the catalog of currently available methods indicates lack of a solution that would allow for simultaneous assessment of the organization in all the areas mentioned above.

It is important to note that the process of combining concepts into hybrid solutions is the one that comes along with both methodological and organizational difficulties. Comprehensive review of lean and green literature was conducted [10]. It identifies six main streams within which both conceptual and empirical research in lean and green was carried out. Recommendations include the search for evaluation methods or indicators.

To provide such a tool enabling efficiency assessment in these three areas, a concept and a prototype of the Human Lean Green auditing tool have been developed. The instrument involves main indicators of the organization efficiency in the areas mentioned above: working environment, process improvement and natural environment [7].

The value of human resources, employees' awareness and their attitude to process development and environmental protection cannot be overestimated. The above assumption became another inspiration for enriching the designed audit tool with aspects related to the creativity of employees and their pro-ecological behaviour. The Lean-Green philosophy was then supplemented with the so-called corporate social responsibility in which people are the central point [23].

The combination of three goals: protecting employees' interests (Human), eliminating waste from processes (Lean) and environmental protection (Green) is the justification for calling it the "Human-Lean-Green" (HLG) approach. The new approach became the basis for the development of a method and the development of a diagnostic tool that allows, on the one hand, to analyze an organization in each of the three afore-mentioned areas, and, on the other hand, to develop a plan to implement necessary solutions and directions of an enterprise's development.

The authors also assumed that actions taken in one of the areas (e.g. lean) affect the value of indicators defined for ecology or work environment.

## 4 Methods

As the literature review revealed, it was necessary to develop a method for a comprehensive assessment of an organization's condition, the one encompassing three main areas of sustainable development. As a response to the perceived methodological gap, the aforementioned Human Lean Green method has been implemented in the form of an audit dedicated to manufacturing and service companies. A starting point to develop the method was to find a common ground that would enable evaluations and comparisons being made among enterprises irrespective of the type of business, their size, legal form and implemented management systems. As a result of the search and analysis concerning the relevance of the adopted methodology, 5 main components of the Ishikawa cause-effect diagram and the related 5M concept were implemented as the evaluation criterion: Manpower, Methods, Machinery, Materials, Management. The concept claims that most of the problems potentially occurring in a company are concentrated within five categories of causes. Therefore, adopting them as evaluation criteria in the adopted method of assessing sustainable development of enterprises allows covering the most important areas of each enterprise's operations (Fig. 1).

The Human Lean Green concept involves adopting well-balanced activities in three areas:

- efficiency of manufacturing and service processes (Lean),
- environmental impact (Green),
- quality of the work environment (Human).

In order to evaluate the actions taken within each HLG area, it is necessary to de-fine criteria for measuring each of them.

The selection of measurement criteria in the Lean area was based on the classification of waste in the production processes proposed by Masaaki Imai [24]: overproduction, inventories, shortages and defects, excessive motion, waiting, excessive transport, overprocessing.

Selecting measurement criteria for the Green area (scrap, energy consumption, water consumption, resource consumption, air pollution) involved the review of popular standards related to environmental reporting, such as SRG, ISO 14000, ISO 26000, AA 1000. The choice is based on the Sustainability Reporting Guidelines (SRG) introduced by the Global Reporting Initiative (GRI) in the USA, considered the industry's most widely used reporting standard and compatible with other social responsibility standards (ISO 26000).

Measurement criteria for the Human domain (no attention to health, poor ergonomics, poor health and safety conditions, lost human potential and declining biodiversity) were chosen based on the assessment of the human workplace environment in terms of verifying work conditions necessary to protect an employee from premature work-related loss of biological capabilities as well as conditions for their regeneration at work.

Based on the above presented criteria, the components of the model defined for the Human Lean Green method are collected and presented in Fig. 1.

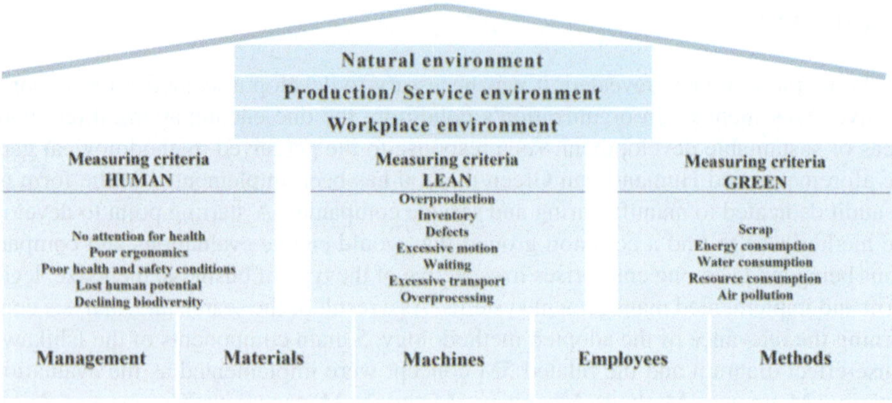


**Fig. 1.** Components of the human lean green method model.

All key measurement values relate to questions in each of the audit areas (HLG) expressed in terms of sentences describing best practices in the respective area as well as quantitative parameters describing e.g. energy consumption, waste-related financial values, etc.

The audit questions were formulated using the best practices observed in modern organizations with respect to the three main pillars of sustainable development [25–28]. In this way, the method aligns with the company's sustainability strategy objectives.

Audits take place at the company's headquarters and begin with the completion of an organization questionnaire (type of activity, legal form, capital structure, implemented management systems). Once the audit area is selected, the survey begins by having employees respond to questions asked (in the form of parameter values). The screening questions in each area were divided into two groups: "Numerical data" (quantitive parameters) and "Good practices" (qualitative parameters with possible answers: "Yes", "Partially to a great extent", "Partially to a moderate extent", "Partially to a small extent", "No" and "No data"). All questions are expressed using sentences depicting best business practices. Where no answers or data were provided, it was considered that above the established threshold for such answers in each of the Human, Lean, Green areas examined, missing data distorted the final audit result making an objective assessment of the company impossible. Figure 2 reflects an audit result snapshot view in the form of a graph showing averaged good practice indicators for HLG areas.

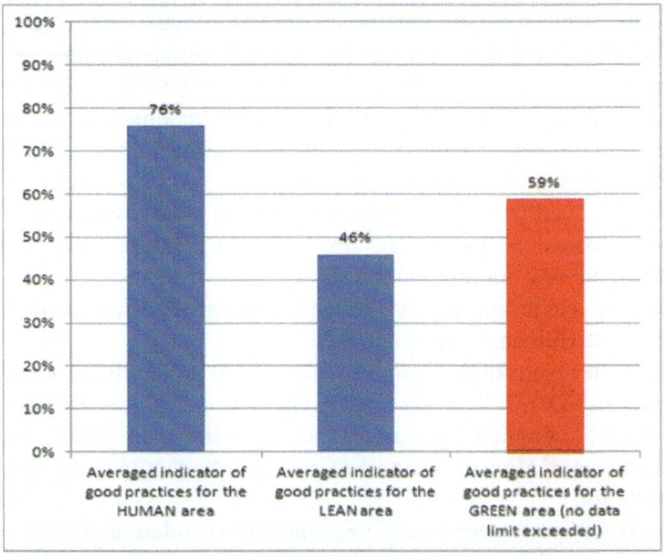


**Fig. 2.** Audit result in the form of a graph of averaged indicators of best practices for the HLG areas.

Primary assessment of the company's condition is made on the basis of measures taken in the form of indicators obtained through answers to screening questions. Indicator values aggregate results for each area by applying a computational model developed as part of the method. They represent compliance with a set of best business practices representing the content of the audit questions based on a defined scale. Each indicator takes values from 0–100.

As mentioned before, the audit tool allows both the analysis of the organization in each of these three areas and the design and implementation plan for the desired solutions and development directions in the company.

Audit report provides recommendations for improvement as audit questions relate to best business practices that are critical to sustainable development. By implementing the activities contained in the post-audit report, the audited organization identifies threats and opportunities on the way to "sustainability" in the areas of Human, Lean and Green.

## 5 Results and Discussion

The audit research was conducted in twenty businesses located in the territory of Poland. These companies declared the introduction of good practices in all areas recognized in the Human Lean Green method; at the same time they agreed to conduct the audit at their headquarters.

Vast majority of the surveyed enterprises are world-class companies providing the benchmark for other companies. The surveyed group included businesses representing manufacturing (industrial processing) and service activities. Regarding the enterprise size, the survey participants represented micro, small, medium, and large companies. A number of the companies audited implemented systems to support organisational management in compliance with ISO standards.

The results obtained from the Human Lean Green audits were analyzed using the Minitab® 19 Statistical Software. The results of the analysis allowed to formulate conclusions from the first audit of enterprises.

The first step was to test normality of the distribution of average audit results in each of the three areas of Human, Lean, and Green.

According to the results of the study, the value of test probability p in each of the three sections, Human $p = 0.077$, Lean $p = 0.643$, Green $p = 0.733$, is greater than the limit value of $p = 0.05$. It can therefore be concluded that the study of the averaged results of audits in the areas of Human, Lean and Green proves that they are characterized by a normal distribution.

Then, a study was performed which concerned the correlation of the averaged results of audits in each of the three areas in relation to the others, i.e. in the following sections: Green vs Lean, Human vs Green and Human vs Lean. The study aimed to determine how a variable in the form of the audit result in one area affects the audit result in another area. The study of the correlation of the averaged audit results in all areas concluded that the higher the audit result in one area, the higher the audit result in another area, i.e. the audit result in one area of Human Lean Green affects the audit result in another area (see Fig. 3).

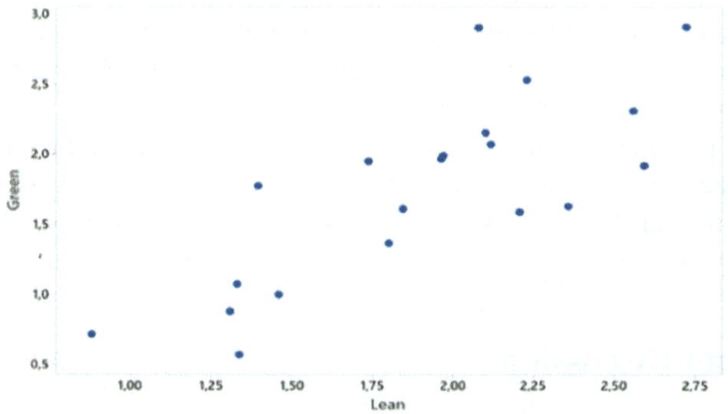


**Fig. 3.** Study of the correlation of averaged audit results in the Lean and Green areas.

For example, by analyzing the regression graph of averaged audit results in the Green and Lean areas based on the R-Sq coefficient, it was shown that the variability of the results in the Green area depends in 61.6% on the results in the Lean area (see Fig. 4).

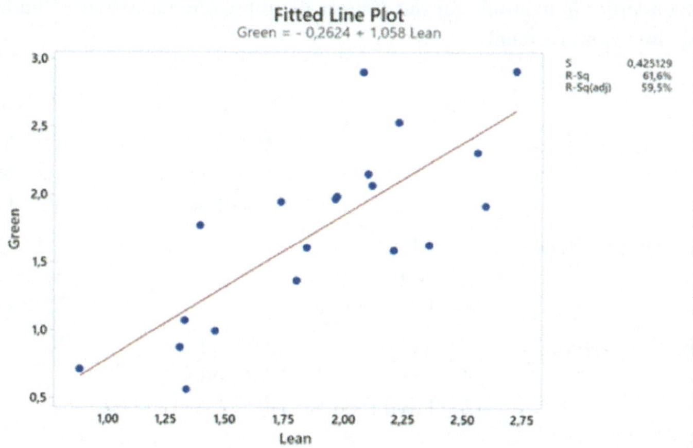


**Fig. 4.** Regression graph of averaged audit results in the Lean and Green areas.

The above study proved not only that the audit result in one area of Human Lean Green influences the audit result in another area, but also indicated the existence of close mutual correlations between the examined areas with a high level of significance, which can also be described by certain functions.

Researchers dealing with the integration of Lean and Green approaches have come to similar conclusions. In general, research results indicate that lean methodology and green activities are largely compatible and have the capacity to successfully operate simultaneously [10, 27], as they benefit from synergies in the areas of waste elimination, lead time, product development, various concepts and techniques for managing people as well as from the supply chain organization and relationships.

Due to the emerging Human Lean Green approach, the results were benchmarked against the results of a study conducted by the Authors in the period prior to the development of the method.

The results of the research preceding the development of the Human Lean Green method, obtained as part of the project called *Lean Green - caring for the environment*, showed that most of the surveyed enterprises (nearly 750 respondents representing enterprises from all over Poland) took steps to improve processes, protect the natural environment and optimize working conditions (including health and safety, workplace ergonomics, work tools). At the same time, the respondents were to comment on the relationships they perceived in practice (by answering: *yes, no* or *I don't know*) between the following areas: Human, Lean and Green. A summary of their opinions is presented in Table 1. For example, the respondents noticed the impact of improving working conditions (Human) on the improvement of processes implemented in an enterprise. The analysis of the results showed that there is a relationship between the three distinguished aspects of an organization's functioning: actions taken in one of them have a positive impact - in the respondents' opinion - on at least one of the remaining areas (Table 1).

**Table 1.** Respondents' indications on the impact of improvement activities undertaken in the areas of Lean, Green and Human.

| | Lean | Green | Human |
|---|---|---|---|
| Process improvement (Lean) | x | 27% YES<br>27% NO<br>46% I don't know | 42% YES<br>23% NO<br>35% I don't know |
| Natural environment (Green) | 27% YES<br>35% NO<br>38% I don't know | x | 27% YES<br>38% NO<br>35% I don't know |
| Working conditions (Human) | 35% YES<br>23% NO<br>42% I don't know | 31% YES<br>27% NO<br>42% I don't know | X |

On the basis of the answers, it was possible to conclude that despite the mutual influence of each of the three analyzed areas, the positive impact of proecological activities and measures related to environmental protection on other aspects of a company's operation is noticed to a negligible extent.

The next step of the main (leading) study was to check the impact of having a standardized management system on the audit results in the surveyed companies. Based on the ANOVA analysis of variance, an analysis was performed for all three areas of Human, Lean and Green. The results of the analysis allowed to conclude that companies with implemented standardized management systems do not obtain a higher result in the Human Lean Green audit.

Another study aimed to check how the size of a company (micro, small, medium, large) affects the result of the Human Lean Green audit. A research hypothesis was formulated that the size of a company affects the audit result. The hypothesis turned out to be true. Large companies achieved the best results in all areas of the Human Lean Green audit.

Another study was to check how the type of a company (manufacturing, service) influences the result of the Human Lean Green audit. A research hypothesis was formulated that the type of a company affects the audit result. The analysis of the variance of the dependence of the average audit result on the type of a company in the Green area shows a statistically significant difference between the individual elements of the assessment. By analyzing the graph of the dependence of the average audit result on the type of a company, it can be concluded that there is a significant statistical difference and the presented results differ significantly from each other. Service companies present a higher average result from audits in the Lean and Green areas than manufacturing companies.

Obtaining the results presented in this paper involved a lot of work, as the presented HLG method also takes some limitations [7]. Implementation of the presented method involves collecting a wide range of information across diverse areas of the company's activity, which entails the potential participation of subject matter experts. While the method can be used on a one-time basis, its relevance becomes apparent in periodic applications. Sustainable development trends within a company can only be identified where the method is applied regularly. Support from an external auditor who will conduct the audit, assisting the company's employees during the audit and providing post-audit recommendations is advisable.

## 6 Conclusions

Based on the results of the research carried out with the use of the Human Lean Green audit tool and the conclusions of the survey conducted as part of the *Lean Green project - caring for the environment*, relationships between three aspects of a company's operation - Human, Lean and Green were demonstrated. Actions introduced in one of the three areas constitute - for the most part - the basis for improvement in the functioning of an enterprise in the remaining areas. Thus, each decision regarding the introduction of specific activities (here: best practices) translates into effects in the remaining, paired areas. The presented results illustrate relationships in the group of the surveyed enterprises.

There are limitations of the proposed model that point the way to future research. Detailed treatment of the perceived interactions and the character of mutual influence should be the subject of further research, which will be possible after collecting more data from audits carried out in subsequent companies representing specific business lines.

The method will undergo further elaboration to incorporate an integrating aspect to the assessment. Future research will investigate ways to identify potential connections between audit areas and shared audit criteria so as to incorporate potential integrated components into a refined version of the method. By doing so, the Human Lean Green methodology will enable more advanced monitoring of sustainable development.

**Acknowledgments.** This research was funded by Swiss-Polish Cooperation Programme (Agreement of Financial Support for Implementation CSR No. 27-SPPW-DPP-2013) and Polish Ministry of Education and Science grant number 0613/SBAD/4710.

## References

1. World Commission on Environment and Development: Report of the World Commission on Environment and Development: Our Common Future. Oxford University Press, Oxford (1987)
2. Lowell Center for Sustainable Production: What is sustainable production? https://www.uml.edu/research/lowell-center/about/sustainable-production-defined.aspx. Accessed 10 Sept 2020
3. Kujawińska, A., Diering, M., Rogalewicz, M., Żywicki, K., Hetman, Ł: Soft modelling-based methodology of raw material waste estimation. In: Burduk, A., Mazurkiewicz, D. (eds.) ISPEM 2017. AISC, vol. 637, pp. 407–417. Springer, Cham (2018). https://doi.org/10.1007/978-3-319-64465-3_39
4. Hallam, C., Contreras, C.: Integrating lean and green management. Manag. Decis. **54**, 2157–2187 (2016)
5. Galeazzo, A., Furlan, A., Vinelli, A.: Lean and green in action: interdependencies and performance of pollution prevention projects. J. Clean. Prod. **85**, 191–200 (2014)

6. Azlina, N., Salleh, M., Kasolang, S., Jaffar, A.: Green lean TQM human resource management practices in Malaysian automotive companies. Int. Sch. Sci. Res. Innov. **6**, 2065–2069 (2012)
7. Starzyńska, B., Bryke, M., Diakun, J.: Human lean green method—a new approach toward auditing manufacturing & service companies. Sustainability **13**(19), 10789 (2021)
8. Bryke, M., Starzyńska, B.: Koncepcja Human Lean Green jako instrument zapewnienia zrównoważonego rozwoju organizacji ukierunkowany na wzrost jej efektywności (Human Lean Green Conception as the instrument of sustainability of organizational development oriented towards the increase of its effectiveness). Pr. Nauk. Uniw. Ekon. Wrocławiu (Res. Pap. Wrocław Univ. Econ.) **377**, 119–136 (2015)
9. Kurdve, M., Zackrisson, M., Wiktorsson, M., Harlin, U.: Lean and green integration into production system models–experiences from Swedish industry. J. Clean. Prod. **85**, 180–190 (2014)
10. Garza-Reyes, J.A.: Lean and green—a systematic review of the state of the art literature. J. Clean. Prod. **102**, 18–29 (2015)
11. Verrier, B., Rose, B., Caillaud, E., Remita, H.: Combining organizational performance with sustainable development issues: the Lean and Green project benchmarking repository. J. Clean. Prod. **85**, 83–93 (2014)
12. Bożek, M., Kujawińska, A., Rogalewicz, M., Diering, M., Gościniak, P., Hamrol, A.: Improvement of catheter quality inspection process. In: 8th International Conference on Manufacturing Science and Education – MSE 2017 "Trends in New Industrial Revolution", vol. 121 (2017). MATEC Web Conference
13. Duarte, S., Cruz-Machado, V.: Investigating lean and green supply chain linkages through a balanced scorecard framework. Int. J. Manag. Sci. Eng. Manag. **10**, 20–29 (2014)
14. Shah, R., Ward, P.T.: Lean manufacturing: context, practice bundles, and performance. J. Oper. Manag. **21**, 129–149 (2003)
15. Shah, R., Ward, P.T.: Defining and developing measures of lean production. J. Oper. Manag. **25**, 785–805 (2007)
16. Lawrence, E., Andrews, D., Ralph, B., France, C.: Applying organizational environmental tools and techniques. Corp. Soc. Responsib. Environ. Manag. **9**, 116–125 (2002)
17. Pipatprapa, A., Huang, H.-H., Huang, C.-H.: The role of quality management & innovativeness on green performance. Corp. Soc. Responsib. Environ. Manag. **24**, 249–260 (2017)
18. Valentine, S.V.: The Green Onion: a corporate environmental strategy framework. Corp. Soc. Responsib. Environ. Manag. **17**, 284–298 (2010)
19. Fleming, J.H., Asplun, J.: Human Sigma. Managing the Employee - Customer Encounter. Gallup Press, New York (2007)
20. Del Brio, J.A., Junquera, B., Ordiz, M.: Human resources in advanced environmental approaches – a case analysis. Int. J. Prod. Res. **46**(21), 6029–6053 (2008)
21. Suárez-Barraza, M.F., Ramis-Pujol, J.: Implementation of Lean-Kaizen in the human resource service process: a case study in a Mexican public service organization. J. Manuf. Technol. Manag. **21**(3), 388–410 (2010)
22. Tortorella, G.L., Fogliatto, F.S.: Method for assessing human resources management practices and organisational learning factors in a company under lean manufacturing implementation. Int. J. Prod. Res. **52**(15), 4623–4645 (2014)
23. ISO 26000:2010 Guidance on Social Responsibility; ISO: Geneva, Switzerland (2010)
24. Imai, M.: Kaizen: The Key to Japan's Competitive Success. McGraw-Hill Education, New York (1986)
25. Veselý, A.: Theory and methodology of best practice research: a critical review of the current state. Cent. Eur. J. Public Policy **2**, 98–117 (2011)
26. Abualfaraa, W., Salonitis, K., Al-Ashaab, A., Ala'Raj, M.: Lean-green manufacturing practices and their link with sustainability: a critical review. Sustainability **12**, 981 (2020)

27. Dües, C.M., Tan, K.H., Lim, M.K.: Green as the new lean: how to use Lean practices as a catalyst to greening your supply chain. J. Clean. Prod. **40**, 93–100 (2013)
28. Starzyńska, B., Bryke, M.: Compilation of the best practices for auditing the sustainable development of organizations. In: Tonkonogyi, V., et al. (eds.) InterPartner 2020. LNME, pp. 833–842. Springer, Cham (2021). https://doi.org/10.1007/978-3-030-68014-5_81

# Industry 4.0 Solutions

# Product Lifecycle Management (PLM) in the Context of Industry 4.0

Jan Duda[1(✉)], Sylwester Oleszek[2], and Krzysztof Santarek[3]

[1] Chair of Production Engineering, Cracow University of Technology, Cracow, Poland
jan.duda@pk.edu.pl
[2] Transition Technologies PSC S.A., Łódź, Poland
[3] Institute of Production Systems Organization, Warsaw University of Technology, Warsaw, Poland

**Abstract.** The article presents the essence, genesis and development trends of PLM systems. Currently, these are advanced tools supporting the work of engineers at all stages of the product lifecycle. They are particularly useful, and even necessary, in the design of products integrating various fields of technology (engineering), i.e. products with mechanical, electrical, electronic, IT components, etc. The development of Industry 4.0 technology creates new requirements, but also opportunities for PLM systems. The article presents the essence, genesis and development trends of PLM systems. I also presents the selected examples of the use of Industry 4.0 technology in PLM systems, including the industrial Internet of Things, virtual and augmented reality and digital twins.

**Keywords:** Product Lifecycle Management (PLM) · Application Lifecycle Management · Service Lifecycle Management · Connected PLM · Industry 4.0

## 1 Introduction

Enterprises are aware of the importance of developing new products and technologies for gaining and maintaining a competitive advantage. The increase in the complexity of products integrating various technologies establishes high demands on engineers - designers and the design process. This applies in particular to the need to consider various aspects of a designed product throughout its lifecycle. Meeting these requirements is facilitated by IT systems of product lifecycle management - PLM (Product Lifecycle Management). Their history dates back to the early 1970's. Initially, they were focused on engineering problems of designing products, mainly mechanical ones. As the applications of these systems grew, problems arose, loosely related to product design process, but more related to the product design management. The example was product information management, especially in the case of carrying out design works in large, dispersed teams. Thus, PDM (Product Data Management) systems appeared which gradually evolved towards PLM systems. Product lifecycle management is the process of managing the entire lifecycle of a product, from the inception of its concept, through design, production, operation, to end-of-life, recycling or disposal. PLM systems integrate people, data, processes and


J. Trojanowska et al. (Eds.): MANUFACTURING 2022, LNME, pp. 171–185, 2022.
https://doi.org/10.1007/978-3-030-99310-8_14

systems by collecting, storing, integrating and ensuring consistency and up-to-date information about the product, delivering it to organizational units of the company and other participants of the product lifecycle. The article presents the essence, genesis and development trends of PLM systems. The development of Industry 4.0 technology creates new requirements, but also creates many opportunities for PLM systems. The article presents examples of the use of Industry 4.0 technology in PLM systems, including the industrial Internet of Things, virtual and augmented reality and digital twins.

## 2 Integration of Computer-Aided Systems in Product Development

The development of PLM (see Fig. 1) systems results from the evolution and continuous assimilation of product-oriented computer solutions: from design (design or computer-aided manufacturing, CAD/CAM systems), to the integration of Enterprise Resource Planning (ERP) systems, Customer Relationship Management (CRM) and Supply Chain Management (SCM) [3].

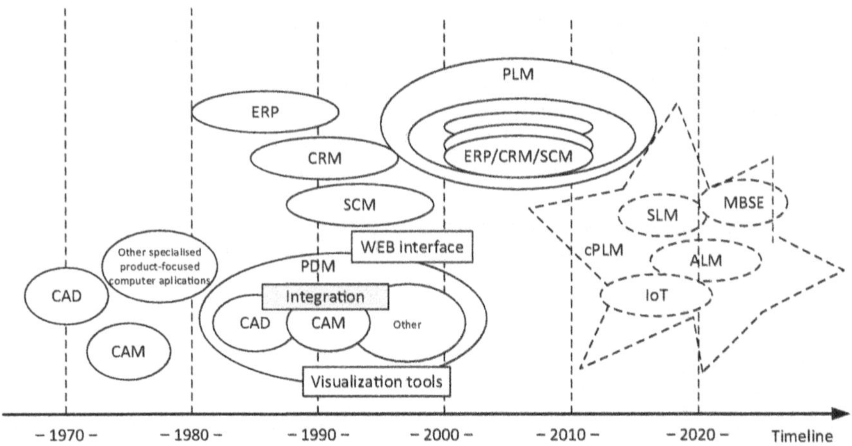


**Fig. 1.** Integration of computer aided systems used in product development. Own study based on [9].

In the 1980s, simultaneously with the development of computer aided design, manufacturing and engineering tools (CAD/CAM/CAE), there appeared PDM (Product Data Management) systems for managing product information created by these tools. These are a special class of workflow and process management systems and were originally intended to record data on the product structure, design and technological documentation and its production processes. They also provided system environment to process and exchange electronic data. In the 1990s, PDM systems based on a web interface appeared, along with more efficient and user-friendly visualization tools. Web-based PDM systems have become available throughout the so-called extended enterprise. Two main limitations made it difficult to further expand PDM systems:

- a limited range of information data managed by early PDM systems, limited to engineering data such as geometric models, BOM, and FEA models,
- working with these systems was not always easy and usually required a strong engineering and technical background.

The core functionality of the early PDM systems was, therefore, to provide users with the required data via a central repository and to ensure product data integrity through continuous updating and controlling the way data was created and modified. The next generations of PDM systems were enriched with new functions such as: change management, document management, workflow management and project management. It enabled the implementation of concurrent engineering (CE) strategies and the improvement of product development processes in the enterprise. The first generation of PDM systems, although very useful and effective in technology-related areas, did not cover non-technology related areas in the company, such as: sales, marketing, and supply chain management, as well as external entities such as customers and suppliers.

Corporate applications such as Enterprise Resource Planning (ERP), Customer Relationship Management (CRM) and Supply Chain Management (SCM) were developed almost in parallel with PDM systems. These solutions, each focused on a specific product lifecycle process, are product information driven. Modern PDM systems enable data exchange between cooperating enterprises. For a detailed discussion of the PDM application, see [19].

Later, PDM systems evolved into PLM product lifecycle management solutions, going beyond the technical aspects of the product, and providing a common platform for creating, organizing, and disseminating product-related information across the extended enterprise.

In this approach, PDM systems are used as a tool that integrates CAx and DFx systems with ERP (Enterprise Resource Planning), CRM (Customer Relationship Management) and SCM (Supply Chain Management) systems.

Unlike PDM systems that focus on data management, PLM systems support knowledge management processes that involve capturing, organizing, and reusing knowledge throughout the product lifecycle. PLM class systems integrate a set of applications supporting product development, which include [2]:

- Product and Portfolio Management (PPM),
- Computer-aided Technologies, CAx,
- Manufacturing Process Management (MPM),
- Product Data Management (PDM).

The development, integration and improvement of systems supporting product development process is still ongoing. Currently, it is a class of the very complex systems that integrate various tools into coherent, inter-organizational PLM solutions.

## 3 Evolution of PLM Systems

Systems supporting the management of product development are evolving towards the concept of Smart Connected PLM and closed loop of the product lifecycle, using, i.a.

Internet of Things (IoT) technologies. Looking from the perspective of the Industry 4.0 concept, the set of applications included in PLM systems is under the constant process of extending by:

- solutions for quality management in the product lifecycle (QMS),
- Model-Based System Engineering (MBSE),
- Application Lifecycle Management (ALM),
- Service Lifecycle Management (SLM),
- modules that allow integration with systems of other classes, including ERP systems (ERP connector).

### 3.1 Model-Based System Engineering (MBSE)

One of the most effective and efficient approaches to carry out design activities is Systems Engineering (SE). One of its implementations is Model-based Systems Engineering (MBSE)[1]. INCOSE[2] defines MBSE as [10]:

"[...] the formalized application of modeling to support system requirements, design, analysis, verification, and validation activities beginning in the conceptual design phase and continuing throughout development and later lifecycle phases."

While using MBSE, computer models support design processes that span multiple technical disciplines, the task of which is to provide both a high degree of clarity and understandability, the freedom to form complex structures, and the connections required when expressing complex information in an accessible and easy-to-understand manner. At the same time, it enables precision, discipline and unambiguity necessary when defining technical means. For this purpose, it is necessary to use a common language that would be understandable to participants of the design process representing different disciplines of science and technology. An example of such a language is SysML (Systems Modeling Language), which provides tools and supports the specification, analysis, design and verification of complex systems [11, 15, 18].

MBSE aims to create a holistic model of the whole system that integrates different sub-models and different modeling activities. A model-based approach that has been used for a long time in some specific fields such as software engineering (e.g. Unified Modeling Language models) and mechanical engineering (e.g. Computer Aided Design models) has replaced the document-based approach as it has many advantages [17]:

- many perspectives on the system model, and thus easier analysis,
- better product quality thanks to consistency,
- the possibility of assessing the correctness and completeness of the product (and its model),
- knowledge re-use through standard information capture,
- possible reduction of cycle times due to standardization and re-use,
- easier maintenance and synchronization of information compared to document-based approaches.

[1] Also referred to in the literature as Model-driven Systems Development (MDSD).

[2] INCOSE – International Council of Systems Engineering.

There are many detailed methodologies to facilitate the use of MBSE. The design process is iterative, which means that the activities influence each other and are refined and improved in subsequent iterations. The tools supporting this method of designing include:

- complete toolkits such as IBM Rational, PTC Windchill Modeler or Enterprise Architect, that are integrated with specialized tools, e.g. for requirements management,
- use case modeling and system architecture tools,
- independent tools supporting specific design tasks, e.g. Eclipse Papyrus for defining the system architecture.

### 3.2 Application Lifecycle Management (ALM) Systems

In the context of the development of mechatronic products, the capabilities of PLM systems turned out to be insufficient, as they do not have functionalities related to the management of the application lifecycle. Instead, ALM systems are used for this purpose. The scope of these ALM class systems includes the following applications [20]:

- agile project management,
- release management,
- requirements management,
- document management,
- integration of software development tools,
- source code management (SCM)-ensuring control over versions and re-visions of the source code of the application and other text files; SCM class systems enable tracking the development of the source code and prevent its modification by more than one person at a time [4],
- test management,
- process management,
- software configuration management,
- change management.

Currently, ALM and PLM class solutions are most often used as stand-alone system environments that must be installed and maintained separately. Selected applications are present both in PLM class systems and in ALM class systems. However, the actual implementation of the available functionalities is different as they serve different purposes, e.g. configuration management or process management applications. Others, such as for managing the source code, are only found in ALM class systems.

For many years, efforts have been made to develop an effective method of integrating both classes of systems in such a way that it is possible to effectively manage product data without the need to duplicate them. In the integrated and uniform ALM-PLM system, all applications should be available in one software environment, the data flow should be undisturbed, and all functionalities should be available based on a unified, common user interface. Product development activities separated into ALM and PLM domains are shown in Fig. 2.

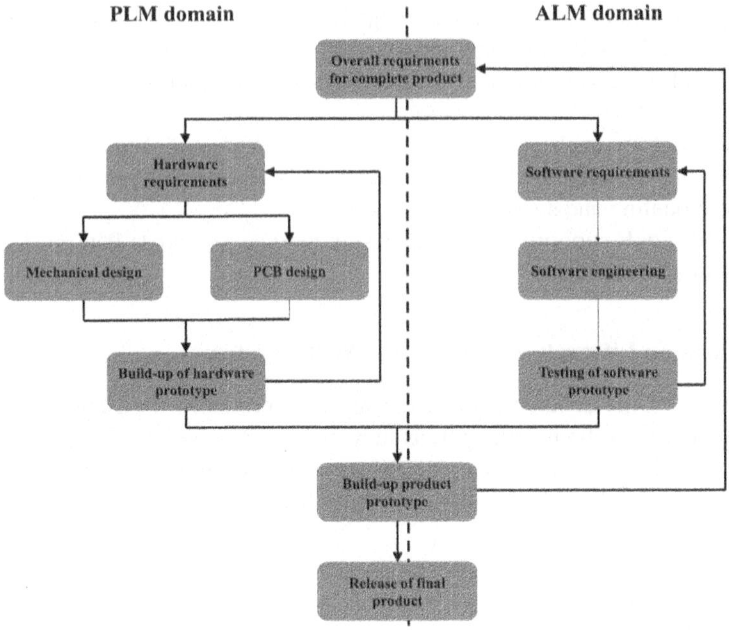


**Fig. 2.** Product development activities separated into ALM and PLM domains. Source: own elaboration based on [20].

### 3.3 Service Lifecycle Management (SLM) Systems

Throughout their lifecycle products are tied to services. A service is here treated as any activity or benefit that does not have a material form that one party (service provider) can offer to the other (recipient). Satisfying the needs of the product buyer also requires continuous support of the product by services. This applies especially to the use (exploitation) stage of the product. Among the services supporting the operation (use) of the product, the following can be mentioned:

- warranty and post-warranty service,
- support for users in the field of self-repair and maintenance of products,
- user training in the use of the product and its operational service,
- remote condition monitoring and diagnostics of products,
- collecting and processing information on products in service and making them available to maintenance services and design engineers, creating a "history" of the product,
- automatic and remote software update (e.g. computer operating systems, application software for smartphones, tablets, watches, cameras, etc.).

Services, like products, have their own lifecycle and are becoming more and more complex, requiring specialist knowledge and high qualifications, special equipment and proper organization. Figure 3 shows a diagram of the service lifecycle. IT tools supporting service management are called service lifecycle management systems - SLM.

The services accompanying products can be broadly classified into two types:

- services whose recipients are product producers, necessary for the production process,
- services aimed at end users of the product.

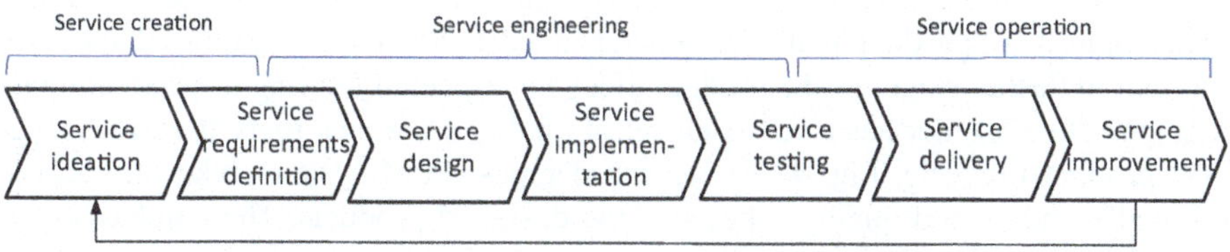


**Fig. 3.** Diagram of the service lifecycle. Source: [21].

In the first case, services related to the maintenance and repair (operational) service of machines and production devices are the most important. Among them there are Computer Maintenance Management Systems (CMMS) supporting the management of machinery and equipment maintenance. They are a growth area for the application of many Industry 4.0 technologies, such as: industrial Internet of things, sensors, artificial intelligence, cloud technologies, big data or virtual or augmented reality technologies. The systems enabling remote supervision and monitoring of the condition of machines and devices are called E-maintenance (or web-based maintenance). Currently, they cover a wide class of systems: from simple web applications monitoring the work of machines and recording key parameters of their work, creating statistics presented in a clear graphic form, machine work and downtime, etc. to advanced solutions that enable forecasting condition of machines and taking preventive actions, counteracting emergency situations. The latter category includes Intelligent Maintenance (smart maintenance) or Predictive Maintenance systems.

The possibilities of systems supporting users (buyers) of products are equally rich. They offer functions similar to those discussed above, but often in a more accessible way, which allows them to be used even by users without specialized education and experience. Usually, services in this form are provided by the manufacturer of the products. Users, especially those of technologically advanced products, can use systems of remote monitoring and diagnostics of the condition of products, support in learning how to use the product. A specific type of services is the collection of data on used products and customer comments from the service network or directly from users and making them available to the manufacturer so it can be used in the process of further product development.

The role of SLM systems increases with the creation and development of product - service systems (PSS). They are a combination of products and accompanying services, which constitute one offer. Product - service systems are an expression of the so-called servitization of the economy (including industry), the development of smart products and the search for new business models by enterprises previously dealing mainly (or to a large extent) with production activity of physical products. The implementation of such services usually requires the use of advanced IT tools that enable the service to be delivered to the client, most often remotely. A good, although not the only example, are telecommunications services addressed to smartphone owners.

### 3.4 Quality Management Systems (QMS)

In general, quality is the degree to which the customer's needs are met. In other words, quality is one of the basic requirements that a product must meet. In the course of designing, the majority of product quality parameters are decided, including those that become important in further stages of the product lifecycle, especially during its production and operation. The designer has a very rich set of quality assurance tools at various stages of the product lifecycle, Fig. 4. To varying degrees, PLM systems take into account the broadly understood quality aspects of the designed products. The emphasis is put mainly on supporting design (engineering) works aimed at developing products that meet previously defined operational and functional requirements. Due to the progress in the development of PLM systems, they more and more often offer quality assurance tools of varying degrees of complexity.

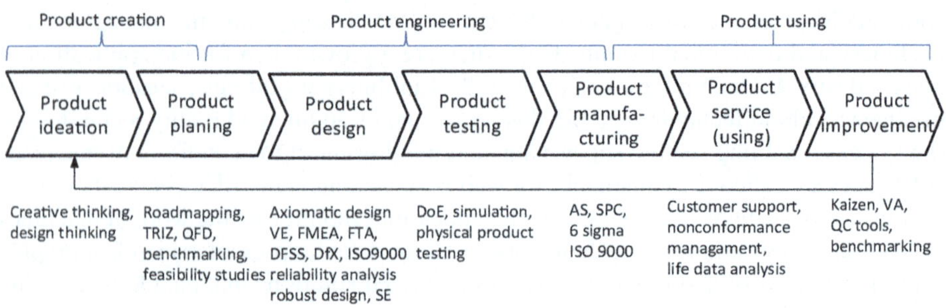


**Fig. 4.** Quality assurance methods and techniques used at various stages of the product lifecycle.

The basic design guideline is to use proven design methodologies, methods and guidelines. They concern in particular the organization (course) of the design process. These include ISO9000 standards, methods derived from systems engineering, and others. The relatively least susceptible stage to computer support is the one of creative search for product ideas (product ideation). Product planning consists of searching for acceptable (feasible) methods of implementing the generated ideas, taking into account market, economic, organizational, technical, environmental and other criteria. Some of the relevant tools are supported by IT and can be integrated with PLM systems (e.g. TRIZ, QFD). When designing a product, the most problems that affect the quality of the final design (result of the design process), the quality of realization (the result of the production process) or the quality of use (quality of use of the product). Designers have many effective quality assurance tools at their disposal: design for six sigma, axiomatic design, failure mode and effect analysis, fault tree analysis, robust design, value engineering, design for manufacturing, for assembly, for testing, for environment, etc. The use of these methods is aimed at, inter alia, early detection of weaknesses in the product structure, improvement of its reliability and resistance to changes in the external operating conditions of the product and user errors during product operation. When designing products with a particularly high degree of complexity and requirements, a lot of attention is paid to functional design, within which, among other things, system architecture it is determined, and thus its structure, decomposition of functions and their allocation to the indicated subsystems. The evaluation of such projects requires the use of advanced modeling techniques and MBSE class tools (see Sect. 3.1). Prior to launching the production, the product design's models, prototypes, product trial series, etc., must be examined and tested. Computer (digital) models of the product can be examined. In many cases, however, it is necessary to conduct research on physical models. The results of these tests are the basis for making changes in the product structure, hence the need to integrate the research and testing phases with PLM systems.

The production phase is a typical example of the use of quality assurance tools such as: acceptance sampling, control charts, 6 sigma, etc. The purpose of using them is to ensure the stability of the production process and product quality parameters. They enable identifying and eliminating the causes of process disruptions and continuous improvements. These tools require the continuous gathering and processing of large amounts of data from various sources, including sensors, measuring devices, etc. These tools are often integrated with PLM and ERP systems.

The exploitation phase is an important source of information about the quality in the actual conditions of the use of the product. The source of this information may be the service and sales network, as well as complaints and comments submitted by their users. This information is a valuable source of knowledge about the actual quality level of products, necessary in the process of improving their design and production methods. The development of PLM systems indicates the growing importance of the issues of product quality assurance at various stages of the lifecycle.

## 4 Influence of Industry 4.0 Concepts and Technologies on the Development of PLM Systems

### 4.1 Connected PLM

Carrying out the process of new product development (NPD) or change management (CM) poses a number of challenges for enterprises. One of them is the need to work in distributed systems. This way of working causes additional workload and excessive use of resources for coordinating, supervising and maintaining data consistency. Technologies related to the Industry 4.0 concept enable significant improvements in this regard. In general, the essence of Industry 4.0 is the creation of intelligent value chains based on dynamic, self-organizing and optimizing socio-technical systems, known as smart factories.

One of the most important principles of the Industry 4.0 concept is interoperability, defined as the ability of employees, robots, production systems and machines to communicate with each other. For this to be possible, common standards, a common integrating platform, sensors and cloud computing are necessary.

The concept of the so-called connected PLM (cPLM), defined as the association of data representing the product structure with other relevant data that enable participants in the product development process to gain easy access to a wide range of information about the product. This enables the product to be tracked throughout its entire lifecycle. Product lifecycle management based on the cPLM concept gives enterprises the opportunity to make better decisions in the early stages of the development process. Thanks to this, it is possible to limit costly changes/rework and, consequently, to improve the quality of the product.

## 4.2 RFLP Requirements Specification

Classic PLM systems (integrated with CAD systems) operate on the basis of the physical representation of the product and mainly focus on the mechanical area. Simply put, it can be said that the product definition provided by these systems refers to what we can see in the physical world. However, in the case of very complex products, consisting of many integrated and cooperating systems from various areas of engineering (mechanics, electronics, IT, etc.), it is necessary to provide support for activities in areas other than mechanical, as well as the formulation and analysis of functional dependencies that can be expressed at a higher level of abstraction. The complexity of these products is not only due to the number of interconnected physical components, but also to the functional relationship between their systems. In other words, advanced products such as an airplane, computer, or car function properly only when all of their systems work together correctly [14].

The development of such a complex product must involve specialists from many different areas of engineering and using tools of different classes. The process of designing such a product must also be carried out properly.

In order to enable the methodical and systematic implementation of this process and to ensure the possibility of carrying out complex activities with the use of tools specific to all required areas of engineering, a concept abbreviated as RFLP (Requirements-Functional-Logical-Physical) has been developed. It is a new way of organizing the product structure definition, according to which the traditional (physical) product definition (P) is supplemented with the requirements concerning the system definition (R), product functionality definition (F) and logical connections (L). Therefore, higher than just physical, levels of abstraction are taken into account [8]. The procedure according to the RFLP methodology is shown in Fig. 5.

One of the main benefits of an environment structured in accordance with the RFLP concept is the linking of requirements to system components at all abstraction levels and the possibility of tracing them throughout the product lifecycle. This allows us to check and understand how the requirements are implemented at each stage and to what extent they are met. In the case of introducing changes to the requirements, it also makes it possible to specify all areas in which the change should be implemented. Due to the fact that the connection provided by RFLP is bi-directional, it is also possible to identify situations in which an element of the system has been modified in an uncontrolled manner and no longer meets a given requirement [1].

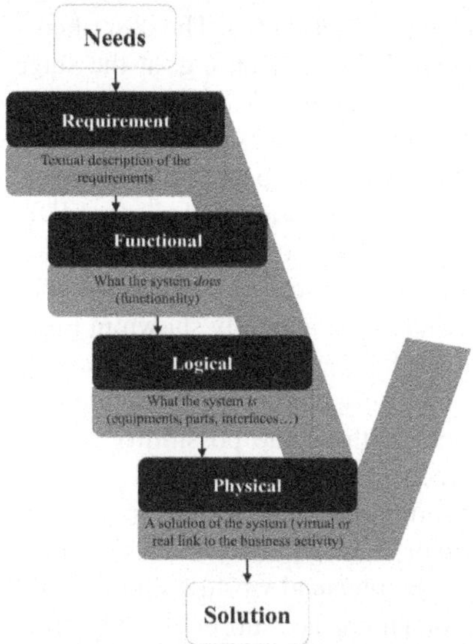


**Fig. 5.** RLFP process flow. Source: own elaboration.

The concepts presented in Fig. 5 are related to the idea of V model and the systematic development of complex mechatronic systems according to the VDI Guidline 2206. The origination and importance of V model of VDI Guidline 2206 has been discussed in [6].

The PLM system operating in accordance with the assumptions of this concept should enable the implementation of the product development process in one integrated system environment, using tools from many different, also distant from each other, areas, such as e.g. mechanics, electrical engineering, electronics, software development, hardware engineering and embedded systems [8]. Nevertheless, the assumptions of the RFLP concept do not impose that the PLM system should be a monolithic solution or that the related data should be stored in one database. Therefore, it is possible to use distributed systems, consisting of components and subsystems of various classes (e.g. PLM, ALM, MBSE, etc.), provided, however, that the related data will be available whenever the need arises [1].

### 4.3 Industry 4.0 Technologies in Product Lifecycle Management

At the core of the Industry 4.0 concept there are technologies such as: PLM (Product Lifecycle Management) systems, Industrial Internet of Things (IIoT), Digital Twin (DT), Augmented Reality (AR) and Virtual Reality (VR), mass customization, cloud computing or cyber-physical systems.

One of the concepts of using AR techniques in the PLM system was proposed by PTC and implemented in the commercial Windchill system. In this process, it is possible to publish data (3D models) stored in the PLM system to a portal managed in a cloud computing environment. Then 3D models are optimized and converted to a multimedia format and can be downloaded using a mobile application and displayed through it in the 1:1 scale compared to the real world [16]. The described solution is fully automated and enables a quick preview of the project, e.g. at the stage of developing the initial concept or at the stage of approving the final project. Due to the fact that the AR portal is managed in a computing cloud, the access to data - after authentication using personal credentials (in the form of a login and a password) - is possible from any location, i.e. also for the customer ordering the product. A dedicated IDE (Integrated Development Environment) is also available, thanks to which it is possible to create your own mobile application in which proprietary solutions can be implemented, e.g. additional business logic. The described solution is schematically shown in Fig. 6.

The method proposed by PTC, although fully automated and able to be quickly implemented and used, allows only displaying 3D models for the purpose of their presentation and evaluation, but without the possibility of editing their geometric form directly in the AR environment.

The technology of the Industrial Internet of Things (IIoT) is very closely related to the concept of the virtual world. It is mainly thanks to innovations in the IIoT area it became possible to connect the physical world with the virtual one [5].

The Industrial Internet of Things is a subset of the Internet of Things and focuses on industrial automation, device integration, communication between devices, data flow and predictive analytics [5, 23, 24]. IIoT simplifies the level of complexity of communication between machines and enables the acquisition and analysis of the data obtained from sensors deployed in machines and devices [13].

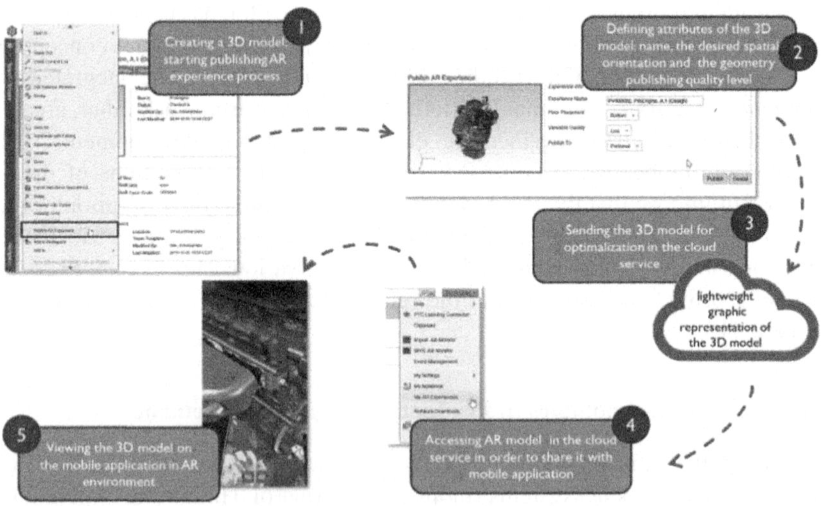


**Fig. 6.** Implementation of the PLM system integration with the AR environment. Source: own elaboration based on [16].

The concept of an intelligent product is based on the physical and informative representation of a product that has a unique identifier, has the ability to effectively communicate with its environment, can record and store information about itself, has a language to display its functions and is able to participate in the decision-making process related to its purpose or arrive at them on its own. The level of intelligence of an intelligent product ranges from the level of just simple data processing to complex proactive behavior [22]. Smart products can be equipped with RFID sensors and built-in data processing units.

The concept of the digital twin and product avatar is based on the concept of the so-called "hybrid world", according to which each material product in the "real world" has its digital representation in virtual reality [22].

The concept of using "twins" goes back to NASA's Apollo program. During the implementation of this program, two identical spacecraft were built. One of them was launched into space, the other remained on Earth and was called its "twin". It was used for many purposes throughout the entire project. In this sense, each type of the prototype that is used to map the real conditions in which another corresponding object or system is located, in order to perform a simulation in real time, can be understood as a twin. The constantly growing computing power and the overall efficiency of the systems and technologies used for simulations make it possible to build more and more perfect, more realistic and more precise models of physical objects. As a result, physical components are replaced with virtual ones. This enables engineers to apply this method at an increasingly early stage in the development cycle, even before the physical components are available. Extending this concept to all stages of the lifecycle leads to the use of a complete digital model of a complex physical system known as the Digital Twin (DT) [7].

A concept related to the digital twin is a virtual twin (VT). While the basis for the development and use of a digital twin are physical models, the concept of a virtual twin is primarily related to the visualization of material objects. In general terms, the implementation of a virtual twin is based on the dynamic mapping of the physical location of real objects in an environment enabling their visualization (e.g. in a virtual world) and ensuring continuous real-time synchronization between a real object and its virtual counterpart [12].

The concepts of a digital and virtual twin are closely related to the concept of a virtual factory. This term is a virtual, complete equivalent of a real object or system (e.g. a production system), which is characterized by continuous synchronization between the virtual and real systems. Thanks to the collected data and connected smart devices, mathematical models and real-time processed data, such a system can be used for many different purposes. Due to real-time synchronization and access to data read directly from the physical counterpart, it is possible to forecast and optimize the real system at every stage of the lifecycle in almost real time. By providing a virtual graphical representation of a real factory, the data obtained from the digital twin can be linked and superimposed on virtual three-dimensional models of machines and devices.

## 5 Summary

PLM systems are becoming an indispensable and standard workshop and work environment for designers, especially for products with a high degree of complexity and high requirements. Like other IT tools, they are constantly evolving as a result of not only increasing requirements for these systems, but also the development of IT technologies. The article presents the essence, genesis and development trends of PLM systems. Particular attention was paid to the need to expand PLM systems with applications supporting product development in all phases of the product life cycle and the use of Industry 4.0 technologies in these systems. The theses of the article are illustrated with examples of the use of selected Industry 4.0 technologies in PLM systems: industrial internet of things, virtual and augmented reality, and digital twins. The article is a fragment of a larger undertaking carried out by the Authors. Its result will be a book entitled "Product Lifecycle Management", which will soon be published by PWE.

## References


1. Chadzynski, P.Z., Brown, B.: Enhancing automated trade studies using MBSE, SysML and PLM. In: INCOSE International Symposium, vol. 28, pp. 1626–1635 (2018). https://doi.org/10.1002/j.2334-5837.2018.00572.x
2. Duda, J.: Managing the Development of Products in a Systemic Perspective. Cracow University of Technology, Cracow (2016)
3. Ameri, F., Dutta, D.: Product lifecycle management: closing the knowledge loops. Comput. Aided Des. Appl. **2**(5), 577–590 (2013)
4. Fisher, A., et al.: Model lifecycle management for MBSE. In: INCOSE International Symposium (2014). https://doi.org/10.1002/j.2334-5837.2014.tb03145.x
5. Gilchrist, A.: Middleware industrial internet of things platforms. In: Industry 4.0, pp. 153–160. Apress, Berkeley (2016). https://doi.org/10.1007/978-1-4842-2047-4_10
6. Graessler, I., Hentze, J.: The new V-Model of VDI 2206 and its validation. Automatisierungstechnik **68**(5), 312–324 (2020). https://doi.org/10.1515/auto-2020-0015
7. Hehenberger, P., Bradley, D. (eds.): Mechatronic Futures. Springer, Cham (2016). https://doi.org/10.1007/978-3-319-32156-1
8. Horvath, L., Rudas, I.J.: Systems engineering methods for multidyscyplinary product definitions. In: IEEE 12th International Symposium on Intelligent Systems and Informatics (SISY 2014), pp. 293–298 (2014). https://doi.org/10.1109/SISY.2014.6923604
9. Cao, H., Folan, P.: Product life cycle: the evolution of a paradigm and literature review from 1950–2009. Prod. Plan. Control **23**(8), 1–22 (2011)
10. INCOSE, International Council on Systems Engineering: Systems Engineering Handbook. A Guide for System Life Cycle Processes and Activities. Wiley, Hoboken (2015)
11. IT-Consulting: SysML - what is worth reading and keeping on the shelf (2020). https://it-consulting.pl/autoinstalator/wordpress/2020/05/25/sysml-co-warto-przeczytac-i-miec-na-polce/. Data dostępu: 18 Apr 2021
12. Konchenko, A., Chatar, C., Doronichev, S., Saetern, D., Bruns, J.: Oilfield virtual twin. In: SPE Annual Technical (2020)
13. Kumar, A., Nayyar, A.: si$^3$-industry: a sustainable, intelligent, innovative, internet-of-things industry. In: Nayyar, A., Kumar, A. (eds.) A Roadmap to Industry 4.0: Smart Production, Sharp Business and Sustainable Development. ASTI, pp. 1–21. Springer, Cham (2020). https://doi.org/10.1007/978-3-030-14544-6_1
14. Li, T., Lawson, C.P., Lockett, H.: Using Requirement-Functional-Logical-Physical models to support early assembly process planning for complex aircraft systems integration. J. Manuf. Syst. **54**, 242–257 (2020). https://doi.org/10.1016/j.jmsy.2020.01.001
15. Pieńkowska, M.: SysML Language in Requirements Engineering, Studia Ekonomiczne/Uniwersytet Ekonomiczny w Katowicach, nr 128 Wyzwania w rozwoju podstaw metodycznych projektowania systemów informatycznych zarządzania, pp. 91–102 (2013)
16. PTC: Using Augmented Reality Experience from Windchill, Windchill Help Center – 12.0.0.0 (2021). https://support.ptc.com/help/wnc/r12.0.0.0/en/#page/Windchill_Help_Center%2FARPublishAbout.html%23. Accessed 17 Oct 2021

17. Schmidt, M., Stark, R.: Model-Based Systems Engineering (MBSE) as computer-supported approach for cooperative systems development. In: Proceedings of 18th European Conference on Computer-Supported Cooperative Work, Siegen, Germany (2020)
18. Soares, M.: Model-based framework for influencing consumer products conceptual de-signs, pod red. In: Karwowski, W., Soares, M., Stanton, N. (eds.) Human Factors and Ergonomics in Consumer Product Design: Methods and Techniques, pp. 63–82. CRC Press, Boca Raton (2011)
19. Stark, J.: Product Lifecycle Management: 21st Century Paradigm for Product Realisation. Springer, London (2011). https://doi.org/10.1007/978-0-85729-546-0
20. Stark, J.: Product Lifecycle Management (Volume 4): The Case Studies. Springer, Heidelberg (2019). https://doi.org/10.1007/978-3-030-16134-7
21. Wiesner, S., Freitag, M., Westphal, I., Thoben, K.-D.: Interactions between service and product lifecycle management. Procedia CIRP **30**(2015), 36–41 (2015)
22. Wuest, T., Hribernik, K., Thoben, K.-D.: Accessing servitization potential of PLM data by applying the product avatar concept. Prod. Plan. Control Manag. Oper. **26**(14–15), 1198–1218 (2015). https://doi.org/10.1080/09537287.2015.1033494
23. Xu, R., Yang, L., Yang, S.H.: Architecture design of Internet of Things in logistics management for emergency response. In: 2013 IEEE International Conference on Green Computing and Communications and IEEE Internet of Things and IEEE Cyber, Physical and Social Computing, pp. 395–402 (2013)
24. Yan, B., Huang, G.: Supply chain information transmission based on RFID and Internet of Things. In: 2009 ISECS International Colloquium on Computing, Communication, Control, and Management (2009)

# Product Co-design Supported by Industry 4.0 in Customized Manufacturing

Wiesław Urban, Elżbieta Krawczyk-Dembicka(✉), and Krzysztof Łukaszewicz

Bialystok University of Technology, Wiejska 45A Street, 15-351 Bialystok, Poland
e.dembicka@pb.edu.pl

**Abstract.** In an economy faced with new economic and civilisation challenges, the product development process still plays an invariably important part. Increasingly often it is an area of close cooperation of enterprises with customers who expect a customized product and want to have a significant influence on the process of creation. The analysis of the existing mechanisms of co-design with customers will help determine the potential of further development of such activity and indicate the tools that boost its efficiency. The research involved case studies of two small and medium-sized enterprises, the operation of which includes co-design with the customer, the product development process was shown, with the indication of the manufacturer's and customer's roles. The process of creation of the production line for animal food packing was analysed, as well as the process of developing a printable cardboard packaging. The Industry 4.0 technologies meeting the co-design needs were presented. This issue requires further empirical study, extended over products which are not yet customized to a large extent.

**Keywords:** Co-design · Industry 4.0 · Product development process

## 1 Introduction

A gradual evolution of consumption patterns is closely connected with the technological changes that manufacturing is undergoing. Currently, a trend to re-orient enterprises towards the so-called mass customization of production [1, 2] is noticeable. The mass customization of products to the needs of customers is a manifestation of the 'Long Tail' strategy which consists in offering and efficient delivery of non-standard products, adapted to specific needs of customers, sometimes at higher price, which reflects extra effort and costs covered by the manufacturer. This strategy is proving to be effective to such extent, that it is included in business strategies of a growing number of companies. Anderson [3] makes a hypothesis that this strategy is the future of business, while Fischer et al. [4] and Imgrund et al. [5] stress the importance of the improvement potential integrated in the strategy.

The adoption of a strategy characterized by 'mass customization' and adaptation of products to individual requirements of customers requires more pressure on customer interaction and service. Increasingly often, the research results indicate that the boundaries between sectors and the very nature of 'product' are becoming blurred.


J. Trojanowska et al. (Eds.): MANUFACTURING 2022, LNME, pp. 186–199, 2022.
https://doi.org/10.1007/978-3-030-99310-8_15

One of the reasons for this blur should be searched for in the Industry 4.0 solutions implemented in the manufacturing and service sectors. The customers' expectations shift from buying/consuming products to adopting sophisticated solutions which solve context problems and generate value in use. Instead of paying for products, business customers increasingly often want to gain value inherently offered by using the product, thus consuming it as a service [6, 7]. The 'Product as a Service' model assumes a change in the approach not only to production, but also to entire product design and life cycle. It stresses the constant interaction with the customer in the long term, not only a one-off sale and warranty service.

Depending on application, co-design takes on different meaning, but it is always a process which actively engages the customer in product development [8–10]. It is a process which ensures the delivery of products more tailored to the preferences and needs of customers. One of the factors that has a positive impact on new product development is the involvement of first-tier suppliers in co-design as soon as on the very early stage of product development. Despite the fact that the concept of co-design is very widespread in service provision, the modern design routines pertaining both to process and product in many industrial sectors only to a small systemic and systematic extent take into consideration co-design with the customer. There are selected areas, wherein, to some extent, the end user of the product is involved, however this is not sufficient, developed methodologies are absent, as are developed theoretical frameworks, there are many indications that the co-design concept still has potential for broader application. Therefore, scientific research must define such deficiencies, pinpoint potential benefits of co-design, provide guidelines for broader application of co-design, including with the use of emerging new technologies.

The purpose of this study is then to assess the currently used methods of product design by virtue of co-design with the customer, and then to determine the potential of co-design development with involving the customer through the use of technologies referred to as Industry 4.0. However, empirical investigations will be primarily devoted to the comprehensive understanding the current role of customers in product development process, the potentiality of Industry 4.0 implementation will be determined by a conceptual work. The Industry 4.0 technologies were treated in the study as a mean allowing the customer to participate directly in the product development process and to influence the course of this process at any place and time. The research is based on case studies of industrial products, in the preparation of which co-design is employed. It allows the identification of the customers' participation in product development process, it is also a starting point for concept works related to the possibility to develop co-design using emerging technologies.

### 1.1 Co-design, a Key Point of Co-creation

Co-creation is a concept widely discussed in service research but it is relevant not only to pure service operation but wider in other business kinds. Value creation along with value co-creation with customers is the core of business activities on competitive markets [11]. Vargo and Lusch [12] argue that all business activity types are in fact service centric, collaborative activities of co-creation between parties are the in the heart of market exchange. The concept of co-creation is mainly on business mindset, the traditional approach which might be summarized in a question 'what can we do for you?', co-creation concept climes to be replaced by another question 'what can we do together?' [13].

Authors name customers participating in services processes as co-developer, co-designer or co-producer [14, 15]. Service studies underline service encounter processes are to develop successful co-creation opportunities during interactions between service providers and customers [16]. Authors state that even the responsibility for the process outcome meant as quality and added value are shared with participating customers [17]. However, the main issue referred to co-creation in services is gathering knowledge from customers along with prevision of their expectations, service providers are listening costumers so that customizing its actions [16, 18]. The study by Chang et al. [19] taking advantage of knowledge based management concept demonstrate several conditions necessary to occur for effective co-creation of service innovations, companies need adjusted tools for involving customers also user-friendly co-creation platform for coordination customers involvement.

Co-design is a specific part of co-creation where the customer is involved in design processes. However, this notion has different connotations, e.g. in architecture design and software design the user involvement is called as collaborative design [20]. Co-design is perceived as an opportunity users can contribute with a unique and latent needs to viable concepts [21]. Hurley et al. [22] underline that co-design this is a method enabling users to contribute with their unique knowledge and skills to new service ideas design. Customer/user ideas emerged during co-design make the starting point for new value propositions in new service products. Authors tested a co-design procedure composed of six steps including sensitizing of participants and facilitation leading to open discussion [22].

On the other hand, scholars see co-design area having still limited insights with regard to its benefits for the innovations [23]. Moreover, co-design being a pretty popular concept in service literature has almost no references in the manufacturing product/process design discussion. Many basic issues need to be deeply studied when it comes to co-design in manufacturing industry. Authors [24, 25] mention several conditions in terms of practical organization issues when cooperating with customers for successful innovation outcomes, these issues indispensably need studied with reference to manufacturing products design. Customer-driven innovation in any field of business activities need to take seriously into consideration technology advancements utilized for better collaboration during co-design process [26], this might have the greatest practical implications.

### 1.2 Product Development Process in Manufacturing Industries

Literature sources state that manufacturing products development is a long-lasting processes bundle which started from initial triggers, as strategic product plan or requirements analysis, and it ends at a stage of product withdrawal or utilization [27–29]. Among number of product development process a BIG Picture model [29] proposes a kind of holistic and integrated view. It is characteristic od of several paths considering uncertainty and risk, dealing with gathering appropriate information as well as different types of innovations [30]. However, the design activities meant as different kinds of creations and innovation making occur in such steps as definition of product assumptions, product modelling, prototyping and testing, preparing product for production [27, 28, 31]. Liu and Tsai [32] generalizing with four steps of product design process (1) product ideas generating along with the assessment, (2) product concept design and its further development, (3) product testing and trials, (4) launching product manufacturing.

Product development process models presented by the literature show a striving towards limitation of feedback loops between process steps, which make the process costly and lasting longer. This is clearly manifested in the stage-gate process model [33, 34], this approach bases on the belief that formal and structured reviews need to be implemented and exploited to ensure that the design is ready enough to allow the transition from one stage to another. Literature model also underline the knowledge management as a key issue in product development, combining knowledge domains of companies and clients raises capabilities to create successful innovative products [35, 36].

## 2 Methodology

Taking into account the research objectives, the use of the case study method was considered appropriate. The case study is qualitative approach to research which facilitates exploration of an interested phenomenon taking into consideration its context and by exploring a variety of data sources [37]. The case study procedure ensures that investigated issue is not explored from one viewpoint, but rather multiple facets are observed, considered and understood. The case study methodology has very wide application, scholars encourage to employ it also in exploratory manner, for example for getting known and understood some phenomenon in its real state [38], possibly to build on it some theory or a concept. This meets the research expectations set by the purpose of this study, because it is a multi-faceted research and understanding of the nature of co-design in the creation of an industrial product. It further allows to find existing challenges which can be solved by Industry 4.0 technologies exploitation.

The investigation strategy relies upon following the process of product development on the stages the product is being initiated, formed, its final form and structure decided, the technology of the product developed and set up, as well as it manufacturing process established. Special attention is devoted for the customer role and interaction with it during this process. The question which is asked is what is the contribution of the customer to the product development process. The data has been gathered from unstructured interviews with managers, documentation analysis and direct observation by investigators. Several visits were carried out in each of the surveyed companies, and the progress of the existing production processes was observed and analysed. Discussions were conducted with the employees indicated by the company's management board, during which all doubts were clarified and opinions were collected on the solutions proposed by the investigators.

For a case study procedure, intentionally have been chosen manufacturing processes having intensive cooperation with recipients on product development stage, one of them is producer of manufacturing lines for food processing industry, who design, produce and install comprehensive technology solutions for recipients, another one is a packaging maker delivering premium cardboard packaging at which product design and set-up have crucial consequences for the process costs and the quality of output. Two manufacturing cases represent two different manufacturing industries, support the multiple sources of data postulated by case study methodology. They describe the current state.

The use of Industry 4.0 technologies in the analysed case studies is of a conceptual nature. In the current state, identified during the research, the research objects were not equipped with specialized tools enabling, for example, the virtualization of the production process, such as goggles AR/VR, but during the observations, a great potential was noticed for the use of Industry 4.0 technologies at various stages of the process.

## 3 A Case of Industrial Production Line Co-design

Co-design plays an important role in the operation of the analysed company supplying customized technologies to manufacturers in the food industry. The company provides comprehensive services, from technology design to installation of finished product at the customer's premises, including maintenance services over a period use agreed on with the customer. The research production process covers an industrial production line for animal food packing. It was analysed in terms of cooperation between the company and the customer in the scope of new product design. From the point of view of the enterprise (manufacturer), in this case the product is the food packing technology, and from the customer's (buyer's) point of view, is it the production line.

The product development process presented in the flow chart at Fig. 1 starts with the customer placing a request for quotation, wherein product functionality needs are initially defined. Having analysed the request, the manufacturer starts developing technological and structural requirements and then creating a parametric model in the CAD environment, taking into consideration the customer's expectations. A period of intense cooperation between the buyer and manufacturer follows, in which the product is created and all technological and structural requirements are specified. This stage has at least a few iterations, during which the manufacturer presents 3D simulations and consults product functionalities to prepare an offer which is perfectly tailored to the customer's needs. The offer includes detailed information on the product, workload and costs required by the technology. Thus, the manufacturer presents the customer with terms of further cooperation and upon approval, a business agreement is entered into, under which the enterprise starts the implementation of the production process. In absence of approval, negotiations are usually started to introduce modifications in the designed product or the new product design process is interrupted. This stage is important from the enterprise's point of view due to the risk that the customer may resign, which may lead to the incurred costs not being recovered.

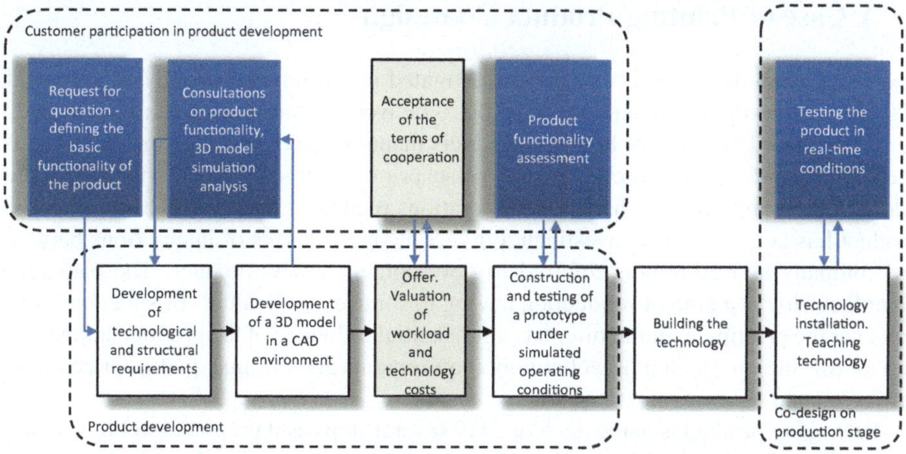


**Fig. 1.** Co-design in development process of industrial production line.

The creation of a production line starts with a prototype, treated also as the first version of technology. Based on the 3D model a real-time model is built, which usually requires certain modifications of the technological and structural requirements are made, impossible to identify on the parametric model. Next, the prototype is tested in operational conditions. The manufacturer carries out a simulation of food packing line in their manufacturing facility, and the results are transferred to the customer. The customer assesses the product functionality and, if necessary, informs of the need to modify the product. This is the last moment to introduce customer's modifications in the technological and structural requirements. The customer's role is therefore to approve the prototype's functionality, which allows the manufacturers to start building the technology. From the enterprise's perspective this is the final stage of product (technology) design, and the initial stage of manufacturing.

Building technology requires considerable creative work from the manufacturer, necessary to solve any previously identified technical problems and to guarantee the proper operation of the industrial line. It also covers product optimization and full adaptation to the customer's requirements, which requires the integrity of activities performed by mechanics, electricians and control engineers. The finished device is installed in the customer's premises and prepared for testing in real-time conditions, which for the customer is equivalent to product implementation. From the manufacturer's view, this marks the beginning of co-design of technology at production stage, when the manufacturer accompanies the customer during tests and assures technical support while 'teaching' the technology. The manufacturer carries out necessary training and adjusts the product to the actual needs of the user, identified during the work in the manufacturing site.

## 4 A Case of Printing Product Co-design

The issue of co-design has been also investigated in a printing company, a SME scale enterprise. Investigated manufacturing process delivers premium, highly ennobled, packaging cardboard product for branded products manufacturers in beauty and food industries. The manufacturing process consists offset printing, die-cutting, hot-stamping, automatic folding and different handmade operations related to finishing. Setting-up a new product has been carefully investigated in a company in terms of interactions between the company and customers referred to development a new product. The packaging manufacturing is a kind of production having to some extent features of subcontracting, however the printing manufacturer has a substantial influence of the product, adjusting it and modifying on account of technology requirements and optimal quality/price effects for recipients.

Packaging product is being developed in several steps as it presented in Fig. 2 below, the block chart presents also actions undertaken by a recipient regarding product design. The beginning of the product design process is the recipient contribution with graphics of the packaging desired, they are delivered to the company on the initial stage form business relations – the offering. Now, before the indication of the production costs along with the price proposal for customer several loops of consultations and creations between both sides take place. The different options of changes are proposed to the client, several further creations from the client sides come back to the printing company. This is a step of "clarifications" for the producer side, however for both sides this step consists of creations, if they are done well both sides benefit, the recipient has maximally exploited potential of the technology using by a manufacturer and obtain high and stable quality at moderate cost, the producer has satisfied customer with minimalised probability of defects and complaints. This first step is the most intensive when it comes to creativeness and new creations, many conceptualizations and technical/technological options are exchanged and considered, and all these are happening before a business agreement occurred between the parties.

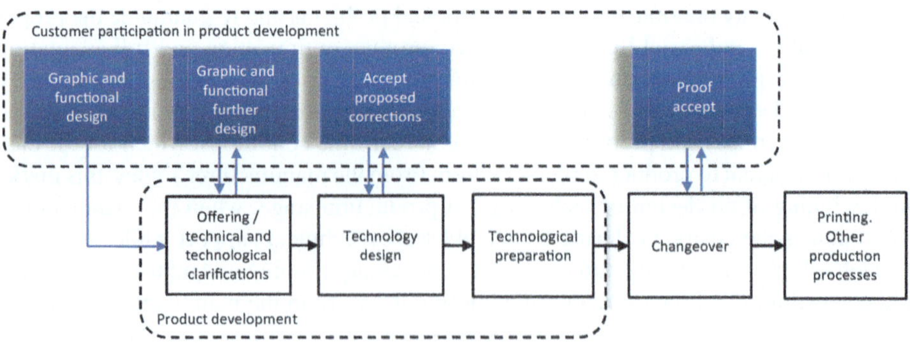


**Fig. 2.** Co-design in development process in printing cardboard packaging production

When business terms are finally accepted preceded by consent as to the assumptions of the product the printing company might start preparing many details referred to the technology process, however still some minor changes in the product itself also occur. The recipient is not interfering in the technology process, it accepts or not some detailed changes prosed by the manufacturer, they appear when the manufacturer preparing the production process, it is done with the cooperation with dies producer subcontractors. Next step is direct preparation to the manufacturing process, usually in this step the customer is not engaged in any way.

Before the production process is started the machines need to be set-up. One of the manufacturing processes of the primary importance is printing, in this process all the visual characteristics of the product are achieved. For many new products, the recipient is asked/expected to be asked to accept the product appearance, colours and fitness. The changeover step is for the manufacturer a production process set up, however for the recipient is still the product design, during this still some variables are changed some options, not ground-breaking, proposed. After it the final version of the product are shown for the customer and it say if it is satisfying or not.

## 5 Discussion: Prospects how Industry 4.0 Technologies Can Support Product Co-design

There are many indications that the current economies of some countries are at the stage of the fourth industrial revolution [39–41]. By definition, the fourth industrial revolution is a concept pertaining to the expansion of automation and robotization, constant optimization of products and processes, accumulation and real-time processing of large amounts of data, as well as to quick adaptation to changes in market conditions. Basically, it aims at full digitalization of production, wherein the equipment and technological systems communicate with each other, and intelligent networks connect machines, processes, systems, products, suppliers and customers. The I4.0 concept may also be characterized by such terms as: decentralization, cooperation, virtualization, real-time assessment of capacity, service orientation and modularity [42].

Report data [43] show that printing, as a customer-oriented industry, will be transforming, as the companies are met with constant trend of mass customization. The report [43] states also that 72% of users are reporting a growing need for quick processing of orders, 61% have noticed a growing demand for short series, and 59% are seeing growing expectations regarding just-in-time services. The attention to consistency information and image in all contact channels and contact points with customer is the fundamental step to achieve success of the proposed transformation.

In the perspective of Industry 4.0, simulation can be evaluated as a supportive tool to follow the reflections gathered from various parameter changes and enables the visualization in decision-making. Therefore, simulation tools can be used with other fundamental technologies of Industry 4.0 [44, 45]. The simulations are complemented by AR/VR technologies which allow the designers and customers to 'see and test' the finished product before it is physically manufactured, which carries high value from the marketing perspective and therefore influences the building of relationship with the customer.

An important quality of VR/AR which makes them different from traditional computer simulation, is the constant presence of the user in the virtual environment. For the discussed case, the VR technology will help place the model of device, created initially in the CAD system, in any virtual environment, next to other virtual objects. User immersion in the virtual world may depend on the simulation scenario, as well as on the applied peripheral devices. The user's influence on the environment may be modified to a large extent, also directly during the simulation. Apart from influencing the senses of sight and hearing of the user, the simulation in virtual reality may also use touch feedback by means of haptic devices presented in the literature on the topic [46, 47]. Such an approach allows the effective introduction of changes in the design and facilitates the contact with customer. The AR technology, based on the virtual object being visualized in the real spatial context, while the view of the object depends solely on the user's position against the location of the designed object [48], may be used for the final presentation of the design and verification of design requirements.

Söderman [49] described the use of the VR technology to assess products with potential customers. He concluded that the user's understanding of the representation of the design depends both on the user's knowledge of the product (previous experience in using the analysed product), as well as their knowledge of product representation (previous experience in the type of design representation used, e.g. 2D/3D sketch, VR, AR etc.). It suggests that full value of the VR technology will be realized only when the users, or designers, customers or consumers, gain a reasonable level of knowledge of these technologies [50].

Many researchers studied the impact the type of design representation has on various aspects of the design process. As for creating ideas, both Häggman et al. [51] suggested that the use of CAD tools at an early design stage may at times lead to premature limitations in the exploration of design space, which results in lower innovativeness of ideas.

A vital factor that may limit the generation of ideas is the time, effort or cost required to create design representation. Viswanathan and Linsey [52] discovered, that designers demonstrate a tendency to favour ideas they invested more time and effort in, even if they are less innovative or effective than other concepts.

Both product preparation processes analysed above, wherein the customer's participation was tracked, show explicitly that probably the most important factor of co-design is good communication between the parties, manufacturer and the customer. Communication challenge does not come down to a comprehensible and responsive exchange of messages, e.g. through e-mail, in the case of analysed processes it is about communicating visual impressions, including those impressions that are connected with the sizes and geometry of objects, as well as interaction facilitating creative team work which yields original solutions.

The AR/VR technologies presented above are of fundamental importance in this case. These technologies, connected with on-line team work environments and real-time data exchange, may have a significant impact on co-design. Such a solution, assuming its high functionality, may take co-design with customer to a totally new level. The number of interaction cycles, joint creative works and corrections of product/technology at various stages of development will be probably larger and limited primarily by the degree of readiness to involve the customer in achieving a fully satisfying product. The number of interaction cycles and team work engaging both participants in co-design does not determine success, it is the added value manifested by the adjustment, functions and optimization of products that determines efficiency. It should be expected that these technologies will considerably boost these measures. Moreover, the reduction of interaction costs through the elimination of personal presence of the customer's representatives at the manufacturer's facility, or vice versa, with be of considerable importance. For example, El-Jarn and Southern [26] present an interesting solution that fits the concept presented herein, consisting in the integration of software for 3D design with VR glasses and simulation tools as a means to study the prototypes with active participation of customers.

One of the major directions of changes in customer service process which will surely be, in the foreseeable future, implemented by entities operating similarly to the second of the analysed cases, is the use of AI-based tools which will support the customers in independent preparation of products, facilitating, e.g. smart tips and suggesting own 'automatic' hints as regards printing products. AI-based solutions will help learning from experiences of individual orders and then will intelligently guide through the preparation process and will help/substitute customers in many detailed jobs.

## 6 Conclusions

In the traditional product development process, manufacturers always work in conditions of limited amount of information and transparency of actual expectations and perceptions of the designed product by future users. Therefore, apart from handling various types of research, the manufacturers are forced to fill the concept gaps with their ideas of what is most important for potential customers, i.e. specific features of products, cost level etc. Without adequate information from customers, both in quantitative and qualitative aspect, the team working on the product may develop solutions that are in a totally different area than expected and approved by the future user. The co-design concept, which has gained considerable popularity in literature, is a direct answer to the basic problem occurring in each new product development process. Involving the customer in various stages of the process does not eliminate the problem, but helps contain it considerably, provided the organization has the possibility and skills to involve the customer properly. Despite the fact that in the industrial sector the customer is involved in product development, the scientific literature does not give much attention to this issue. An absence of co-design conceptualization is visible in the area of industrial products, and as this study shows, it is an important aspect and demonstrates enormous potential for the future. Therefore, further theorygenic works, conceptual works, as well as designing organizational and process solutions may be regarded as necessary. It seems that using co-design in the service perspective, after necessary adaptations, may give new potential when designing a product and taking into consideration its servitization, i.e. the entire ecosystem of product-related services.

The research using the case study method was conducted in entities operating on B2B markets, where high involvement of customers in the whole product development process is high. This helped thoroughly trace the co-creation of new product. In the analysed cases, the customer's presence is seen almost at each stage of the new product design and manufacturing process which, on one hand, ensures better product customization, and on other helps avoid costly mistakes. The application of co-design for customized products also distributes the product liability to the manufacturer and the customer. At the same time, it helps speed up the product design process thanks to shortening the decision-making processes.

The Industry 4.0 concept may open a totally new potential for co-design. It is estimated that is should primarily lead to the improvement of the product's value in use for the future users, and there are many indications that it may also reduce the costs of product development process. However, it requires equipping enterprises with mobile devices for industrial applications (e.g. tablets resistant to the working environment, AR/VR glasses/goggles), which will be adapted to work in often difficult conditions, and at the same time will ensure comfort and safety for employees or customers who use them. Trends towards hardware miniaturization should facilitate this. The challenge, especially in the initial period of use, will be to 'learn' how to work with devices and integrate them with the systems used by enterprises, but in the long run it will improve the activities undertaken and increase the quality of manufactured products. It may be then expected that co-design supported by the Industry 4.0 technologies will lead to the improvement of product competitiveness. It should also be stressed that customer's involvement in the analysed cases extends over product design. The customer is actively engaged as far as the production launch stage, or even production implementation, therefore the notion of co-development would be more adequate for these areas.

The use of Industry 4.0 with co-design demonstrates a very high potential of benefits. The presented discussion shows that technologies not only improve the efficiency and effectiveness of co-design in sectors where it is already in use, but they may also help expand such activities in sectors where they are not used yet. However, Industry 4.0 is still an area where concepts, experimental works, unique and prototypical applications intertwine, and where no widely empirically tested and stable solutions are available. Hence the strong need for research and implementation works in this area, which would take into account the application of these solutions in the area of broadly understood co-design of industrial products. The development of procedures and routine actions, as well as building organizational knowledge to facilitate the proper flow of co-design using the Industry 4.0 technology is a separate issue.

It seems that attention should also be focused on the global experience related to the COVID-19 pandemic which forces employees to keep social distance and reduce direct contact. Undoubtedly it hinders the 'face-to-face' co-design process [53]. In this aspect, the discussed Industry 4.0 technologies bring particularly useful solutions. The digitalization of process and intensification of remote contacts allow not only not to restrict direct contact, but, paradoxically, allow to use these technologies faster and to a broader scale to involve the customers more in the product development process, as well as including in the process those customers who would not decide or be able to participate in a 'face-to-face' co-design.

**Acknowledgments.** This research was funded by a grant from the Ministry of Education and Science received by the Bialystok University of Technology, grant number WZ/WIZ-INZ/3/2019.

## References


1. Dou, R., Huang, R., Nan, G., Liu, J.: Less diversity but higher satisfaction: an intelligent product configuration method for type-decreased mass customization. Comput. Ind. Eng. **142**(C), 1–13 (2020). 106336. https://doi.org/10.1016/j.cie.2020.106336
2. Hamrol, A., Gawlik, J., Sładek, J.: Mechanical engineering in Industry 4.0. Manage. Prod. Eng. Rev. **10**(3), 14–28 (2019). https://doi.org/10.24425/mper.2019.129595
3. Anderson, C.: The Long Tail: why the Future of Business is Selling Less of More. Hyperion, New York (2006)
4. Fischer, M., Imgrund, F., Janiesch, C.: Prioritizing and organizing the modeling of the long tail of business processes. In: Proceedings of the 28th European Conference on Information Systems (ECIS2020), An Online AIS Conference, pp. 1–15, 15–17 June 2020. https://aisel.aisnet.org/ecis2020_rp/18
5. Imgrund, F., Fischer, M., Janiesch, C., Winkelmann, A.: Managing the long tail of business processes. In: Proceedings of the 25th European Conference on Information Systems (ECIS2017), Guimarães, Portugal, 5–10 June 2017. https://aisel.aisnet.org/ecis2017_rp/39
6. Cusumano, M.A., Kahl, S.J., Suarez, F.F.: Services, industry evolution, and the competitive strategies of product firms. Strateg. Manage. J. **36**(4), 559–575 (2015). https://doi.org/10.1002/smj.2235
7. Gebauer, H., Paiola, M., Saccani, N., Rapaccini, M.: Digital servitization: crossing the perspectives of digitization and servitization. Ind. Mark. Manage. **93**, 382–388 (2021). https://doi.org/10.1016/j.indmarman.2020.05.011
8. Ulrich, P.V., Jo Anderson-Connell, L., Wu, W.: Consumer co-design of apparel for mass customization. J. Fash. Mark. Manage. **7**(4), 398–412 (2003). https://doi.org/10.1108/13612020310496985
9. Antikainen, M., Mäkipää, M., Ahonen, M.: Motivating and supporting in open innovation. Eur. J. Innov. Manage. **13**(1), 100–115 (2010)
10. Son, J., Sadachar, A., Manchiraju, S., Fiore, A.M., Niehm, L.S.: Consumer adoption of online collaborative customer co-design. J. Res. Interact. Mark. **6**(3), 180–197 (2012). https://doi.org/10.1108/17505931211274660
11. Prahalad, C.K., Ramaswamy, V.: The Future of Competition: Co-creating Unique Value with Customers. Harvard Business Scholl Press, Boston (2004)
12. Vargo, S.L., Lusch, R.F.: From goods to service(s): divergences and convergences of logics. Ind. Mark. Manage. **37**, 254–259 (2008)
13. Wind, J., Rangaswamy, A.: customization: the next revolution in mass customization. J. Interact. Mark. **15**(1), 13–32 (2001). https://doi.org/10.1002/1520-6653(200124)15:1%3c13::AID-DIR1001%3e3.0.CO;2-%23
14. Grönroos, C., Ravald, A.: Service as business logic: implications for value creation and marketing. J. Serv. Manage. **22**(1), 5–22 (2011). https://doi.org/10.1108/09564231111106893
15. Tronvoll, B.: Customer complaint behaviour from the perspective of the service-dominant logic of marketing. Manage. Serv. Qual. **17**(6), 601–620 (2007)
16. Payne, A.F., Storbacka, K., Frow, P.: Managing the co-creation of value. J. Acad. Mark. Sci. **36**, 83–96 (2008)
17. Eichentopf, T., Kleinaltenkamp, M., van Stiphout, J.: Modelling customer process activities in interactive value creation. J. Serv. Manage. **22**(5), 650–663 (2011)
18. Edvardsson, B., Ng, G., Zhi Min, C., Firth, R., Yi, D.: Does service-dominant design result in a better service system? J. Serv. Manage. **22**(4), 540–556 (2011)
19. Chang, Y.-C., Chiu, W.-H., Wang, J.-H., Teng, M.-J.: Customer involvement in the new process innovation: antecedents, mediation and performance. Eur. J. Innov. Manage Vol. ahead-of-print No. ahead-of-print, (2021). https://doi.org/10.1108/EJIM-09-2019-0268

20. Ulrich, P.V., Anderson-Connell, L.J., Wu, W.: Consumer co-design of apparel for mass customization. J. Fash. Mark. Manage. **7**(4), 398–412 (2003)
21. Trischler, J., Kristensson, P., Scott, D.: Team diversity and its management in a co-design team. J. Serv. Manage. **29**(1), 120–145 (2018). https://doi.org/10.1108/JOSM-10-2016-0283
22. Hurley, E., Trischler, J., Dietrich, T.: Exploring the application of co-design to transformative service research. J. Serv. Mark. **32**(6), 715–727 (2018). https://doi.org/10.1108/JSM-09-2017-0321
23. Ostrom, A.L., Parasuraman, A., Bowen, D.E., Patrício, L., Voss, C.A.: Service research priorities in a rapidly changing context. J. Serv. Res. **18**(2), 127–159 (2015)
24. Trischler, J., Pervan, S.J., Kelly, S.J., Scott, D.R.: The value of codesign: the effect of customer involvement in service design teams. J. Serv. Res. **21**(1), 75–100 (2018)
25. Storey, C., Larbig, C.: Absorbing customer knowledge: how customer involvement enables service design success. J. Serv. Res. **21**(1), 101–118 (2018)
26. El-Jarn, H., Southern, G.: Can co-creation in extended reality technologies facilitate the design process? J. Work-Appl. Manage. **12**(2), 191–205 (2020). https://doi.org/10.1108/JWAM-04-2020-0022
27. Amaral, D., Rozenfeld, H., de Araujob, C.: A case study about the product development process evaluation. In: Loureiro, G., Curran, R. (eds.) Complex Systems Concurrent Engineering, pp. 211–218. Springer, London (2007). https://doi.org/10.1007/978-1-84628-976-7_24
28. Santos, K., Loures, E., Piechnicki, F., Canciglieri, O.: Opportunities assessment of product development process in industry 4.0. Procedia Manuf. **11**, 1358–1365 (2017). https://doi.org/10.1016/j.promfg.2017.07.265
29. Lercher, H.: Big picture the innovation model. Anzeigen und Marketing Kleine Zeitung GmbH & Co KG Verlagsort, Graz, Austria (2019). https://ssrn.com/abstract=2929258. Accessed 24 Mar 2020
30. Aristodemou, L., Tietze, F., O'Leary, E., Shaw, M.: A literature review on technology development process (TDP) models. In: Centre for Technology Management Working Paper Series, pp. 1–31, Cambridge University Press, Cambridge, UK (2019). https://doi.org/10.17863/CAM.35692
31. Rossetti, G.H., Giraudo, F., Murer, P., Arcusin, L.: Comparative analysis of product development process management models. Am. J. Ind. Eng. **2**(1), 5–9 (2014). https://doi.org/10.12691/ajie-2-1-2
32. Liu, P., Tsai, C.: Research on the Influences of new product design and new product development process management on new product development performance in Taiwan's industries. Asian J. Qual. **10**(1), 89–106 (2009). https://doi.org/10.1108/15982680980000629
33. Cooper, R.G.: Stage-gate systems: a new tool for managing new products. Bus. Horiz. **33**(3), 44–54 (1990). https://doi.org/10.1016/0007-6813(90)90040-I
34. Cooper, R.G.: What's next? After Stage-Gate. Res. Technol. Manage. **57**(1), 20–31 (2014). https://doi.org/10.5437/08956308X5606963
35. Magnusson, P.R.: Exploring the contributions of involving ordinary users in ideation of technology-based services. J. Prod. Innov. Manage. **26**(5), 578–593 (2009). https://doi.org/10.1111/j.1540-5885.2009.00684.x
36. Sukhov, A., Magnusson, P.R., Netz, J.: What is an idea for innovation? World scientific book chapters. In: Kristensson, P., Magnusson, P., Witell, L. (eds.) Service Innovation for Sustainable Business Stimulating, Realizing and Capturing the Value from Service Innovation, chapter 3, pp. 29–47, World Scientific Publishing Co. Pte. Ltd. (2019). https://doi.org/10.1142/9789813273382_0003
37. Baxter, P.E., Jack, S.M.: Qualitative case study methodology: study design and implementation for novice researchers. Qual. Rep. **13**(4), 544–559 (2008). https://doi.org/10.46743/2160-3715/2008.1573

38. Rowley, J.: Using case studies in research. Manage. Res. News **25**(1), 16–27 (2002). https://doi.org/10.1108/01409170210782990
39. Dalenogare, L.S., Benitez, G.B., Ayala, N.F., Frank, A.G.: Int. J. Prod. Econ. **204**, 383–394 (2018). https://doi.org/10.1016/j.ijpe.2018.08.019
40. Liao, Y., Deschamps, F., de Freitas, E., Loures, R., Ramos, L.F.P.: Past, present and future of Industry 4.0 - a systematic literature review and research agenda proposal. Int. J. Prod. Res. **55**(12), 3609–3629 (2017). https://doi.org/10.1080/00207543.2017.1308576
41. Schwab, K.: The Fourth Industrial Revolution. Crown Publishing Group, New York, p. 192 (2017)
42. Hermann, M., Pentek, T., Otto, B.: Design principles for industrie 4.0 scenarios: a literature review. Working Paper No. 01/2015 (2015). http://www.thiagobranquinho.com/wp-content/uploads/2016/11/Design-Principles-for-Industrie-4_0-Scenarios.pdf
43. FESPA: 2018 print census reveals strategic responses to escalating demand. FESPA Staff (2018). https://www.fespa.com/en/news-media/press-releases/2018-print-census-reveals-strategic-responses-to-escalating-demand. Accessed 11 Oct 2021
44. Salkin, C., Oner, M., Ustundag, A., Cevikcan, E.: A conceptual framework for industry 4.0. In: Ustundag, A., Cevikcan, E. (eds.) Industry 4.0: Managing The Digital Transformation, pp. 3–23. Springer, Cham (2018). https://doi.org/10.1007/978-3-319-57870-5_1
45. Niekurzak, M., Kubińska-Jabcoń E.: Production line modelling in accordance with the industry 4.0 concept as an element of process management in the iron and steel industry. Manage. Prod. Eng. Rev. **12**(4), 3–12 (2021). https://doi.org/10.24425/mper.2021.139990
46. Buń, P., Grajewski, D., Górski, F., Wichniarek, R.: Application of haptic devices in virtual and hybrid design. Inżynieria Maszyn **18**(1), 18–29 (2013)
47. Sinclair, M., et al.: Three haptic shape-feedback controllers for virtual reality. In: 2018 IEEE Conference on Virtual Reality and 3D User Interfaces (VR), pp. 777–778, Reutlingen (2018). https://doi.org/10.1109/VR.2018.8446399
48. Skórska, H.: Augmented reality systems and their applications. Przegląd Mechaniczny **1**, 48–52 (2017). https://doi.org/10.15199/148.2017.7-8.7
49. Söderman, M.: Virtual reality in product evaluations with potential customers: an exploratory study comparing virtual reality with conventional product representations. J. Eng. Des. **16**(3), 311–328 (2005)
50. Cascini, G., et al.: Exploring the use of AR technology for co-creative product and packaging design. Comput. Ind. **123** (2020). https://doi.org/10.1016/j.compind.2020.103308
51. Häggman, A., Tsai, G., Elsen, C., Honda, T., Yang, M.C.: Connections between the design tool, design attributes, and user preferences in early stage design. ASME. J. Mech. Des. **137**(7), 071408 (2015). https://doi.org/10.1115/1.4030181
52. Viswanathan, V., Linsey, J.: Design fixation in physical modeling: an investigation on the role of sunk cost. In: Proceedings of the ASME 2011 International Design Engineering Technical Conferences and Computers and Information in Engineering Conference, vol. 9: 23rd International Conference on Design Theory and Methodology; 16th Design for Manufacturing and the Life Cycle Conference, pp. 119–130, Washington, DC, USA, ASME, 28–31 August 2011. https://doi.org/10.1115/DETC2011-47862
53. Davis, A., Wallace, N., Langley, J., Gwilt, I.: Low-contact co design: considering more flexible spatiotemporal models for the co-design workshop. Strateg. Des. Res. J. **14**(01), 124–137 (2021). https://doi.org/10.4013/sdrj.2021.141.11

# Design of Customized Sea Container House in the Context of Industry 4.0 - A Case Study

Jacek Frątczak[1] and Justyna Patalas-Maliszewska[2](✉)

[1] Sanpol Sp. z o.o, Sulechowska 27a, 65-119 Zielona Góra, Poland
[2] University of Zielona Góra, Licealna 9, 65-417 Zielona Góra, Poland
j.patalas-maliszewska@iim.uz.zgora.pl

**Abstract.** One of the most important challenges for the success of manufacturers, according to the concept Industry 4.0 (I4.0), is to improve the efficiency and flexibility of production processes. The type of mass production customization consists in personalizing an offer of an enterprise on a large scale, which is possible due to the dynamic development of production technology and in-depth knowledge of customers' needs and preferences. Designing a new product requires not only knowledge of the technology itself or product organization, but also the ability to involve a customer in the design process of a new product. Therefore this article proposes a model of the customized product design process using virtual reality (VR). VR is one of the I4.0 technologies, which is an immersion in a completely virtual environment. The main contribution of this work is an evaluation of this approach on an actual case study, on the process of manufacturing modular homes from prefabricated HQ 40-foot shipping.

**Keywords:** Design · Customized product · Industry 4.0 · A Case Study

## 1 Introduction

The term Industry 4.0 (I4.0) includes not only the computerization of manufacturing and the implementation of new network technologies in order to improve the communication paradigm [1], but also the transformation of current production systems into Smart Manufacturing Systems (SMS) through automation, deployment of cyber-physical systems (CPS), cloud computing capability, robotics, big data, artificial intelligence, and the Internet of Things (IoT) [2]. It is worth mentioning that the concept of I4.0 is also associated with decentralization, interoperability, and virtualization of real-life manufacturing processes, which means that smart factories can effectively use the paradigm of personalised manufacturing [3]. The flexibility of process reconfiguration in order to meet individual customer needs is an important parameter in this regard.

The type of mass production customization consists in personalizing an offer of an enterprise on a large scale, which is possible due to the dynamic development of production technology and in-depth knowledge of customers' needs and preferences. It is somehow a variation of the differentiation strategy, which assumes that every consumer is unique. This type of production is used in order to optimally meet the needs of buyers


J. Trojanowska et al. (Eds.): MANUFACTURING 2022, LNME, pp. 200–208, 2022.
https://doi.org/10.1007/978-3-030-99310-8_16

by including them in the product design process. Product customization is particularly useful in the construction industry, which is a reason why in this article we conducted research on the manufacturing process of Customized Sea Container House, specifically the design stage. De-signing a new product requires not only knowledge of the technology itself or product organization, but also the ability to involve a customer in the design process of a new product.

The use of technologies based on virtual objects at a design stage of a personalized product is significantly beneficial due to the fact that it makes it possible to eliminate errors and reduce costs related to product testing [4]. Augmented reality (AR) is a technology that enables the integration of a virtual environment and its objects with a real environment by interacting with its users using digitally generated images, sounds, graphics, and Global Positioning System (GPS) data [5, 6]. Virtual reality (VR) is a system that allows users to interactively explore and simulate the behaviour of a CPS. Through technology, VR and AR aim to expand an individual's sensory environment [7]. VR solutions that are available on the market enable us to prepare immersive and interactive environments with a high degree of realism. This approach prompts customers to make a quicker decision and engages them in the design process. Another important advantage of VR technology is the usage of small real areas to visualize large spaces and objects, such as interiors or building lumps [8].

This paper deals with problems related to mass-customized production of houses from prefabrication, and especially with the design stage within the production process of the new product, due to the use of modern technologies, namely VR in the context of Industry 4.0. Therefore, this article elaborates on the process of designing a new product in close cooperation with the customer using VR technology. This approach was verified in a real-life case study: the production of the sea container house from the prefabrication of used HQ 40-foot transport containers.

## 2 Model of the Customized Product Design Process Using VR

In the mass customisation process, in the design stage and presentation of products whose design requires customer acceptance, companies use Industry 4.0 technologies, which include visualization, i.e. augmented reality (AR), which is associated with graphics, sounds and haptic feedback; virtual reality (VR), which is an immersion in a completely virtual environment; and mixed reality (MR), which is an intermediate solution. Other areas in which elements of virtual reality are used are logistical processes, including production site planning and material flow simulation; construction processes, for example, physico-chemical simulation, processes related to traffic management, such as machine diagnostics, service, fast parameters reading; and production processes, for example, work according to manuals, job training, insight into production planning, etc. [8].

The model of the new product design process using visual-spatial technology requires establishing boundary conditions for a range of products (i.e. immutable physical parameters, such as dimensions of the workspace), creating libraries containing virtual models of variable components and their material characteristics (e.g. texture, colour, etc.). The next step is to define product groups that will determine the selection of 3D objects from the libraries, according to the customer's preferences or price range. In addition, it is necessary to define the dependencies and interactions which can occur between objects and selected functions. All of these steps described above facilitate creating a presentation of a product and prepare us for the production of its final version. Depending on the technologies used, it is necessary to select an appropriate infrastructure enabling immersion in virtual reality. For example, equipment such as glasses, manipulators, cameras, and software dedicated to the end-user (Fig. 1).

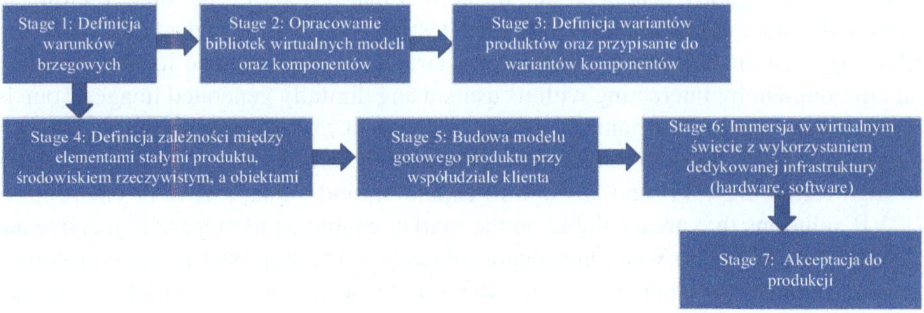


**Fig. 1.** Model of the customized product design process using VR.

The usage of virtual models during the design process is becoming more and more popular in supporting building information modelling (BIM) and architecture presentation [9, 10]. Wang et al. [11] describe five ways of presenting real estate using this method, i.e. one-dimensional text, e.g. advertisements; 2D floor plan visualization; architectural model; animation of movement inside the building; and a model showing the functionality of the building. The product can be adjusted to the customer according to a completely individual design or as a proposal of certain options [12]. This ability to adjust the product to the customer gives the company a competitive advantage in modern times of changing trends, especially when the offered product is innovative [13]. In the case of prefabricated buildings, the trend is based particularly on reducing construction time and minimalizing costs at the same time. In order to produce a complex and also personalized product, it is recommended to use computing solutions, such as virtual reality system for designing of finished product in an intelligent factory, working according to the I4.0 requirements [14].

## 3 A Case Study

The usage of VR technology in the production of the sea container houses fits into the concept of I4.0 as it allows for effective changes in a project and im-proves communication with a client, as well as lets the company present the final product and verify design assumptions. The process of manufacturing modular homes from prefabricated HQ 40-foot shipping containers (Fig. 2) is an exam-ple of an implementation of a concept of mass customization using VR.

**Fig. 2.** A customized product: house from prefabricated HQ 40-foot sea containers.

The production process of a customized home from prefabricated HQ 40-foot sea containers involves four main stages:

- Stage 1: Designing a new product:
  - Step 1: choosing the size variant: one, three, four-module variant,
  - Step 2: selection of interior organization, layout of structural and partition walls,
  - Step 3: choosing the option of interior design along with their spatial location,
  - Step 4: selection of the variant of the organization of internal installations of the new product: electricity, water and sewage, heating.
- Stage 2: Preparation of the raw material (container) for the next stages of prefabrication:
  - Step 1: mechanical cleaning of rust and other contaminants,
  - Step 2: chemical cleaning of biological and chemical contaminants,
  - Step 3: cutting holes in structural elements for windows and doors,
- Stage 3: Structural stabilization of the prefabricated element:
  - Step 1: stabilization of the construction by welding closed steel profiles in the places of the holes from Stage 1,
  - Step 2: welding of external and internal wall cladding carriers, as well as construction of partition walls,
  - Step 3: coating all steel elements with a chemical anti-corrosion repellent.
- Stage 4: Assembly, reinforcement and installation works:
  - Step 1: installing window and door frames (without wings),

- Step 2: installing internal systems reinforcements, such as an electrical installation, water-supply system, and sewer installation,
- Step 3: installing the thermal insulation using the spray method (PUR foam).

In the analysed manufacturing process, the first step is the design stage of a new product, for which we implement the model of a customized product design process using VR technology (Fig. 1).

## 4 Research Results

According to the stages presented in a model of the customized product design process (Fig. 1), the process of designing a home from prefabricated marine HQ 40-foot structures using VR is carried out as follows:

- Stage 1: Establishing the boundary conditions.
  The step is to determine which real items will be referenced by virtual objects. It is assumed that one sea container will constitute a single actual accommodation unit. It is partially covered with PUR foam insulation and finishing material. At this stage, the basic restrictions related to the modification of the container in accordance with the art of engineering is declared, e.g. determining the maximum area and number of the window opening on the longer and shorter wall, etc. For advertising and demonstration purposes, the facility is located at the headquarters of the design studio. It is also provide an insight into the structure of the container body. Using VR the container is equipped with a number of cameras supporting movement in a completely modeled space. A finite set of variants and settings for residential units will be proposed. At this stage, the visualization wizard architecture and the customer-visible interface is build.
- Stage 2: Creating virtual model and component libraries.
  A team of designers and graphic designers collects ready-made 3D solid models and builds their own. For this purpose, the CAD Inventor software is used. In the first step, standard solutions for containers are designed to visualize door and window openings, staircase cut-outs and chimneys. Subsequent activities include the preparation of groups of 3D components for interior fittings, finishes along with the specification of their attributes: style or price level.
- Stage 3: Product variant definition and component variant assignment.
  In order to improve the customer's choices and the functioning of the product visualization wizard, it is assumed that the container houses will have default variants depending on the customer's budget (basic, comfort, premium). In addition, each version will offer 3 interior styles (scandinavian, industrial, vintage). Component libraries described with appropriate attributes (stage 2) will be associated with the selected products. Depending on the price range agreed with the client, the definition of his style, the starting point is visualization and then the modification of the selected project.

- Stage 4: Defining the dependencies between a product's constant elements, the real world, and objects.
  The interior of the facility is equipped with markers or a grid to which the possibility of displaying/inserting virtual objects (e.g. cabinets, basic household appliances, etc.) is assigned. In addition, at this stage, the interaction between the selected branch of products and the actual customer movements or indications resulting from the manipulator will be programmed. For example, grabbing the handle of a virtual door should visualize another interior of a room (e.g. bathroom) or the entire residential module. However, indicating a specific component will display its attributes, unit price (e.g. price of a selected object, possible sizes of a set of furniture). Moreover, individual areas of the housing modules will be associated with the display of a given family of objects, e.g. markers on the wall of the container activate the options for inserting/moving the window opening. During the definition of dependencies and relationships with objects, restrictions resulting from construction art (stage 1) will be respected.
- Stage 5: Constructing the final product model in collaboration with the customer.
  Below is an example of the operation of the configurator application of the finished product variant (in this case a single-module version house – Fig. 3a, 3b. 3c):

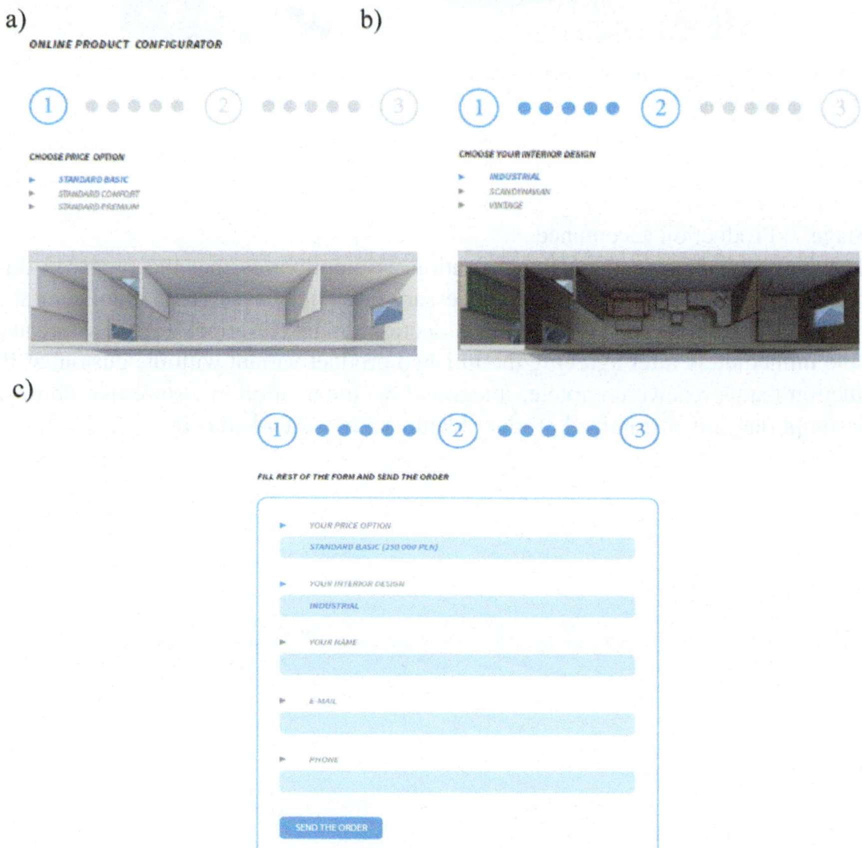


**Fig. 3.** Configurator application of the finished product: house from prefabricated HQ 40-foot sea containers, variant A: a) product configuration, b) choice of price level, b) pre-order form.

- Stage 6: Immersion in VR through the use of dedicated infrastructure.
  VR applications allow to visualize and confirm customer choices without the need to incur additional costs, for example with the creation of real prototypes. In the analysed case study the software is SketchUp 3D design and modeling program is used (Fig. 4).

**Fig. 4.** Use of the sketchup program in the design of a 3D model of a house from prefabricated HQ 40-foot sea containers.

- Stage 7: Production acceptance.
  The purpose of using the VR application assumptions is that the prepared documentation is equipped with all the necessary drawings based on an architectural 3D model. Construction details, floor plans, sections, main elevations and views are available immediately after agreeing the finished product variant with the customer. Production teams receive complete, integrated documentation to ensure that no design assumptions and customized product features are overlooked (Fig. 5).

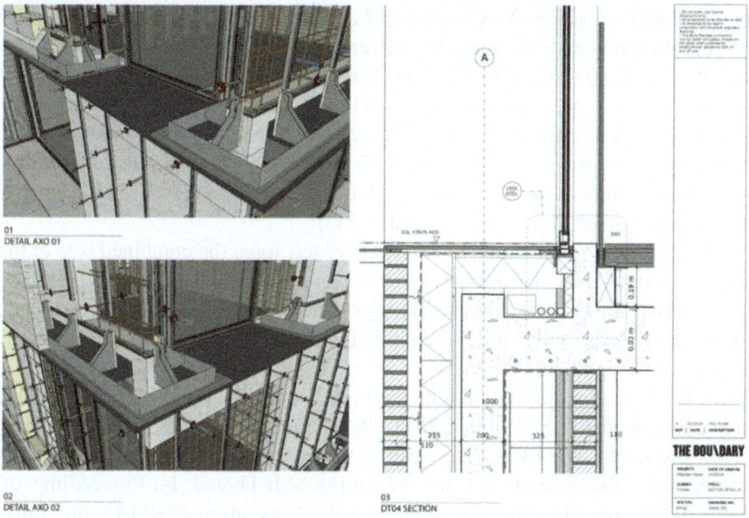


**Fig. 5.** An example of the structural decomposition of a building designed on the basis of 3D technology with the use of VR – . Source sketchup.com.

Next, the next three stages of the production process of a customized home from prefabricated HQ 40-foot sea containers can be realised. Further work, on the production process of a customized home should be extended throughout definition of the methodology for assessing the effectiveness of the production process in line with the mass customisation concept under condition of the unexpected changes in the environment and in customer demands.

## 5 Conclusions

Modern prefabrication in the Industry 4.0 trend is one of the most important construction trends of the future, which has many applications in the field of housing construction. The research results illustrate the current work of the customized product: a customized home from prefabricated HQ 40-foot sea containers design process using VR. By proposing a model of the customized product design process which addresses the influence of customer activity on the production process of a customized product using a configurator application of the finished product, this study contributes knowledge transfer applications support within an organization and customer. The observations within a real-life case study, proposed in this paper, can be a good benchmark for how managers can improve their production process of the customized product at the design stage in line with the Industry 4.0 requirements.

## References

1. Urban, W., Łukaszewicz, K., Krawczyk-Dembicka, E.: Application of industry 4.0 to the product development process in project-type production. Energies **13**(21), 5553 (2020). https://doi.org/10.3390/en13215553

2. Kamble, S.S., Gunasekaran, A., Gawankar, S.A.: Sustainable industry 4.0 framework: a systematic literature review identifying the current trends and future perspectives. Process Saf. Environ. Prot. **117**, 408–425 (2018). https://doi.org/10.1016/j.psep.2018.05.009
3. Ferrero, R., Collotta, M., Bueno-Delgado, M.V., Chen, H.-C.: Smart management energy systems in industry 4.0. Energies **13**(2), 382 (2020). https://doi.org/10.3390/en13020382
4. Bachvarov, A., Maleshkov, S., Häfner, P., Katicic, J.: Design-by-the-customer through virtual reality. In: Proceedings of the 4-th International Conference on Advanced Research and Rapid Prototyping, pp. 561–566, Leiria, Portugal (2009)
5. Daassi, M., Debbabi, S.: Intention to reuse AR-based apps: the combined role of the sense of immersion, product presence and perceived realism. Inf. Manage. **58**(4), 103453 (2021)
6. Kurt, Y., Öztürk, H.: The effect of mobile augmented reality application developed for injections on the knowledge and skill levels of nursing students: an experimental controlled study. Nurse Educ. Today **103**, 104955 (2021)
7. Gandolifi, E.: Virtual reality and augmented reality (2018). Accessed 12 Oct 2021. https://www.researchgate.net/publication/324571346_Virtual_Reality_and_Augmented_Reality
8. Śliwa, M., Babirecki, W.: XR technologies in support of knowledge management in production companies – the concept of using AR in the SMED area. In: Proceedings of the 35th International Business Information Management Association (IBIMA), pp. 15623–15637, Seville, Spain (2020)
9. Wang, J., Wang, X., Shou, W., Bo, X.: Integrating BIM and augmented reality for interactive architectural visualisation. Constr. Innov. **14**(4), 453–476 (2014)
10. Schranz, C., Urban, H., Gerger, A.: Potentials of augmented reality in a BIM based building submission process. J. Inf. Technol. Constr. (ITcon) **26**(24), 441–457 (2021)
11. Calderon-Hernandez, C., Brioso, X.: Lean, BIM and augmented reality applied in the design and construction phase: a literature review. Int. J. Innov. Manage. Technol. **9**(1), 60–63 (2018)
12. Fogliatto, F.S., da Silveira, G.J., Borenstein, D.: The mass customization decade: an updated review of the literature. Int. J. Prod. Econ. **138**(1), 14–25 (2012)
13. Ellena, T., Mustafa, H., Subic, A., Yen Pang, T.: A design framework for the mass customisation of custom-fit bicycle helmet models. Int. J. Ind. Ergon. **64**, 122–133 (2018). https://doi.org/10.1016/j.ergon.2018.01.005
14. Żywicki, K., Zawadzki, P., Górski, F.: Virtual reality production training system in the scope of intelligent factory. In: Burduk, A., Mazurkiewicz, D. (eds.) ISPEM 2017. AISC, vol. 637, pp. 450–458. Springer, Cham (2018). https://doi.org/10.1007/978-3-319-64465-3_43

# Digital Maturity of the Enterprise as an Assessment of its Ability to Function in Industry 4.0

Alla Polyanska(✉), Svitlana Savchuk, Ivanna Zapukhliak, Yaroslav Zaiachuk, and Irina Stankovska

Ivano-Frankivsk National Technical University of Oil and Gas, Ivano-Frankivsk 76019, Ukraine
parvs@ukr.net

**Abstract.** The article investigates the digital maturity of energy companies through an assessment of their ability to operate in conditions of digital era. It is substantiated that the assessment of digital maturity is a necessary condition for the digital transformation of the enterprise. The purpose of the article is to assess the digital maturity of enterprises by applying the methodology of digital maturity state definition on the example of energy enterprises using fuzzy logic methods. Based on the analysis of existing models of digital maturity assessment and identification of their shortcomings, the author's model of digital maturity assessment of the enterprise is developed. Proposed model includes such descriptors of enterprise management system: strategy, personnel, organizational culture, technologies, structure, marketing. The indicators of digital technology used at energy enterprises of Ukraine were analysed on the base of static data. Using the questionnaire method, information on digitalization was collected at Ukrainian energy enterprises. Conclusions about the state of digitalization at the selected enterprises was made based on the processing of the survey data. The model of enterprise's digital maturity assessment was applied using the provisions of fuzzy set theory and interval scales. There were calculated the levels of digital maturity of Ukrainian energy enterprises by determining the integrated level of digital maturity and maturity level of its individual components. The obtained results of energy enterprises digital maturity assessment made it possible to orientate on specific measures for the digital transformation of energy enterprises.

**Keywords:** Digital maturity · Digital transformation · Assessment model · Energy enterprises · Fuzzy logic method

## 1 An Introduction

Klaus Schwab, founder and chairman of the World Economic Forum, in 2016 published his vision for future development on the margin of a technological revolution that will significantly change our life as well as habits. The digital transformation will change the focus on resources and capabilities that we were not able to experience and explore before and will be characterized by the growth of innovation, which will achieve significant


J. Trojanowska et al. (Eds.): MANUFACTURING 2022, LNME, pp. 209–227, 2022.
https://doi.org/10.1007/978-3-030-99310-8_17

improvements in efficiency, productivity and cost reduction; ensure the growth of data and opportunities of their use for new technologies with the involvement of different levels of developers - users - customers and will promote the development in many fields; will develop artificial intelligence and ways of its application in various spheres of activity [1]. The conditions in which digital changes are taking place in the world differ both at the level of countries and enterprises. Globalization opens up new opportunities for joint decisions, which are made to solve economic problems both at the level of individual enterprises and within the framework of a strategic partnership. Here there is a need to study the parameters of readiness and ability to implement these solutions under conditions of significant environmental influences. The concepts of readiness and capability in management are associated with the maturity of the organization, which means that it is ready for a certain model of action to better perform tasks and identifies necessary improvements by developing a system of criteria that will allow to compare the results with competitors, strategic partners or customers [2]. Digital maturity is considered as a measure of an organization's ability to create value through the possibility to predict success for companies launching a digital transformation [3].

Assessing the digital maturity of the enterprise allows it to determine its ability to achieve better results and occupy a stable position compared to competitors. This is especially true in an era when the digital environment is subject to constant unpredictable changes. The benefits of digitalization may be obvious, but relatively few domestic enterprises are fully realizing their digital potential, losing many opportunities and competitive advantages [4]. This is often due to the complexity of the digital transformation process itself, i.e., the company cannot assess its current capabilities and choose the most effective way to implement them.

It worth to note that the processes of digital transformation have already acquired some practice. In particular, the European Research Centre for Information Systems (ERCIS) [5] considers activities that an enterprise must take to create the primary preconditions for digital transformation, namely:

- to assess the current level of digital maturity and existing development skills in this area,
- to determine the level of current and desired digital maturity of the business,
- to identify the action plan needed to achieve the desired level of digital maturity.

Assessing maturity as a standardized structure for determining the overall capabilities and needs of the enterprise helps to prioritize those areas that have the greatest impact on business and can provide maximum effect in the digitization [6].

Today, most experts describe digital maturity as the level at which an organization is able to optimize its online presence using data analysis to predict future customer needs and improve digital performance accordingly [7]. In a study by Capgemini and the MIT Digital Business Centre [8], digital maturity is considered as a function of two dimensions, namely:

- digital intensity - is the amount of investment of the enterprise in digitalization; measured by coordinated areas of digital transformation (for instance, digital project portfolio),

– the intensity of digital transformation management or digital support of the enterprise, measured by the presence of digital vision, the effectiveness of digital strategy, etc.

We consider these dimensions interdependently and complementary, as digital intensity depends on the intensity of digital transformation management, and the last depends on investment in appropriate management measures and tools. Despite the fact that the digital transformation of the enterprise is realized through different processes, in organizations, digital change can occur differently, in particular, either evolutionarily or revolutionary. It mainly depends on the current digital maturity of the organization. That is why understanding the state of digital maturity is the first logical step towards digitalization.

In this context the purpose of the article is to substantiate the relevance of digital maturity of enterprises to assess its ability to operate in Industry 4.0, considering a methodological approach to determining the state of digital maturity on the example of energy enterprises using fuzzy logic methods. The achieving of declared article purpose is based on the consequent performance of such steps: an underlining the actuality of digital maturity investigation in the time of INDUSTRY.4 revolution; backgrounding the method of fuzzy sets to determine the state of digital maturity in energy companies as appropriate math tool that uses qualitative results of questionnaire assessments; the addressing the relative questionnaire for getting relevant information before proceeding to a direct analysis of the development and usage of digital technologies; conducting the analysis of energy companies for preconditions determination of digital transformation; analysing of got results for determining the level of digital maturity of Ukrainian energy enterprises; discussing of other assessment methodologies on digital maturity investigation; conclusions and perspectives of future research's on mentioned field.

## 2 Research Methods

The task of the enterprise's digital maturity assessing involves solving problems in the field of expertise, so it is worthwhile to use methods of fuzzy logic, which provides an opportunity to set and mathematically solve even such problems for which there are no full statistics, or when among the informative factors there are only qualitative indicators, while providing the opportunity to adapt economic and mathematical models to changing economic conditions [9, 10]. In addition, fuzzy set methods should be used in cases where the system operates under uncertainty and is characterized by the lack of an accurate mathematical model by which is possible to describe the functioning of the system [11]. It should be noted that this technique has been effectively used in the case of assessing the state of corporate culture in enterprises [12] and to assess the state of readiness of enterprises for change [13]. All cases involve the processing of analytical material, which is obtained by conducting qualitative, expert evaluations. Therefore, we consider it appropriate to use the method of fuzzy sets to determine the state of digital maturity in energy companies.

To solve this problem, the expression of the fuzzy products rule is as follows:

$$(i) : \Omega; \Pi; A \Rightarrow B, S, F, Y \qquad (1)$$

where ($i$) is the name of the fuzzy product; $\Omega$ - scope of fuzzy products; $\Pi$ - the condition of application of the core of fuzzy products; $A \Rightarrow B$ - the core of fuzzy products; $S$ - method of determining the quantitative value of the truth of the core; $F$ - coefficient of certainty of fuzzy products; $Y$ - postcondition of fuzzy products.

By fuzzy products we mean expression (1) except that the condition of the core (antecedent) A and the conclusion of the core (consequent) B are fuzzy linguistic expressions.

The main feature of fuzzy rules used in fuzzy inference systems is that the conditions and inferences of individual fuzzy rules are formulated in the form of fuzzy linguistic expressions of one of the following types in relation to the values of certain linguistic variables, namely:

- expression "$\beta \in \alpha$", where $\beta$ - the name of the linguistic variable; $\alpha$ - its value, which corresponds to a separate linguistic term from the basic term set $\Upsilon$ of a linguistic variable $\beta$,
- the expression "$\beta \in \nabla\alpha$", where -$\nabla$ modifier, which corresponds to such words as: "Very", "More or less", "Much more" and others.

Based on the above, we propose a model for assessing digital maturity, which will be as close as possible to the current realities of the energy companies' operation in Ukraine. This model involves determining the level of digital maturity in 6 areas of enterprise's activity, such as: strategy, staff, structure, technology, marketing, organizational culture. According to the study of enterprises' digitalization as well as the analysis of attributes of digital maturity in different models and characteristics of digital transformation, there were identified the list of relevant questions, answers to which help to determine the state of digital maturity. This approach will provide the assess of digital maturity in each of the enterprise's activities.

The model is developed in the form of a questionnaire and consists of 61 questions on 6 components: "Strategy" (13 questions), "Personnel" (15 questions), "Organizational culture" (10 questions), "Technologies" (10 questions), "Marketing" (6 questions), "Structure" (7 questions). For each question/statement there are 5 possible answers, from which you must choose one.

Analysing the internal environment of the enterprise in terms of assessing the state of its digital maturity, experts do not use quantitative values of various indicators, but make linguistic assessments of quality (linguistic variables): very good, good, satisfactory, bad, very bad. We use a fuzzy-sets model to assess the level of digital maturity of the enterprise, which is presented in the form of a hierarchical relationship between input variables (X1-X61), integrated characteristics of six elements of the enterprise: strategy, staff, technology, organizational culture, structure, marketing and output variable - integrated indicator of digital maturity of the enterprise (Fig. 1).

The fuzzy inference system converts the values of input process variables into output variables based on the use of fuzzy product rules. To do this, fuzzy inference systems must contain a base of fuzzy product rules and implement fuzzy inference based on conditions presented in the form of fuzzy linguistic expressions. The basis of fuzzy systems design is the construction of a knowledge base (BS) using imaging methods and knowledge retrieval. In the knowledge base, the expert indicates his answer in each situation in the form of products IF… THAT…, many of which make up the knowledge base. The expert is a qualified specialist who knows the subject area.

In the design of the expert system, considerable effort and time are spent on the development of the database: the accumulation of knowledge, determining the model of their presentation, structuring, filling the knowledge base and keeping it up to date.

The set of product rules is a priori knowledge of the process, which is obtained by experts based on their experience and intuition using logical connections AND, OR, NO and the implications of "IF… THAT…". A set of rules for designing a fuzzy system for assessing the level of digital maturity of the enterprise has been formed separately.

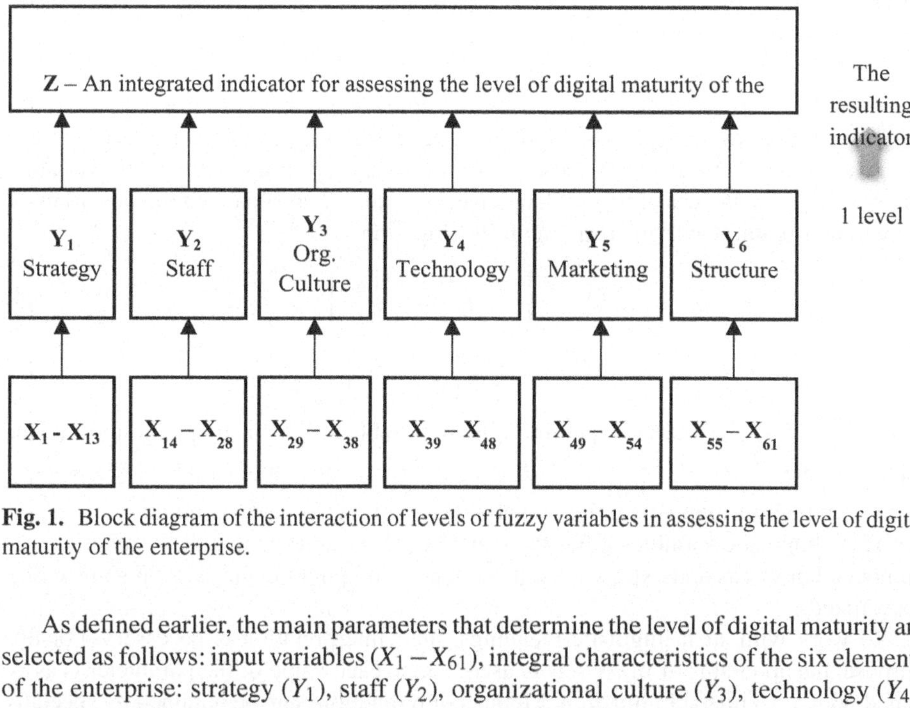


**Fig. 1.** Block diagram of the interaction of levels of fuzzy variables in assessing the level of digital maturity of the enterprise.

As defined earlier, the main parameters that determine the level of digital maturity are selected as follows: input variables ($X_1 - X_{61}$), integral characteristics of the six elements of the enterprise: strategy ($Y_1$), staff ($Y_2$), organizational culture ($Y_3$), technology ($Y_4$), marketing ($Y_5$), structure ($Y_6$). The output variable is an integral indicator of the digital maturity of the enterprise (Z).

There is a certain functional connection between these parameters:

$$Y_k = f\left(X_i, \ldots, X_j\right), i \in (1, 61), j \in (1, 61), k = \overline{1, 6},$$
$$Z = f(Y_1, Y_2, \ldots, Y_6),$$

where $k$ - the number of elements of the enterprise.

Areas of change of parameters that characterize the level of digital maturity:

$$X_i \in \left[x_{i,\min}; x_{i,\max}\right], \tag{2}$$

Similarly for the original variable $Z$ we will have:

$$Z \in \left[\left(z_i\right)_{(\min)}; \left(z_i\right)_{(\max)}\right], \tag{3}$$

where the indices *min*, *max* indicate the lower and upper values of the input and output variables.

Based on the analysis and research, the ranges of these parameters change are established (Table 1).

**Table 1.** Areas for changing parameters.

| Parameter | Change range |
|---|---|
| $X_i$ | [0;1] |
| $Y_k$ | [0;1] |
| $Z$ | [0;1] |

In the article we consider the input and output variables as linguistic variables that are given on universal sets (2) and (3). We assign the meaning of linguistic variables using words or sentences of natural language, i.e. terms. To evaluate linguistic variables, we will use qualitative terms from the following term sets:

$$L_i^{(f)} = \left\{ l_i^{(1)}, l_i^{(2)}, \ldots, l_i^{(q_f)} \right\}, \tag{4}$$

$$Z = \left\{ z^{(1)}, z^{(2)}, \ldots, z^{(s)} \right\}, \tag{5}$$

where $L_i^{(f)}$, $Z$ - term sets of input and output variables; $l_i^{(p)}$ - *p*- linguistic term of the input variable $x_i^{(f)}, p = \overline{1, q_f}; z_i^{(s)}$ - *s* - linguistic term of the output variable ї $z_i$, $s = \overline{1, b_i}$; $q_f, b_i$ - number of linguistic terms of variables $x_i^{(f)}$ і $z_i$; $f-$ variable number.

Thus, linguistic variables differ from numerical variables in that their values are not numbers, but words and expressions of ordinary colloquial language or, in some cases, specialized.

To work with such linguistic meanings, their interpretation is necessary. For this purpose, the apparatus of fuzzy sets is used, where each value of the parameter is associated with a membership function. Membership functions can be obtained by specially developed methods, namely the most common: the method of static information processing and the method of pairwise comparisons. In addition, the standard membership functions can be used [14].

According to the definitions, given by experts in fuzzy logic method, fuzzy sets, which have to be used in solving practical problems, are unimodal and normal. One of the unimodal methods of approximation of unimodal normal fuzzy sets is approximation by means of functions of (L-R) type [15]. However, the use of such functions is associated with additional investigations for selecting unknown parameters based on expert information.

Therefore, there is a target of constructing the membership function in terms of the minimum of the original data, which include:

- parameter name $x_i^{(f)}, f = \overline{1, d}, i = \overline{1, m}$,

- range $\left[\overline{x}_i^{(f)} \quad \underline{x}_i^{(f)}\right]$ of parameter change $x_i^{(f)}$,
- the number of terms $q_f$ used to evaluate the parameters $x_i^{(f)}$,
- the name of each linguistic term.

There are no specific recommendations in the literature on determining the number of terms. Their number is chosen from the interval $7 \pm 2$ [16]. A further increase in the number of terms leads to the complication of the model without increasing its accuracy. Considering the above, the task is to choose the number of terms of parameters that characterize the level of digital maturity of the enterprise.

The number of terms can be determined in two ways [16]:

- setting of some universal sets, and then the transition to real values based on the use of expert knowledge,
- determining the number of terms, based on the range of change of some value and the quantization interval $\Delta X_i$:

$$r = \frac{X_i}{\Delta X_i}, \quad \Delta X_i = X_{i+1}^* - X_i^*, \tag{6}$$

where $X_i^*$ is the level of quantization.

In the article the first method was used. The results of the calculations are given in Table 2.

**Table 2.** The number of terms required to fuzzify the parameters.

| Parameter | The required number of terms |
|---|---|
| $X_i$ | 5 |
| $Y_k$ | 4 |
| $Z$ | 4 |

The membership functions of inputs $X_i$, $Y_k$ and output $Z$ parameters are formed using several methods [16]: a method based on statistical processing of expert's statements; the method of pairwise comparisons performed by one expert; fuzzy clustering method. The following sets are given: set of terms $L = \{L, ML, M, MH, H\}$ for input parameters, $Y = \{O, B, E, I\}$ - for group parameters and - $Z = \{GI, A, TK, S\}$ for output parameter, where $L$ - "Very bad"; $ML$- "Bad";$M$ - "Satisfactory"; $MH$- "Good"; $H$- "Very good";$O$ - "Observers"; $B$ - "Beginners"; $E$- "Experienced"; $I$ - "Innovators"; $GI$- "Observers"; $A$- "Beginners"; $TK$ - "Experienced"; $S$- "Innovators".

In the article, for the input parameters and the output parameter, the function that based on the results of statistical processing of statements of many experts as well as the method of pairwise comparisons performed by one expert [13] were used. The following basic functions were used to approximate the obtained dependences:

– for terms $x_1$, $y_1$ and $z_1$ - Z-shaped function (L-Function):

$$f_L(h,a,b) = \begin{cases} 1, h \le a, \\ 1 - 2\cdot\left(\frac{h-a}{b-a}\right)^2, a < h \le \frac{a+b}{2}, \\ 2\cdot\left(\frac{b-h}{b-a}\right)^2, \frac{a+b}{2} < h \le b, \\ 0, h > b; \end{cases} \quad (7)$$

– for terms $x_5$, $y_4$ and $z_4$ - S-Function:

$$f_s(h,a,b) = \begin{cases} 0, h \le a, \\ 2\cdot\left(\frac{h-a}{b-a}\right)^2, a < h \le \frac{a+b}{2}, \\ 1 - 2\cdot\left(\frac{b-h}{b-a}\right)^2, \frac{a+b}{2} < h < b, \\ 1, h \ge b, \end{cases} \quad (8)$$

– for all other terms - trapezoidal function (Trapezoidal function):

$$f_t(h,a,b,c,d) = \begin{cases} 0, h \le a, \\ \frac{h-a}{b-a}, a < h \le b, \\ 1, b < h \le c, \\ \dfrac{d-h}{d-c}, c < h \le d, \\ 0, h > d, \end{cases} \quad (9)$$

where $a, b, c, d$ - some numerical parameters that can take arbitrary values, while $a < b \le c \le d$.

## 3 Questionnaire Results Description

Before proceeding to a direct analysis of the development and implementation of digital technologies in energy companies, it is advisable to assess their current state of digitalization (Table 3).

**Table 3.** Use of digital technologies in energy companies

| Indicators | Units of measnt | Years | | | Growth rate | |
|---|---|---|---|---|---|---|
| | | 2017 | 2018 | 2019 | Chain | Base |
| Number of energy companies using computers | Units | 647 | 706 | 714 | 101 | 110 |
| The share of energy companies that used computers | % | 95,5 | 93,9 | 93,0 | – | – |

*(continued)*

**Table 3.** (*continued*)

| Indicators | Units of measnt | Years | | | Growth rate | |
|---|---|---|---|---|---|---|
| | | 2017 | 2018 | 2019 | Chain | Base |
| Number of enterprises that had access to the Internet | Units | 644 | 701 | 709 | 101 | 110 |
| Number of enterprises that had specialists in the field of ICT | Units | 238 | 251 | 241 | 96 | 101 |
| Number of businesses that had a website | Units | 275 | 317 | 330 | 104 | 120 |
| The number of enterprises that used social. network | Units | 155 | 174 | 179 | 102 | 115 |
| Number of enterprises that bought cloud computing services | Units | 77 | 73 | 89 | 121 | 115 |
| Number of enterprises engaged in 3D printing | Units | – | 9 | 8 | 88 | – |
| The volume of sold products (goods, services) obtained from trade through websites or applications | Units | – | 441913 | 338606 | 76 | – |
| The share of sold products (goods, services) obtained from trade through websites or applications | % | – | 0,1 | 0,1 | – | – |

Source: formed on the base [17], data are given for enterprises supplying electricity, gas, steam and air conditioning

Table 3 demonstrates positive changes of current state on digitalization at energy enterprises during 2017–2019, namely: the number of energy companies that used computers have grown on 10.4%; the number of energy companies that had access to the Internet – on 10.1%; ones that had a website – on 20%; ones that used social networks is growing – on 15, 5%; ones that bought cloud computing services – on 15.6%. Ukrainian energy companies start using 3D printing technology, although their number is quite small. Thus, digital technologies in the energy field are developing in Ukraine.

Using the questionnaire method, information on digitalization was collected at energy enterprises of Ukraine. The results of survey data processing have allowed getting the following conclusions about the state of digitalization at selected enterprises (Fig. 2). In most of the surveyed enterprises (45%) the formation of a digital development strategy is at a satisfactory level and only in 9% it is very well developed.

Estimates of the measures to overcome resistance to changes in the digitalization process were similarly divided, in particular, there are enterprises with a very poor level of this indicator (9%). Cybersecurity is positioned at a good level in 45% and at a very good level in 27% of the surveyed enterprises. The level of use of SMM technologies in the work of enterprises is weak: 36% of respondents indicate a poor level of this indicator.

In the most enterprises there is no position or department for digitalization in the organisational structure (36%). The implementation of measures for the digital literacy development, digital skills and staff skills (45%) is sufficiently ensured. According to the survey, there is potential not only for the implementation of digital measures in energy companies, but also for improving the management system taking into account the impact of digitalization on its components.

The results of 96 respondents survey at the investigated enterprises have allowed to obtain the membership functions of all system parameters (Table 4) on the base of formulas (7–9).

On the basis of the averaged membership functions of the terms of input and output parameters, the correspondences between the membership functions $\mu_{li}$ and the control rules $P_{li}$ according to Zade were created.

The value of the digital maturity level was determined due to the results of the synthesized structure of the Mamdani type in the MATLAB system (Fig. 3). Using the developed fuzzy structure of Mamdani type, we consider an example of its use at calculation of digital maturity of the enterprise on the basis of the data collected by a questionnaire method.

## 4 Research Results

As a result of solving the problem of fuzzy inference about the state of digital maturity, the corresponding values are obtained. It is proposed to consider the importance of the three enterprises of regional electricity distribution. For the first enterprise, the value of the state of digital maturity corresponds to the number 0.254 (Table 5), which indicates that the value of the integrated indicator of enterprise's digital maturity level is equal to = 0.254, so the level of "Beginners".

The assessment of digital maturity level for this company allows us to conclude that the company has "Beginners" level of digital maturity, which indicates that the company really appreciates the importance of using digital technologies and their advantages over analogy technologies, but very slowly implements digital technology. The company devotes enough time and resources to develop a digital strategy and establish a structure of information communications in the company. However, digitalization in the field of marketing and operations is not sufficiently developed, strategic planning needs to be intensified in the conditions of digital economy development. In this regard, there is a need to develop specific measures to improve the management system in the context of digitalization of the enterprise.

At the same way the results of other enterprises questionnaires were processed and determined the integral indicative level of digital maturity for the other two enterprises.

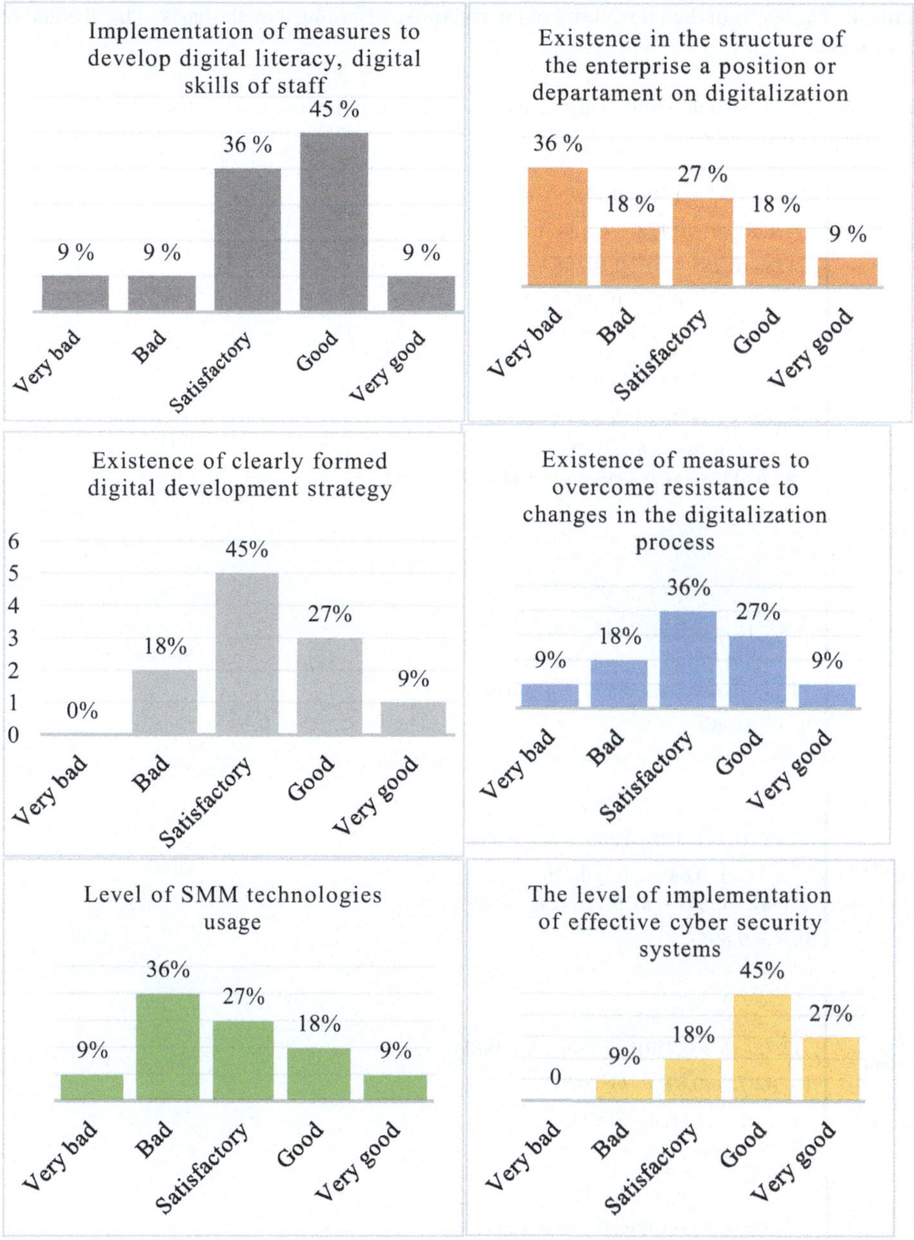


**Fig. 2.** The state of digitalization at energy enterprises of Ukraine. Source: formed on the basis of survey data.

**Table 4.** The levels of digital maturity of the enterprise determined on the basis of the method of fuzzy logic.

| Analytical expression of the levels of digital maturity of the enterprise based on the method of fuzzy logic | Interpretation of digital maturity levels |
|---|---|
| $x_1(H) = \begin{cases} 1, h \leq 0.128, \\ 1 - 128 \cdot (h - 0.128)^2, 0.128 < h \leq 0.192, \\ 128 \cdot (0.255 - h)^2, 0.192 < h \leq 0.255, \\ 0, h > 0.255. \end{cases}$ | Very bad |
| $x_2(H) = \begin{cases} 0, h \leq 0.145, \\ 10 \cdot (h - 0.145), 0.145 < h \leq 0.245, \\ 1, 0.245 < h \leq 0.325, \\ 9.8 \cdot (0.427 - h), 0.325 < h \leq 0.427, \\ 0, h > 0.427. \end{cases}$ | Bad |
| $x_3(H) = \begin{cases} 0, h \leq 0.318, \\ 8.93 \cdot (h - 0.318), 0.318 < h \leq 0.430, \\ 1, 0.430 < h \leq 0.590, \\ 11.11 \cdot (0.680 - h), 0.590 < h \leq 0.680, \\ 0, h > 0.680. \end{cases}$ | Satisfactory |
| $x_4(H) = \begin{cases} 0, h \leq 0.560, \\ 7.93 \cdot (h - 0.560), 0.560 < h \leq 0.686, \\ 1, 0.686 < h \leq 0.776, \\ 10.989 \cdot (0.867 - h), 0.776 < h \leq 0.867, \\ 0, h > 0.867. \end{cases}$ | Good |
| $x_5(H) = \begin{cases} 0, h \leq 0.740, \\ 42.72 \cdot (h - 0.740)^2, 0.740 < h \leq 0.817, \\ 1 - 42.72 \cdot (0.893 - h)^2, 0.817 < h \leq 893, \\ 1, h \geq 0.893. \end{cases}$ | Very good |
| $y_1(X) = \begin{cases} 1, x \leq 0.138, \\ 1 - 125 \cdot (x - 0.138)^2, 0.138 < x \leq 0.248, \\ 125 \cdot (0.359 - x)^2, 0.248 < x \leq 0.359, \\ 0, x > 0.359. \end{cases}$ | Observers |
| $y_2(X) = \begin{cases} 0, x \leq 0.198, \\ 5 \cdot (x - 0.198), 0.198 < x \leq 0.322, \\ 1, 0.322 < x \leq 0.398, \\ 5 \cdot (0.5453 - x), 0.398 < x \leq 0.5453, \\ 0, x > 0.5453. \end{cases}$ | Beginners |

*(continued)*

**Table 4.** (*continued*)

| Analytical expression of the levels of digital maturity of the enterprise based on the method of fuzzy logic | Interpretation of digital maturity levels |
|---|---|
| $y_3(X) = \begin{cases} 0, x \leq 0.396, \\ 5 \cdot (x - 0.396), 0.396 < x \leq 0.5807, \\ 1, 0.5807 < x \leq 0.662, \\ 5 \cdot (0.821 - x), 0.662 < x \leq 0.821, \\ 0, x > 0.821. \end{cases}$ | Experienced |
| $y_4(X) = \begin{cases} 0, x \leq 0.672, \\ 25 \cdot (x - 0.7722)^2, 0.672 < x \leq 0.7722, \\ 1 - 25 \cdot (0.8724 - h)^2, 0.7722 < x \leq 0.8724, \\ 1, x \geq 0.8724. \end{cases}$ | Innovators |
| $z_1(Y) = \begin{cases} 1, y \leq 0.138, \\ 1 - 25 \cdot (y - 0.138)^2, 0.138 < y \leq 0.223, \\ 25 \cdot (0.3053 - y)^2, 0.223 < y \leq 0.3053, \\ 0, y > 0.3053. \end{cases}$ | Observers |
| $z_2(Y) = \begin{cases} 0, y \leq 0.15, \\ 5 \cdot (y - 0.15), 0.15 < y \leq 0.322, \\ 1, 0.322 < y \leq 0.383, \\ 5 \cdot (0.5579 - y), 0.383 < y \leq 0.5579, \\ 0, y > 0.5579. \end{cases}$ | Beginners |
| $z_3(Y) = \begin{cases} 0, y \leq 0.391, \\ 5 \cdot (y - 0.391), 0.391 < y \leq 0.559, \\ 1, 0.559 < y \leq 0.64, \\ 5 \cdot (0.8053 - y), 0.64 < y \leq 0.8053, \\ 0, y > 0.8053. \end{cases}$ | Experienced |
| $z_4(Y) = \begin{cases} 0, y \leq 0.6465, \\ 25 \cdot (y - 0.6465)^2, 0.6465 < y \leq 0.7313, \\ 1 - 25 \cdot (0.816 - y)^2, 0.7313 < y \leq 0.816, \\ 1, y \geq 0.816. \end{cases}$ | Innovators |

Source: [19]

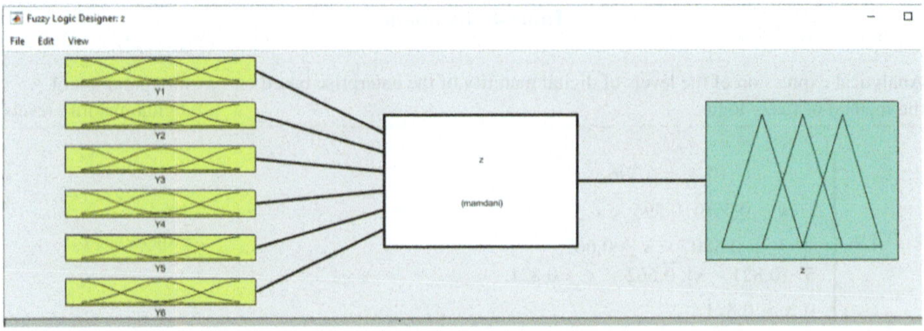


**Fig. 3.** Mamdani type structure [19].

**Table 5.** The results of digital maturity level assessment for the enterprise 1.

| Elements of the enterprise | Values for each subsystem | | General significance | |
|---|---|---|---|---|
| | Estimated value | Characteristics by a criterion scale | Estimated value | Characteristics by a criterion scale |
| Strategy | 0,273 | Beginners | 0,254 | Beginners |
| Personnel | 0,505 | Experienced | | |
| Organizational culture | 0,741 | Experienced | | |
| Technology | 0,453 | Beginners | | Accelerate current digital efforts |
| Marketing | 0,229 | Observers | | |
| Structure | 0,453 | Beginners | | |

The results of the calculation indicate that enterprise 2 has, as well, the "Beginners" level of digital maturity. The estimated value is slightly higher (0.354) in comparison with the enterprise 1, which indicates a superficial digitalization of the enterprise 2, that does not cover all activities. The company has a sufficient level of organizational culture, so it's expected that digital changes will be well received with minimal resistance (Table 6).

**Table 6.** The results of digital maturity level assessment for the enterprise 2.

| Elements of the enterprise | Values for each subsystem | | General significance | |
|---|---|---|---|---|
| | Estimated value | Characteristics by a criterion scale | Estimated value | Estimated value |
| Strategy | 0,298 | Beginners | 0,354 | "Beginners |
| Personnel | 0,397 | Beginners | | |
| Organizational culture | 0,549 | Experienced | | |
| Technology | 0,353 | Experienced | | Accelerate current digital efforts |
| Marketing | 0,489 | Beginners | | |
| Structure | 0,705 | Experienced | | |

The development of digital strategy element deserves more attention as well as its integration with the overall development strategy of enterprise. Another important step of digital maturity level increasing is to the definition and formation of digital competence and digital literacy of staff, as well as to assess the prospects for the use of digital marketing tools.

Compared with other surveyed enterprises, enterprise 3 has a higher "Experienced" level of digital maturity. The company uses the opportunities provided by digital technologies, has a digital strategy and invests in this area. Some digitization measures can serve as an example for other businesses. However, efforts to improve the technological and marketing components need to be intensified (Table 7).

**Table 7.** The results of digital maturity level assessment for the enterprise 3.

| Elements of the enterprise | Values for each subsystem | | General significance | |
|---|---|---|---|---|
| | Estimated value | Characteristics by a criterion scale | Estimated value | Estimated value |
| Strategy | 0,518 | Experienced | 0,598 | Experienced |
| Personnel | 0,397 | Beginners | | |
| Organizational culture | 0,619 | Experienced | | Take the experience of advanced companies |
| Technology | 0,253 | Beginners | | |
| Marketing | 0,282 | Beginners | | |
| Structure | 0,409 | Beginners | | |

Thus, the result of the digital maturity assessment is the identification of the current state of digitalization and the selection of those areas of activities that require the development of specific measures to improve level of digital maturity, taking into account the potential of the enterprise and trends in competitors, countries, economies and more. Scientists support the idea that the digital economy will allow the usage of new management methods, including blockchain, big data, expert systems and general decentralization in all areas and spheres of human life. These methods are promising and in the near future, taking into account foreign experience, will become widespread in all industrial arears. For energy companies, new management methods create both opportunities and can pose significant threats. They also do not always fit into the framework of existing management technologies, in particular strategic and project management, which, according to scientists, are not always suitable for the digital economy [18].

Studies have shown that many companies don't transform their digital development strategy due to the imperfection or lack of a digital maturity. Therefore, one of the promising areas of continuing the study of digital maturity is to find ways to increase its level, in particular by developing measures to improve the competencies of management and employees in the field of digital experience, as well as strengthening psychological readiness for change.

## 5 Discussions

We would like to admit that there were conducted other researches by Ukrainian and foreign scientists in the field of digital maturity investigation and digital transformation. So, there isn't universe approach how to make progress in digital world. The analysis of all enterprises in comparison with the manufacturing enterprises on digital maturity in Central and Eastern European (CEE) countries worth to be mentioned. The digital maturity assessment at the enterprises of CEE country indicated that the dimensions of Internet of things, big data analytics and artificial intelligence are the most important criteria for assessing the digital maturity of enterprises [20]. Noteworthy is the example of assessing the enterprise's digital maturity, which identifies 4 levels and highlights such important areas as technology, organization, and social dimensions and considers corporate Strategy [21]. Another research points out such levels of digital maturity: preparation, computerization, connectivity, transparency, predictability, adaptability which depend on the specific criteria applied to both the developing and implementing systems of the company [22]. The existing research on digitalization theme propose the elements of digital transformation – aspects that could be used for further investigation of digital maturity of company. One of such examples identify technology and actor as the two dimensions of digital transformation that elaborates on the predominant contextual concepts within these dimensions [23, 24]. McKinsey Global Survey has confirmed the prior importance of technology for the future companies' capabilities in the post pandemic era [25]. Deloitte's digital capabilities are assessed in 5 key areas: consumers, strategy, technology, manufacturing, organizational structure and culture. The five main dimensions are divided into 28 sub-dimensions, which in turn are broken down into 179 indicators that measure digital maturity [26]. The German National Academy of Sciences and Technology has developed the Industry Maturity Index 4.0 Acatech, which covers four key areas of digital transformation: resources, information systems, culture and organizational structure [27]. The Model of Centre for Digital Business and Capgemini Consulting conducts analysis of more than 400 large companies from different industries identified three key areas of digital transformation, bases on such arears: transforming customer experience, transforming operational processes and transforming business models [28]. As the result of such a great experience of digital maturity investigation we can discuss about peculiarity of enterprise management in the context of digitalization that is projecting on digital alternatives existing in management systems, methods, techniques, etc. and requires creating a suitable environment for digital changes. Many of opportunities provided by digitalization (development of digital technologies, automation, new channels of promotion, etc.) require proper management and coordination of actions for digitalization of the enterprise.

## 6 Conclusions

The model of enterprise's digital maturity assessment was proposed using the fuzzy logic methodological approach. As the result of model implementation, the levels of digital maturity of energy enterprises were calculated by determining the integrated level of digital maturity and maturity level of its individual components for considered enterprises. There was substantiated that the digital maturity of energy companies is reduced by the following factors: the lack of a software document for planning activities for the business digitalization; lack of digital literacy training programs and activities; low speed of introduction of digital technologies in the production process; lack of SMM-strategy, etc. The obtained results of the energy enterprises digital maturity assessment make it possible to determine the reasons for their insufficient level, which could serve as a basis for the development of measures of digital improvement, taking into consideration the mentioned in the article descriptors of enterprise management system: strategy, personnel, organizational culture, technologies, structure, marketing.

Studies of energy companies' digitalization and determining the level of their digital maturity allow concluding that the introduction of digital technologies in the energy companies activities is characterized by different levels and directions of digital transformation. This, in turn, requires the development of a theoretical and methodological basis for the implementation of recommendations that will allow these companies to identify and implement the necessary changes, improving their management system taking into account the requirements of the digital economy. This is demanded by the spread of digitalization trends in Ukraine and in the global energy environment (development of Smart Grid, "smart metering" of electricity, etc.), as well as internal problems of industrial enterprises (high wear of electrical networks and equipment, insufficient cybersecurity, lack of or low efficiency of digital development strategy, etc.).

The further research requires the construction of the road map visualizing the plan-scenario of the enterprise's digital transformation, taking into account alternative ways and possibilities of digital changes. Thus, the development of a road map for digital transformation is an important tool of strategic planning makes it possible to visualize the sequence and intensity of the necessary measures for the digital transformation of the enterprise.

## References

1. Schwab, K.: The fourth industrial revolution: what it means, how to respond. https://www.weforum.org/agenda/2016/01/. Accessed 11 Sep 2021
2. Looy, A.V., De Backer, M., Poels, G.: Which maturity is being measured? A classification of business process maturity models. In: Proceedings of the 5th SIKS/BENAIS Conference on Enterprise Information Systems (EIS-2010), vol. 662, pp. 7–16. Eindhoven, The Netherlands (2010)
3. Digital Maturity Model Achieving digital maturity to drive growth. https://www2.deloitte.com/content/dam/Deloitte/global/Documents/Technology-Media-Telecommunications/deloitte-digital-maturity-model.pdf. Accessed 11 Sep 2021
4. Polyanska, A.S., Savchuk, S.V., Revtiuk, Y.: Assessment of digital maturity of enterprise as a precondition of digital transformation. Sustainable economic development and advancing education excellence in the era of global pandemic. In: Proceedings of the 36th International Business Information Management Association Conference, 4–5 November 2020, pp. 6032–6044. Granada, Spain (2020)
5. Maturity Model and Best Practice Skill Development for Digital Transformation. SAP SE. https://news.sap.com/2017/09/maturity-model-from-sap-and-ercis-helps-companies-develop-skills-for-digital-transformation. Accessed 12 Sep 2021
6. Savchuk, S.: On the issue of assessing the digital maturity of the enterprise in the context of digital transformation. Sci. Bull. IFNTUOG. Ser. Econ. Manage. Oil Gas Ind. **1**(21), 78–85 (2020)
7. What is digital maturity? https://mercury-ecommerce.com/resources/what-is-digital-maturity. Accessed 12 Sep 2021
8. Digital transformation: a road map for billion-dollar organisations. https://www.capgemini.com/wpcontent/uploads/2017/07/DigitalTransformation_ARoad-MapforBillion-DollarOrganizations.pdf. Accessed 11 Sep 2021
9. Matviychuk, A.: Modeling and analysis of economic systems based on the theory of fuzzy logic. Author's ref. for science degree of doctor of econ. sciences with special. 08.00.11 - mathematical methods, models and information technologies in economics. K. 36 p (2007)
10. Perevozova, I., Dalyak, N.: Application of fuzzy set theory to assess the reliability of the business client of the enterprise. Bull. Eastern Eur. Univ. Econ. Manage. Ser. Econ. Manage. **2**(25), 78–90 (2018)
11. Shpak, N., Odrekhivskyi, M., Doroshkevych, K., Sroka, W.: Simulation of innovative systems under industry 4.0 conditions. Soc. Sci. **8**, 202 (2019)
12. Polyanska, A., Zapukhliak, I., Diuk, O.: Culture of organization in conditions of changes as an ability of efficient transformations: the case of gas transportation companies in Ukraine. Oeconomia Copernicana **10**(3), 561–580 (2019)
13. Zapukhliaka, I., Zaiachuk, Y., Polyanska, A., Kinash, I.: Applying fuzzy logic to assessment of enterprise readiness for changes. Manage. Sci. Lett. **9**, 2277–2290 (2019)
14. Rotshtein, O.: Soft Computing in Biology: Factorial Analysis and Diagnostics: Monograph. UNIVERSUM-Vinnytsia.144 p (2008)
15. Sementsov, G.: The method of choosing the number of terms for a fuzzy description of the basic variables in the F-transformation of parameters and indicators of the drilling process. Bull. Khmelnytsky Natl. Univ. **1**(1), 30–35 (2005)
16. Kondratenko, N., Kuzemko, S.: Fuzzy logic systems using fuzzy sets of higher types. Syst. Res. Inf. Technol. **4**, 106–113 (2006)
17. The use of information and communication technologies in enterprises. State Statistics Service of Ukraine. http://www.ukrstat.gov.ua/operativ/operativ2018/zv/ikt/arh_ikt_u.html. Accessed 16 Sep 2021

18. Digitalisation and Energy. IEA. https://www.iea.org/reports/digitalisation-and-energy. Accessed 15 Sep 2021
19. Savchuk, S.: Managerial improvement on utility companies in the context of the digital economy. Doctoral thesis paper in the specialty 073 – management. The preparation of the thesis paper was carried out at Ivano-Frankivsk National Technical University of Oil and Gas. Defense of a thesis paper will take place at Ivano-Frankivsk National Technical University of Oil and Gas (2021)
20. Brodny, J., Tutak, M.: Assessing the level of digital maturity of enterprises in the central and eastern European countries using the MCDM and Shannon's entropy methods. PLoS ONE **16**(7), 1–38 (2021). https://doi.org/10.1371/journal.pone.0253965
21. Siedler, C., et al.: Maturity model for determining digitalization levels within different product lifecycle phases. Prod. Eng. Res. Devel. **15**(3–4), 431–450 (2021). https://doi.org/10.1007/s11740-021-01044-4
22. Ilin, I., Levaniuk, D., Dubgorn, A.: Assessment of digital maturity of enterprises. In: Murgul, V., Pukhkal, V. (eds.) EMMFT 2019. AISC, vol. 1259, pp. 167–177. Springer, Cham (2021). https://doi.org/10.1007/978-3-030-57453-6_14
23. Nadkarni, S., Prügl, R.: Digital transformation: a review, synthesis and opportunities for future research. Manage. Rev. Quart. **71**(2), 233–341 (2020). https://doi.org/10.1007/s11301-020-00185-7
24. Bodrov, V., Lazebnyk, L., Hurochkina, V., Lisova, R.: Conceptual scheme of digital transformation of business model of industrial enterprises. Int. J. Recent Technol. Eng. **8**(3C), 107–113 (2019)
25. The new digital edge: Rethinking strategy for the postpandemic era.. McKinsey Global Survey. https://www.mckinsey.com/business-functions/mckinsey-digital/our-insights/the-new-digitaledge-rethinking-strategy-for-the-postpandemic-era. Accessed 20 Sep 2021
26. Digital Maturity Model. Achieving Digital Maturity to Drive Grow. https://www2.deloitte.com/content/dam/Deloitte/global/Documents/Technology-MediaTelecommunications/deloitte-digital-maturitymodel.pdf. Accessed 15 Sep 2021
27. Shue, G., Anderl, R., Gauzemeier, J., Ten Hompel, M., Walster, W., et al.: Industry maturity index 4.0 – managing the digital transformation of companies (Acatech RESEARCH), Munich: Herbert Utz Verlag (2017)
28. Westerman, G., Bonnet, D., McAfee, A.: The nine elements of digital transformation. MIT Sloan Manage. Rev. **55**(3), 1–6 (2014)

# The Experimental SMART Manufacturing System in SmartTechLab

Jakub Demčák[1(✉)], Natalia Lishchenko[1], Ivan Pavlenko[2], Ján Piteľ[1], and Kamil Židek[1]

[1] Faculty of Manufacturing Technologies with a Seat in Prešov, Department of Industrial Engineering and Informatics, Technical University of Košice, Bayerova 1, 08001 Prešov, Slovakia

jakub.demcak@tuke.sk

[2] Faculty of Technical Systems and Energy Efficient Technologies, Department of General Mechanics and Machine Dynamics, Sumy State University, 2 Rymskogo-Korsakova Street, Sumy 40007, Ukraine

**Abstract.** The laboratory SmartTechLab as a key of experimental SMART manufacturing system based on Industry 4.0 concept is described in the paper. This laboratory focuses on four research fields: assembly, identification, digitalization and online monitoring. The laboratory is designed also to improve theoretical and practical knowledge of students and young researchers. The development of the assembly is due to use of modern industrial and collaborative robots on the assembly line, designed to move objects and assemble products. The field of identification is supported by RFID technology and two machine vision measuring stations. KEYENCE devices - industrial cameras and laser profilometer are used as machine vision technology. Digital twins, OPC server, HP server, IoT Data and Cloud platforms are the basic elements of digitalization research field. The quality control system in 3D printing for assessment of the product accuracy and surface quality is based on camera and laser displacement sensor.

**Keywords:** SMART Manufacturing · Industry 4.0 · Automation · Digitalization · Quality control

## 1 Introduction and Related Works

The assembly workplace named SmartTechLab (Fig. 1) is a laboratory used for teaching and research in the field of digitalization, automation and implementation of robots into manufacturing regarding the Industry 4.0 concept [1–4]. SmartTechLab is located at the Faculty of Manufacturing Technologies of Technical University of Košice with a seat in Prešov, Slovakia. The laboratory was established to improve the research activity of university staff and practical knowledge of its students in the teaching-learning process. The faculty students gain knowledge in the automatization field, which increases their practical skills.

Today, several types of research and projects that enrich the activities of the faculty take place in the laboratory. These activities provide for the development of the faculty


J. Trojanowska et al. (Eds.): MANUFACTURING 2022, LNME, pp. 228–238, 2022.
https://doi.org/10.1007/978-3-030-99310-8_18

and the improvement of the education process mainly at the Department of Industrial Engineering and Informatics.

**Fig. 1.** Photo of SmartTechLab.

The following parts of the paper describe the SmartTechLab, its constituent parts, its principles of operation, as well as what devices and technologies support the Industry 4.0 concept.

Similar issues (SMART Lab) are dealt with by laboratory in Kielce University of Technology – Poland, or for example by Laboratory for Industry 4.0 - Smart Mini Factory from Free University of Bozen-Bolzano – Italy, or Graz University of Technology - Austria. The authors of the works [1, 3, 4] explore the concept of Industry 4.0 and SMART manufacturing, which presents a considerable challenge for the production and service sectors. These papers define and discuss the smart manufacturing system and states it current implementation status and analyzes the gap between current manufacturing system and the predicted future smart manufacturing system. There are presented the challenges of implementation, opportunities and the future directions for smart manufacturing system. Key technologies and their possible applications to Industry 4.0 smart manufacturing systems are reviewed. In papers [2, 5–8] authors deal with assembly. The main task was the identification and control of parts with the synchronization of all data into the digital twin model. This area is open to new research, methodology development and definition of basic requirements, because real applications in production processes are currently still limited. Based on the obtained knowledge from the research on diagnostics of errors at the component surface by vision recognition systems using machine learning algorithms they have started to use a convolutional neural network (CNN) for the recognition of standardized industrial parts. In papers [9, 10] a visual detection platform is built, and several products produced are matched with cloud storage data through experiments and there is proposed a new approach to creating these data fully automated based on a virtual 3D model of the standardized parts. In papers [11, 12] methodology monitors the FDM process and correlates the theoretical 3D model with the manufactured one is proposed. The reconstructed and theoretical point cloud of the test specimen are building and their correlation are carried out. Numerical estimate is absence.

## 2 Assembly

The assembly line (Fig. 2) consists of the conveyor belts, which are in a closed-loop. The transfer of the objects from one belt to another is ensured by automatic pneumatic pistons, which are extended after touching the object based on information from the inductive sensor using a pneumatic control system.

Three robots (Fig. 3) are used in the laboratory – the ABB Yumi collaborative robot (Fig. 3 - left) for assisted assembly supported by mixed reality [5], the Scara Mitsubishi (Fig. 3 - middle), which transfers objects between the rotary table and the conveyor belt. The third robot in the input storage is DOBOT Magician (Fig. 3 - right). Dobot operates the input and output space with components and moves them to the conveyor belt. These robots work quickly, accurately and efficiently and can replace in this way the assembly workers in the production process.

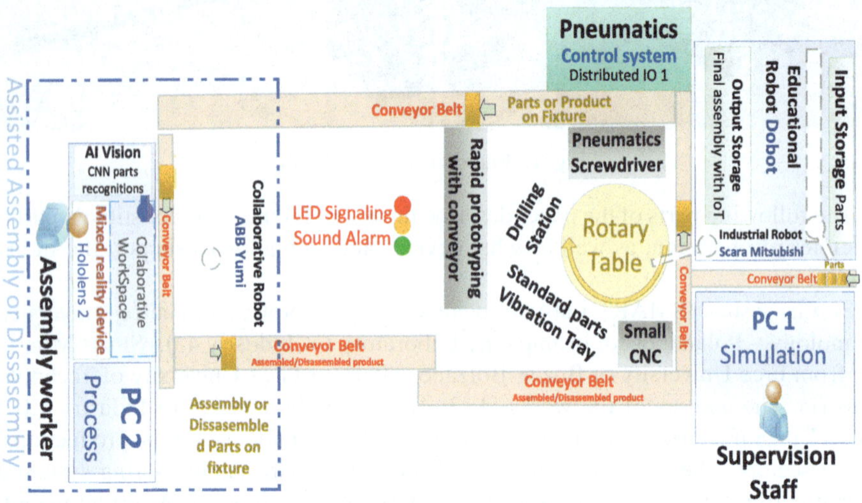


**Fig. 2.** Scheme of the assembly part.

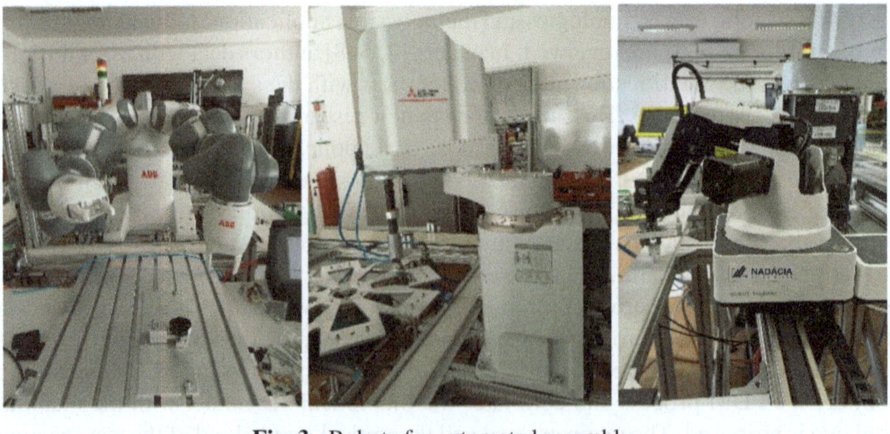

**Fig. 3.** Robots for automated assembly.

PLC-based control system S7-1500 (Fig. 4) is used to control and synchronize data from all integrated industrial technologies (vision systems, industrial and collaborative robots, pneumatics valves, RFID, converters for motors, rotary table, CNC, Rapid prototyping). Rapid prototyping is a group of techniques used to quickly fabricate a scale model of a physical part or assembly using three-dimensional computer aided design data. Construction of the part or assembly is usually done using 3D printing or "additive layer manufacturing" technology. The laboratory can be also controlled via Amazon voice assistant Alexa.

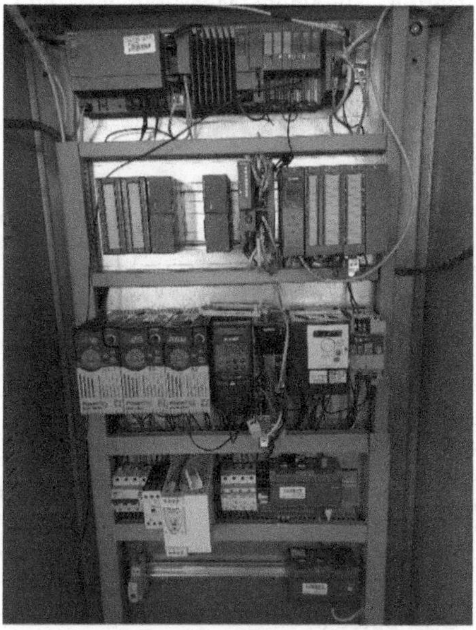

**Fig. 4.** Control system.

## 3 Identification

In the field of identification (Fig. 5), SmartTechLab includes RFID technology (Radio-frequency identification uses electromagnetic fields to automatically identify and track tags attached to objects) and machine vision systems. Siemens devices are used for radio-frequency identification. This technology recognizes in which part of the line the components are located for effective tracking of their movement. The principle of this technology is that the antenna transmits and receives a signal that captures an RFID tag oriented on the identification object. The signal is transmitted employing radio frequency waves and is processed using a reading device.

KEYENCE devices - industrial cameras (CA-H048MX and CA-H048CX) and laser profilometer (LJ-X8020) are used as machine vision technology. This system aims to replace the human in the process of inspecting objects [9, 10]. Product control is thus faster, more accurate and more reliable. The principle of the machine vision technology is the automatic scanning of products using lasers, cameras and additional lighting. The whole system is controlled by a control unit (Fig. 6 – right) connected to a computer, where various identification tools can be selected in the software.

The software for Keyence CV-X Series camera and LJ-X series profilometer offers several categories of tools for inspection: 3D Presence/Comparison, Height, Presence/Absence, Flaw Detection, Alignment, Measurements and Dimensions, Count, ID, Graphic Display and Mathematical Operations.

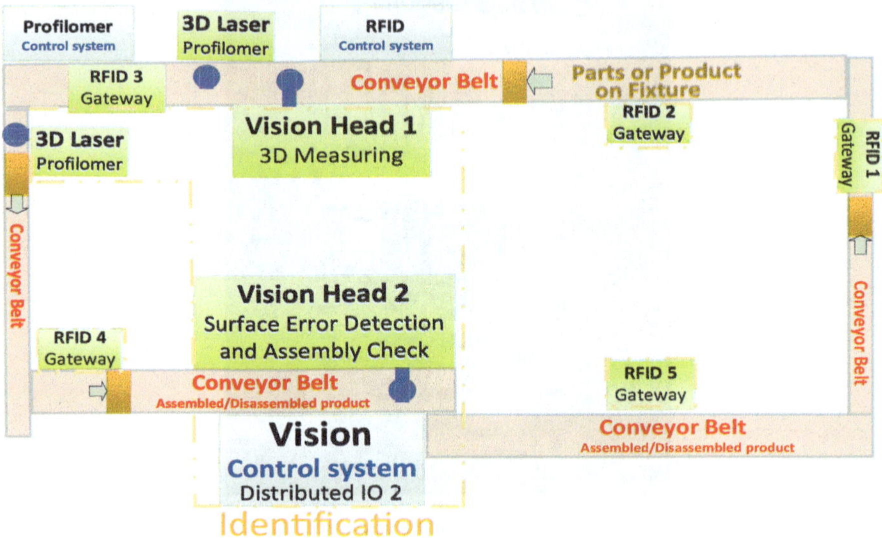


**Fig. 5.** Scheme of the identification part.

There are two machine vision measuring stations in the laboratory. One of them (Fig. 6 – left) is used for 3D scanning (the process of analyzing a real-world object or environment to collect data on its shape. The collected data can then be used to construct digital 3D models) of objects using white structured lighting (CA-DQP12X). This light illuminates the object from eight directions, allowing the camera system to evaluate the object's height and other parameters for 3D detection based on light reflection (see Fig. 6 - middle).

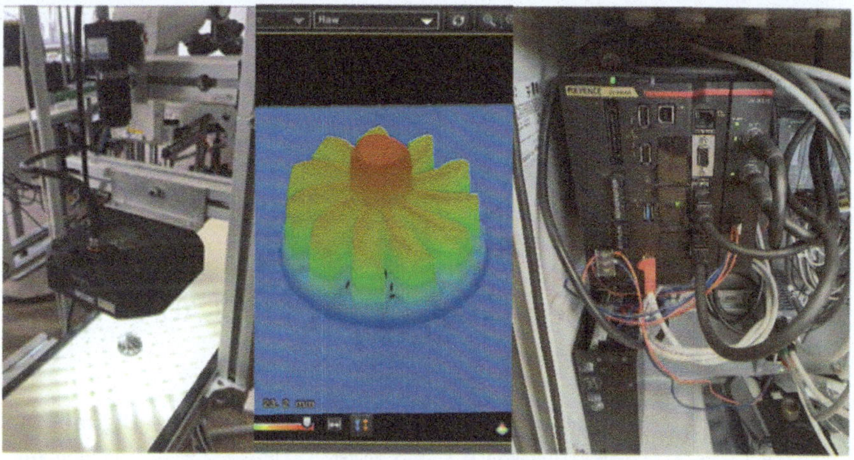

Fig. 6. Machine vision - 3D identification.

The second station uses multispectral lighting (CA-DRM10X), which offers eight different light wavelengths. It can detect different surface flaws with different colors of lighting: UV, blue, orange, red, infrared, far-red, green and white (Fig. 7).

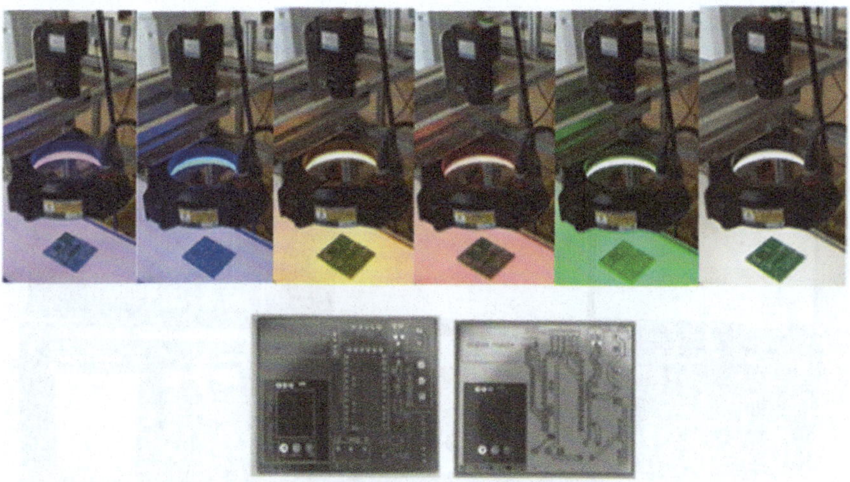

Fig. 7. Machine vision - multispectrum identification.

Figure 7 (bottom) shows the identification of the same object, at the same time using different type of light (white and infrared).

## 4 Knowledge Extraction and Digitalization

The knowledge extraction and digitalization scheme can be seen in the Fig. 8. Data from the automatic assembly line are displayed on 6 LCD monitors (Fig. 9), which are used to visualize all the necessary information. Workers thus have control over the entire assembly process in real-time with all relevant data.

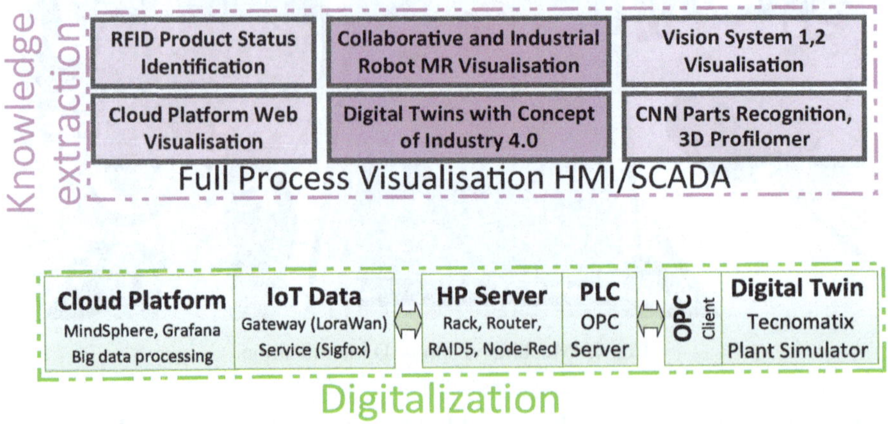


**Fig. 8.** Scheme of knowledge extraction and digitalization part.

Two servers are used for data collection, the first collects data from the PLC to the MindSphere platform, and the second provides data collection from sensors, identification technologies and IoT devices to the open-source Grafana platform with the Influx DB database.

**Fig. 9.** Data visualization.

## 5 Quality Control

The concept of the quality control system for 3D printing was designed as it is presented in Fig. 10. This system is divided into consistent in time three parts: preparation stages, manufacturing stage, and post-manufacturing stage. Control phase at each stage is carried out by the user (preparation stages), on-line monitoring system, including cameras, laser displacement sensor (manufacturing stage) and camera, profilometer, 3D scanner and Scanning Electron Microscope (post-manufacturing stage).

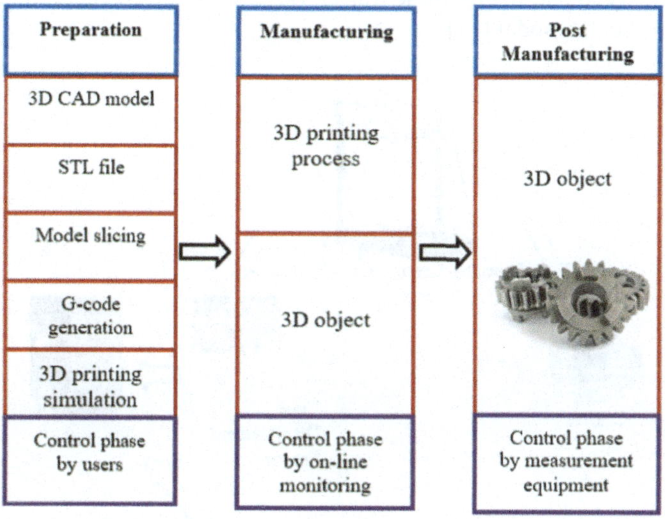


**Fig. 10.** Quality control system.

The result of preparation stages is the virtual 3D object with theoretical point cloud, the result of manufacturing stage is the real 3D object and data base (real point cloud) characterizing the quality of the object and the change in the size of the object as the layers are formed. The result of post-manufacturing stage is data base characterizing the final quality of the object (geometry, surface finish, surface topography, materials properties). It is known that the object quality is determined by its geometry, surface finish, surface topography, materials properties [13–15]. The final values of each of these indicators are formed in the process of printing each layer of the object. Therefore, to ensure the quality of the 3D object, it is necessary to develop an online monitoring system at manufacturing stage, at the same time, pay special attention to the diagnostics of the printing of the 1st layer, since it predetermines the operational characteristics of the printed object.

The system architecture consists of two parts including the hardware and software part. As an example of the quality control in 3D printing the system has been proposed according to the Fig. 11. The hardware part is made up of 3D printer (Creality CR-10 MAX, Shenzhen Creality 3D Technology Co., Ltd), Laser Displacement Sensor (ILD 1420–10, Micro-Epsilon), camera, and PC (Fig. 11).

The 3D printer Creality CR-10 MAX realizes the FDM molding technology for PLA filament (diameter 1.75 mm). FDM is Fused deposition modeling, also known as fused filament fabrication (FFF). In FDM, an object is built by selectively depositing melted material in a predetermined path, layer by layer. After creating a 3D model in modeling programs such as SolidWorks or Autodesk Inventor, it is imported to the SW Cura, which transforms 3D models into 3D printing commands. The outcome of this stage is the Gcode that contains X, Y and Z coordinates to which the print head and build platform have to move for building the 3D model. Then the 3D printing process is simulated using the Gcode. In the next stage a structure containing every path of the 3D printer is created and finally these trajectories are sampled generating a theoretical point cloud of the 3D model [11].

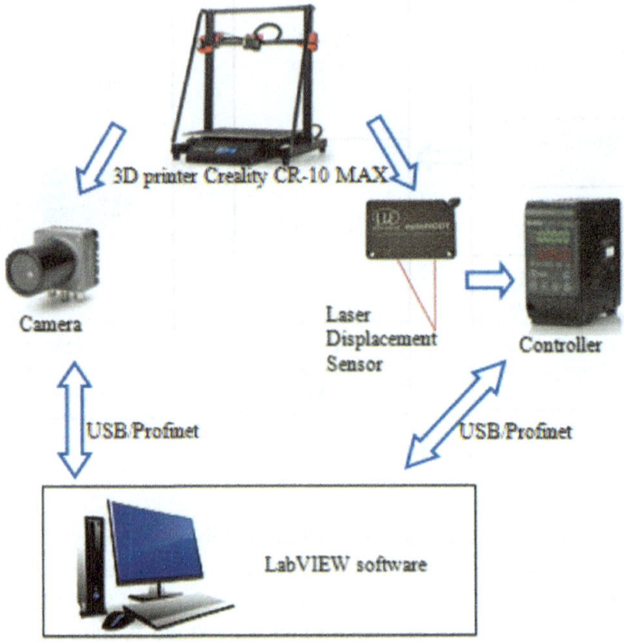


**Fig. 11.** Scheme of 3D printing monitoring system.

The fundamental measurement principle of Laser Displacement Sensor is based on triangulation. For example, for ILD1420–10 the measuring rate is 4 kHz, repeatability 0.5 $\mu$m, measurement range 10 mm, linearity $<\pm 8$ $\mu$m, light spot diameter $45 \times 40$ $\mu$m. The Laser Displacement Sensor is designed to measure displacement, distance and thickness of surfaces. In 3D printing process with this sensor it is possible to determine the change along the axis of the coordinate characterizing the position of the filament relative to the plane of the table. This sensor is connected to PC with LabVIEW via controller compatible with USB/Profinet.

In 3D printing it is possible appearance of distortion, forced by changing the temperature or speed of filament feeding; cracks as result of lack of adhesion between neighboring layers; holes; poor bonding quality between the filament and build platform or inter-layers; shrinkage; warpage and product position error. For real-time monitoring it is possible to choose the following types of cameras: parallel Digital Camera or

Camera Link Cameras. National Instruments (NI) hardware and software supports cameras that complies with the USB 3 Vision bus. The NI Vision Development Module is supported by LabVIEW and is intended for image processing and analysis. In Vision Development Module there are hundreds of image processing algorithms and machine vision functions of enhance images, locate features, identify objects, and measure parts. In 3D printing based on FDM the filament is deposited layer by layer and the stacked layers are observed from the top and the side, so it is most advisable to choose side location of cameras.

## 6 Conclusion

This paper introduced the SMART workplace called SmartTechLab. This laboratory is built regarding the Industry 4.0 concept. The workplace was disassembled into key parts - Assembly, Identification, Digitalization and Quality Control in 3D printing. The modern technologies, sensors and devises at the SmartTechLab were summarized. All systems work together to create an experimental automated workplace for state-of-the-art research and the teaching-learning process.

The developed concept of a quality control system for 3D printing is divided into consistent in time three parts: preparation stages, manufacturing stage, and post-manufacturing stage. At the manufacturing stage, it is necessary pay special attention to the diagnostics of the printing of the 1st layer, since it predetermines the operational characteristics of the printed object. For this purpose, a promising direction is the use of appropriate software and hardware containing the camera, laser displacement sensor and LabVIEW.

The future direction of activities in SmartTechLab is focused on the completion of the digital twin model of the workplace, the transfer of data from the actual assembly to this model and design and realization of the real-time monitoring system of 3D printing.

**Acknowledgements.** This work was supported by the Slovak Research and Development Agency under the contract No. APVV-19-0590 and also by the projects VEGA 1/0700/20, 055TUKE-4/2020 granted by the Ministry of Education, Science, Research and Sport of the Slovak Republic.

## References

1. Matt, D.T., Modrák, V., Zsifkovits, H. (eds.): Industry 4.0 for SMEs. Springer, Cham (2020). https://doi.org/10.1007/978-3-030-25425-4
2. Židek, K., Piteľ, J., Adámek, M., Lazorík, P., Hošovský, A.: Digital twin of experimental smart manufacturing assembly system for industry 4.0 concept. Sustainability **12**(9), 3658 (2020). https://doi.org/10.3390/su12093658
3. Phuyal, S., Bista, D., Bista, R.: Challenges, opportunities and future directions of smart manufacturing: a state of art review. Sustain. Futures (2020). https://doi.org/10.1016/j.sftr.2020.100023
4. Zheng, P., et al.: Smart manufacturing systems for Industry 4.0: conceptual framework, scenarios, and future perspectives. Front. Mech. Eng. **13**(2), 137–150 (2018). https://doi.org/10.1007/s11465-018-0499-5

5. Židek, K., et al.: CNN training using 3D virtual models for assisted assembly with mixed reality and collaborative robots. Appl. Sci **11**, 4269 (2021). https://doi.org/10.3390/app11094269
6. Varela, L., Putnik, G.D., Manupati, V., Rajyalakshmi, G., Trojanowska, J., Machado, J.: Collaborative manufacturing based on cloud, and on other i4.0 oriented principles and technologies: a systematic literature review and reflections. Manage. Prod. Eng. Rev. **9**(3), 90–99 (2018). https://doi.org/10.24425/119538
7. Trojanowski, P.: Comparative analysis of the impact of road infrastructure development on road safety–a case study. Sci. J. Marit. Univ. Szczecin 23–28 (2020). https://doi.org/10.17402/436
8. Lazar, I., Husar, J.: Validation of the serviceability of the manufacturing system using simulation. J. Effi. Responsib. Educ. Sci. **5**(4), 252–261 (2012). https://doi.org/10.7160/eriesj.2012.050407
9. Zheng, S.: Design of intelligent manufacturing product identification and detection system based on machine vision. In: Xu, Z., Parizi, R.M., Hammoudeh, M., Loyola-González, O. (eds.) CSIA 2020. AISC, vol. 1146, pp. 258–265. Springer, Cham (2020). https://doi.org/10.1007/978-3-030-43306-2_37
10. Židek, K., Lazorík, P., Piteľ, J., Pavlenko, I., Hošovský, A.: Automated training of convolutional networks by virtual 3d models for parts recognition in assembly process. In: Trojanowska, J., Ciszak, O., Machado, J.M., Pavlenko, I. (eds.) MANUFACTURING 2019. LNME, pp. 287–297. Springer, Cham (2019). https://doi.org/10.1007/978-3-030-18715-6_24
11. Kopsacheilis, C., Charalampous, P., Kostavelis, J., Tzovaras, D.: In situ visual quality control in 3D printing. In: Proceedings of the 15$^{th}$ International Joint Conference on Computer Vision, Imaging and Computer Graphics Theory and Applications – IVAPP, pp. 317–324 (2020). https://doi.org/10.5220/0009329803170324
12. Starzynska, B., Szajkowska, K., Diering, M., Rocha, A., Reis, L.P.: A study of raters agreement in quality inspection with the participation of hearing disabled employees. In: Hamrol, A., Ciszak, O., Legutko, S., Jurczyk, M. (eds.) Advances in Manufacturing. LNME, pp. 881–888. Springer, Cham (2018). https://doi.org/10.1007/978-3-319-68619-6_85
13. Kaščak, J., et al.: Design of an atypical construction of equipment for additive manufacturing with a conceptual solution of a printhead intended for the use of recycled plastic materials. Appl. Sci. **11**, 2928 (2021). https://doi.org/10.3390/app11072928
14. Ivchenko, O., et al.: Method for an effective selection of tools and cutting conditions during precise turning of non-alloy quality steel C45. Materials **15**, 505 (2022). https://doi.org/10.3390/ma15020505
15. Wojciechowski, S., Twardowski, P.: Tool life and process dynamics in high speed ball end milling of hardened steel. Procedia CIRP **1**, 289–294 (2012)

# AR/VR/MR Applications in Manufacturing

# Conceptual Use of Augmented Reality in the Maintenance of Manufacturing Facilities

Jakub Kaščak[1], Jozef Husár[2(✉)], Lucia Knapčíková[2], Justyna Trojanowska[3], and Vitalii Ivanov[4]

[1] Faculty of Manufacturing Technologies with a Seat in Prešov, Department of Computer Aidded Manufacturing Technology, The Technical University of Košice, Šturova 31, 08001 Prešov, Slovak Republic

[2] Faculty of Manufacturing Technologies with a Seat in Prešov, Department of Industrial Engineering and Informatics, The Technical University of Košice, Bayerova 1, 08001 Prešov, Slovak Republic

jozef.husar@tuke.sk

[3] Faculty of Mechanical Engineering, Department of Production Engineering, Poznan University of Technology, Piotrowo Street 3, 61-138 Poznan, Poland

[4] Faculty of Technical Systems and Energy Efficient Technologies, Sumy State University, Rymskogo-Korsakova Street 2, Sumy 40007, Ukraine

**Abstract.** The presented article deals with the possibility of using augmented reality (AR) technology in the field of mechanical engineering. In several chapters, it describes the possibilities of using a specialized software package. It allows the use of CAD models assemblies processed into sequences, which serve as a template for creating AR experiences. The first chapter informs about the possible benefits of the presented solution, focuses on the selection of equipment and software that can be used during problem-solving. It also describes the actual work in PTC's specialized software package, the addition, and the importance of different parts of the sequence to streamline selected maintenance steps of the Emco Mill 55. The second chapter focuses on the part of the software package specializing in AR. It describes the creation of both parts of the basis for the AR experience, i.e., its 2D and 3D parts. By 2D part, we mean the user interface displayed by the output device in the user's field of view. The 3D part displays an interactive AR sequence connecting the 3D models displayed in the AR, with individual sequence steps and basic CAD functions. The results described at the end of the publication represent a comprehensive way of applying AR technology in this area. At the same time, this solution is one of the ways to reduce the qualification requirements for new employees in the future and the time required for their successful training and education.

**Keywords:** Augmented reality · Manufacturing · Machining · Maintenance · Internet of things · IoT

## 1 Introduction

AR technology is becoming more and more popular nowadays, also due to the interest in various industries. This interest, resp. the demand for these technologies results in the


J. Trojanowska et al. (Eds.): MANUFACTURING 2022, LNME, pp. 241–252, 2022.
https://doi.org/10.1007/978-3-030-99310-8_19

development of new specialized software and hardware solutions, the use and availability of which is diametrically different compared to the past. It mainly benefits areas where the integration of these technologies is problematic, it is out of target interest, or financially demanding [1].

The potential of this technology, whether in the field of IoT and e.g. visual display of the parameters of production equipment with the interpretation of data into the field of view of the employee through the AR. Other potential areas of use include increasing the effectiveness of education and staff training, or direct integration into the educational process in schools with a technical focus during the teaching of subjects requiring the ability to read technical documentation. [2] Its use is also possible in other areas, for example in the publication published by R. Costa et al. "Intelligent mixed reality for the creation of ambient assisted living" presenting the possibility of using the means of mixed reality in care in the concept of solving an electronically oriented health care monitoring system. As can be seen, mixed reality technologies, in particular AR technology, have a wide range of uses [3].

This publication will introduce its possible use in the field of manufacturing, more precisely it will offer the concept of a possible solution to a situation in which assistance is required, for a new or insufficiently qualified worker in the process of simulated service intervention [4].

### 1.1 Technical Illustration Software as a Part of AR Sequence

Regarding the methodology of the simulation solution designed to present the possibilities of implementing the concept of using AR in maintenance operations, there are several model situations to choose from. There is an opportunity to use AR in positioning jigs and fixtures in the machining process, as described in the publication dealing with the determination of the position of the workpiece and fixture with the help of AR. [5] Another possibility is the use of interactive models in the creation of technical documentation [6], or user training in increasing the effectiveness of manufacturing workshops [7].

In this case, the model situation involves the need to replace the milling mandrel and remove the electric motor case cover to perform an unspecified inspection. The sequence thus contains the already mentioned spindle drive model, shown in Fig. 1b. It is then imported into the technical documentation software, in which the individual parts of the sequence are arranged chronologically and the necessary effects and warnings are added. [8] The publication then describes the creation of the UI of the output device and the unification of all elements of the AR sequence into one unit. The expected result is therefore to be the model of the spindle drive displayed in the field of view of the user, who gradually plays back the individual steps of the sequence and thus performs necessary steps for a service intervention. At the same time, the UI should provide sufficient space for the interactivity of the CAD model provided by predefined functions, which will be available within the UI and will allow more effective control over the sequence [9].

As already mentioned, the equipment selected for the application of this technology is from the Emco mill 55 milling machine, shown in Fig. 1a. Specifically, one part of it, the spindle drive. As part of creating the sequence, it was necessary to make a dimensionally and geometrically accurate copy of the sequenced device. This CAD model is visible in Fig. 1b. and it is a complex set at a scale of 1: 1, containing all the components of real equipment.

a) b)

| **Milling spindle drive** | |
|---|---|
| Motor power 3 phase asynchronous motor | 0,75 kW |
| Speed range | 150 – 3500 rpm |
| Max. torque | 3.7 Nm (2.72 ft/lbs) |
| Speed with optional engraving spindle | 14000 min$^{-1}$ |
| **Feed drive** | |
| 3 phase step motor in X / Y / Z axis with resolution | 0,5 μm / 0,5 μm / 0,5 μm |
| Rapid motion speed and feed rate X / Y / Z | 0 – 2 m / min (0 – 78.74 ipm) |
| Average positioning variation in | 8 μm / 8 μm / 8 μm |

**Fig. 1.** Emco mill 55 and its spindle drive, selected for the application of AR technology.

Individual steps will show the sequence preparation, i.e. creation of an accompanying ARE for the worker. This sequence will then be used to guide the user by the replacement of the milling mandrel and intervention in the electrical box of the device motor [10].

## 1.2 Sequence Creation

The presented ARE should also include a sequence that shows the individual steps of service actions in the field of view of the user. At the same time, it is necessary to achieve the interactivity of the used CAD models with the user. If only animation was used as a form of the instructional video, this video would simply play in the user's field of view and its interaction would be limited to the basic controls used in the video files. In this case, however, we want to achieve the playback sequence of the layout of specific parts of the device, the possibility of its rotation, resizing, zooming, and the implementation of visual notifications and warnings. These should alert the user to important sections of the sequence or imminent danger [11].

To create such a sequence, we use the program PTC Creo illustrate. This software is originally intended for the processing of technical documentation, the creation of assembly and disassembly plans, manuals, or presentation animations consisting of CAD assemblies. In our case, this software is an essential part of the whole process. Thanks to its compressed PVZ output format. By simple extraction of the PVI format, we obtain vector data related to the movement and position of individual CAD models. We can then assign them to individual models within the AR sequence. This process is necessary to optimize the sequences for the output hardware. Which is characterized by diverse and lower performance, which would be difficult to work with complex animations [12].

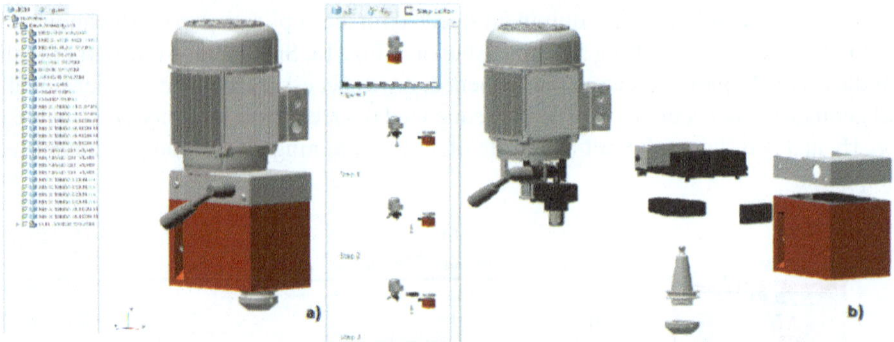

**Fig. 2.** Example of using software for creating technical illustrations.

Figure 2a shows the CAD model of the spindle drive imported into the program environment, along with all assembly elements. As can be seen, this is an identical environment to various commercially used CAD software. One of the differences is the ability to create sequences that are combined with common CAD software features. Figure 2b, thus, shows an example of a possible arrangement of sequential images when creating a so-called spindle drive disassembly plan. The layout of the images, the number of operations on each image, and the various other types of modifications available to this software can also be included in the content displayed in the AR [13].

### 1.3 Creation of Animation Sequence for AR Experience

As already mentioned, the creation of individual sections of sequences and their complexity depends on the displayed action. Figure 3 shows the introduction to the creation sequence, which instructs the employee in two steps to replace the milling mandrel. As we can see in Fig. 3a, in the first phase, the handle is highlighted and so alerts the user to notify the model. We achieve this by using the "Flash" function and creating the color that should be used to highlight the model. In the second step, visible in Fig. 3b, we add a rotation of the handle in the desired direction. In these two steps, we create the basis of the sequence to release the milling mandrel.

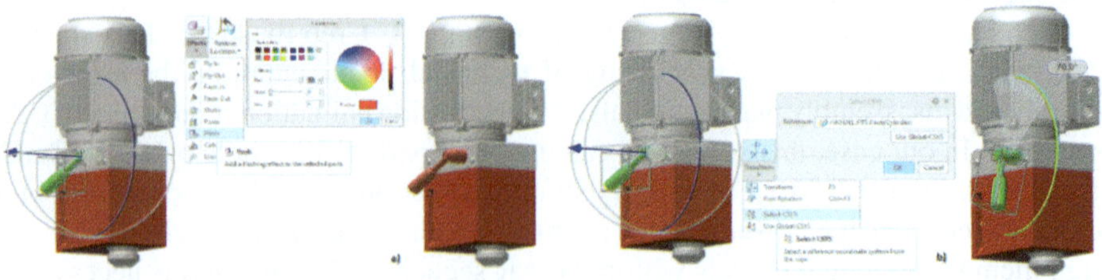

**Fig. 3.** The gradual addition of functions to the sequence for AR experiences.

The next phase of the sequence alerts the user to the release of the milling mandrel. The same highlight function is used as in the first step. The image of the sequence is then extended with a translation and a "fade out" function. By adding these changes, we achieve the display of the user's warning of the release of the milling mandrel, the display of direction and distance to which this component must be extended, and its removal from the user's field of view. The end of the sequence consists of the rotation of the spindle drive assembly and the approach of the distribution part of the electric motor (Fig. 4).

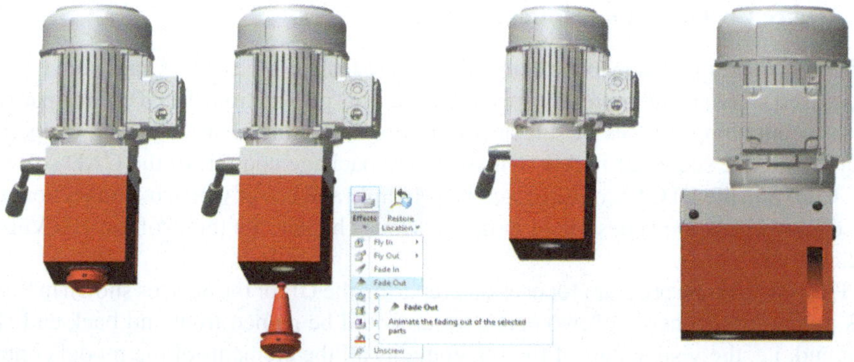


**Fig. 4.** Adding additional parts of the sequence.

The final part of the sequence is the removal of the electric motor case cover. Four functions were used in connection with this action. The first was to highlight the screws that needed to be removed. This was done in the same way as in the previous two steps. Another step was to add an animation of unscrewing the bolts in a preselected direction and removing them along with the case cover. The final part of the sequence consists of displaying the contents of the case and adding alerts, as can be seen in Fig. 5.

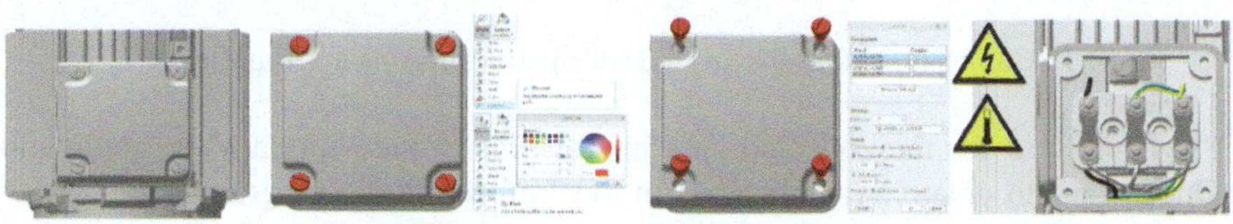

**Fig. 5.** Creation of warning sequence.

After defining all the images in the sequence and adding functions to all the components, the file is easily saved. Its output format is PVZ, it is a compressed format containing vector data and individual CAD models. Due to the hardware performance of the devices for which the sequence is prepared, especially for Android and iOS devices, the file in this format is unsatisfactory. For this reason, only the PVI file containing the vector data is extracted from it. It should be noted that the individual parts of the sequence thus created are repeatable and the position of each component can be reset [14].

## 2 Definition of User Interface Settings for a Specific Type of Device

After successful timing and preparation of the sequence, there are two further steps left. Those are necessary and described in the following chapters and for the implementation of ARE. This chapter introduces two characteristic sequence creation steps dealing with the selection of a suitable device type and UI definition [15].

### 2.1 Device Customization Options

As already mentioned, due to the demand for these technologies, more and more types of specialized software are currently appearing. For this publication, the PTC Creo package is used, containing 3 types of self-operating programs, the result of which cooperates in an interactive sequence in AR. This software package consists of the CAD software PTC Creo 8.0, the PTC Creo Illustrate, the principle and use of which in connection with the solved simulation is described in the previous chapter, the third software is Vuforia studio.

This stand-alone program focuses on preparing the UI for the devices shown in Fig. 6. This preparation consists of two parts, which could be named front and back end. The front end, i.e. the visible part of the UI, contains all the elements of the model controls ensuring the interactivity of the user with the model in the AR. This part is optimized for 3 types of devices [16].

**Fig. 6.** Options for selecting a device before creating a new project.

To be able to see the collected data in real-time in the Vuforia View application, it is necessary to create an application in the Vuforia Studio program. First, we will open Vuforia Studio and choose to create a new project. We will see 3 options for a new project, as can be seen in Fig. 6. We can choose between options for 2D glasses (Fig. 7a), 3D glasses (Fig. 7b), and smart devices (Fig. 7c). For this example, a Mobile - Default option is chosen.

**Fig. 7.** Common devices used in combination with the software used.

The Android device was chosen mainly because of its availability and the possibility of integration into production and the conditions in which this technology has to be used. Headset devices, whether in 2D or 3D, project a non-stop image from the sequence into the user's field of view. Given that it is not yet known its negative impact, respectively, there is no defined percentage of the field of view coverage that would be safe in this case, we choose a device that allows a temporary view, according to the needs of the worker. This device is Lenovo M10 plus, with a resolution of 1900 ×1200 px. This resolution is used as the default setting for the UI.

### 2.2 Vuforia Studio - UI Creation

After naming the project the program's user interface is displayed. The first part of the program, its 2D part, which will be visible on our chosen Android device, is visible in Fig. 8. In this phase of sequence preparation, we adapt the UI by changing its dimensions, adding buttons and functions that can replace standard CAD functions. In Fig. 8a, we can see the design of the interface with buttons for starting the sequence, resetting the position of CAD models, removing warnings, and rotating the whole assembly. Figure 8b, in turn, shows a preview of the interface thus created [17].

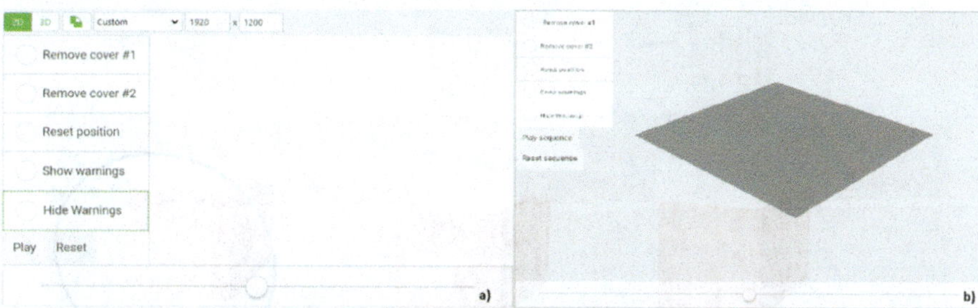


**Fig. 8.** The UI creation process for the android AR application.

At the end of this phase, the final part follows. Thus, the use of the program Vuforia studio on the back-end setting of the sequence, its control, and interaction.

## 3 Definition of AR Maintenance Sequence

The final phase of sequence preparation consists of the back-end settings performed in the Vuforia studio. It is a summary and definition of all the previous steps in a complex application. This chapter covers the individual steps that are necessary to complete this sequence and uses elements from the preparation described in the previous chapters [18].

The phase of these preparations is quite extensive, starting with importing a CAD assembly in STEP format, placing and assigning a market, assigning CAD functions to individual buttons visible on the UI, and importing a PVI file to obtain vector data for timing and model positions within the sequence. In addition, each of these options is editable using javascript expression. Although it is specialized software whose features are user-friendly and intuitive, the process is quite extensive. The following chapter, therefore, describes the necessary ones to finalize the sequence [19].

### 3.1 Assigning a Thingmark to a Specific Model

ThingMark's placement in Vuforia Studio corresponds directly to where AR Experience places the model in Vuforia View. This corresponds to its rotation and placement. We can either rotate the Thingmark in the program, or it is necessary to rotate the printed ThingMark to the desired position so that our application is displayed correctly in the Vuforia View program [20].

Figure 9, shows the possibilities of positioning the marker. As can be seen, there are several variants available. The first that can be seen in Fig. 9a, is the location of the position mark in space. In practice, this means that the position mark would be placed somewhere on the frame of the milling machine and the displayed model would be projected at a specific distance from it. This method is suitable for the use of visualization of interactive CAD models over a 2D technical drawing. But due to the complicated shape of the device in which the sequence is to be run, it is not suitable for this solution. This is because the position marker must always be in the camera's range of the output device, which in this case causes complications in which the position marker is out of focus [21].

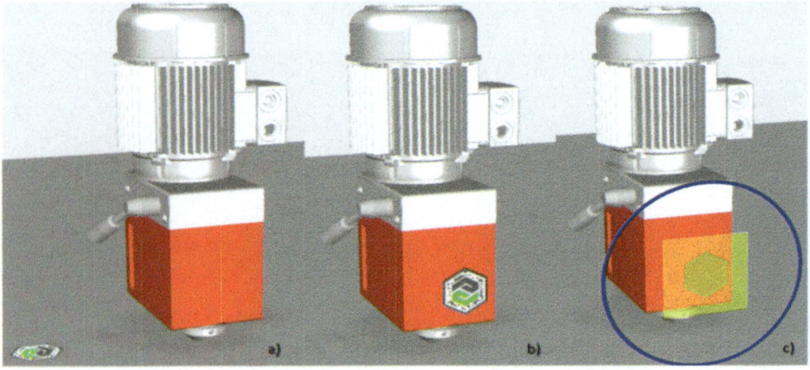


**Fig. 9.** ThingMark placement methods for the selected model.

The locations of the position mark shown in Fig. 9b and Fig. 9c are essentially the same. In the case of the first variant, it is a placement of the position mark directly on the spindle drive. Such a location will provide us with a position mark in a sufficiently visible place and overlap of the CAD model with a real object. Since the CAD model was created on a 1: 1 scale, the sequence placed in this way is ideal. Figure 9c, is essentially the same marker location, except that the position marker is not visible during sequence display, it is more or less an aesthetic matter, without affecting the desired result.

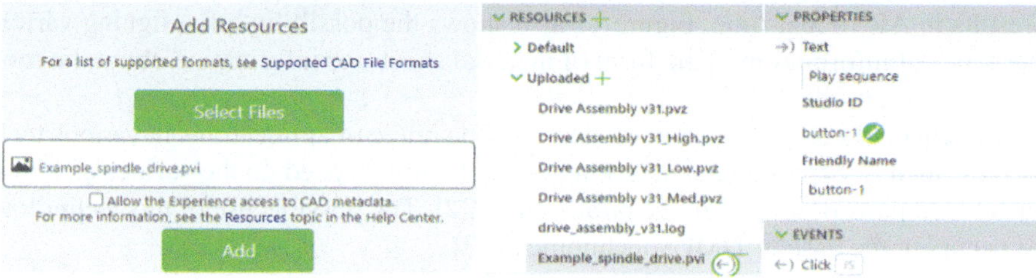


**Fig. 10.** Import and assign of PVI file to the "Play sequence" button.

Figure 10 shows the upload and assignment of the already mentioned PVI file to the "Play sequence" button, which is displayed on the UI. Because it is specialized software, the implementation of these functions is relatively simple. In this way, functions are assigned to all buttons and objects displayed in the AR.

## 4 Results

The results show the location of the spindle drive of the Emco mill 55 in real state and using the sequence for the smart device. As can be seen in Fig. 11, the location of the position mark is on the spindle housing and the CAD model is displayed directly on it. This placement will allow us to correlate the display in the AR and the real world, thus achieving a lower percentage of the employee's field of view. The location is also ideal concerning the course of the sequence, as the individual components are located in the same place.

**Fig. 11.** AR experience launch test on Emco mill 55.

Figure 11 also shows the advanced features of the Vuforia studio software. By adding additional extensions, it is possible to interpret the operating parameters of the device, or the outputs from the diagnostic devices, directly into the field of view of the employee. With this tool, it is, therefore, possible to achieve a form of operational control, where the individual quantities are displayed directly in the field of view together with warnings. This procedure, therefore, creates an ideal environment for the use of other elements of IoT in terms of operational control and predictive diagnostics with the interpretation of results in AR in real-time. Figure 11 also shows the possibility of assigning various other types of information in the form of material sheets, or information about the tools or preparations used.

The sequences created in this way have a wide range of applications in various technologies, such as experimental determinations of cutting speed on the surface character of machined parts described in the publication [22]. The possible use of this technology also occurs in the case of DMLS technology [23].

## 5 Conclusion

The use and relentless demand for AR technology in various industries have created ideal conditions for its use in areas for which it was not entirely ideal in the past. The use of CAD models with advanced functions is very helpful in various branches of the educational process, or training of new employees, even in their "static" form. By integrating various other functions into specialized software, it is now possible to create fully interactive sequences in AR. The use of this software and the definition of individual steps is intuitive due to their specialization and their use is therefore satisfactory even in areas where AR technology is not mainstream.

The publication presents the possibilities of using one of the parts of the Industry 4.0 concept. More specifically, it focuses on creating a virtual training tool for the specific steps that an employee may encounter while working with the manufacturing equipment. It chronologically describes the individual steps of model creation, their integration into individual software up to the final back-end setup. Then present part of the results in the form of verification of the display of the virtual model in the user's field of view. The following steps are currently the subject of research, which promises, the possibility of using and optimizing these steps in various areas.

**Acknowledgments.** This work was supported by the Slovak Research and Development Agency under the contract No. APVV-19-0590, by the projects KEGA 055TUKE-4/2020 granted by the Ministry of Education, Science, Research and Sport of the Slovak Republic.

## References

1. Peraković, D., Periša, M., Cvitić, I., Zorić, P.: Internet of things concept for informing visually impaired persons in smart factory environments. In: Knapčíková, L., Balog, M. (eds.) Industry 4.0: Trends in Management of Intelligent Manufacturing Systems. EICC, pp. 69–86. Springer, Cham (2019). https://doi.org/10.1007/978-3-030-14011-3_7
2. Husár, J., Knapciková, L.: Exploitation of augumented reality in the industry 4.0 concept for the student educational process. In: INTED 2019: the 13th annual International Technology, Education and Development Conference, IATED, pp. 4797–4805 (2019). https://doi.org/10.21125/inted.2019.1195
3. Grabara, J., et al.: Impact of legal standards on logistics management in the context of sustainable development. Acta Logistica **7**(1), 31–37 (2020). https://doi.org/10.22306/al.v7i1.155

4. Kliment, M., et al.: Production efficiency evaluation and products' quality improvement using simulation. Int. J. Simul. Model. **19**(3), 470–481 (2020). https://doi.org/10.2507/IJSIMM19-3-528
5. Ivanov, V., Pavlenko, I., Liaposhchenko, O., Gusak, O., Pavlenko, V.: Determination of contact points between workpiece and fixture elements as a tool for augmented reality in fixture design. Wireless Netw. **27**(3), 1657–1664 (2019). https://doi.org/10.1007/s11276-019-02026-2
6. Costa, R., Neves, J., Novais, P., Machado, J., Lima, L., Alberto, C.: Intelligent mixed reality for the creation of ambient assisted living. In: Neves, J., Santos, M.F., Machado, J.M. (eds.) EPIA 2007. LNCS (LNAI), vol. 4874, pp. 323–331. Springer, Heidelberg (2007). https://doi.org/10.1007/978-3-540-77002-2_27
7. Židek, K., Hosovsky, A., Piteľ, J., Bednár, S.: Recognition of assembly parts by convolutional neural networks. In: Hloch, S., Klichová, D., Krolczyk, G.M., Chattopadhyaya, S., Ruppenthalová, L. (eds.) Advances in Manufacturing Engineering and Materials. LNME, pp. 281–289. Springer, Cham (2019). https://doi.org/10.1007/978-3-319-99353-9_30
8. Straka, M., Rosova, A., et al.: Principles of computer simulation design for the needs of improvement of the raw materials combined transport system. Acta Montanistica Slovaca. **23**, 163–174 (2018)
9. Malindzakova, M., Straka, M., Rosova, A., Kanuchova, M., Trebuna, P.: Modeling the process for incineration of municipal waste, Przemysl Chemiczny **94**(8) (2015). https://doi.org/10.15199/62.2015.8.4
10. Sinay, J., Kotianova, Z., Balazikova, M., Markulik, S., Kotianova, Z.: Measurement of low-frequency noise during CNC machining and its assessment. Measurement. **119** (2018). https://doi.org/10.1016/j.measurement.2018.02.004
11. Luscinski, S., Ivanov, V.: A simulation study of industry 4.0 factor on the ontology on flexibility with using Flexim (R) software. Manag. Prod. Eng. Rev. **11**(3), 74–83 (2020). https://doi.org/10.24425/mper.2020.134934
12. Peraković, D., Periša, M., Sente, R.E.: Information and communication technologies within industry 4.0 concept. In: Ivanov, V., et al. (eds.) DSMIE 2018. LNME, pp. 127–134. Springer, Cham (2019). https://doi.org/10.1007/978-3-319-93587-4_14
13. Židek, K., et al.: Digital twin of experimental smart manufacturing assembly system for industry 4.0 concept. Sustainability **13**(9), 3658 (2020). https://doi.org/10.3390/su12093658
14. Tokorova, M., et al.: Augmented reality as a support tool in machining process. Tem J. **9**(1), 407–411 (2020). https://doi.org/10.18421/TEM91-56
15. Kascak, J., et al.: Implementation of augmented reality into the training and educational process in order to support spatial perception in technical documentation. In: IEEE 6th International Conference on Industrial Engineering and Applications (ICIEA), pp.583–587. IEEE, Tokio (2019). https://doi.org/10.1109/IEA.2019.8715120
16. Buń, P., Trojanowska, J., Ivanov, V., Pavlenko, I.: The use of virtual reality training application to increase the effectiveness of workshops in the field of lean manufacturing. In: 4th International Conference of the Virtual and Augmented Reality in Education, VARE 2018, pp. 65–71 (2018)
17. Luscinski, S.: Digital twinning for smart industry. In: MMS 2018: 3rd EAI International Conference on Management of Manufacturing Systems, vol. 2016, p. 224. (2018). https://doi.org/10.4108/eai.6-11-2018.2279986
18. Husár, J., Knapčíková, L., Hrehová, S.: Augmented reality as a tool of increasing the efficiency of RFID technology. In: Perakovic, D., Knapcikova, L. (eds.) FABULOUS 2021. LNICSSITE, vol. 382, pp. 401–414. Springer, Cham (2021). https://doi.org/10.1007/978-3-030-78459-1_30
19. Knapčíková, L., Husár, J., Behúnová, A., Hrehová, S.: Augmented reality as a tool of increasing of competitiveness of enterprise. In: Perakovic, D., Knapcikova, L. (eds.) FABULOUS

2021. LNICSSITE, vol. 382, pp. 337–349. Springer, Cham (2021). https://doi.org/10.1007/978-3-030-78459-1_25
20. Maščenik, J.: Experimental determination of cutting speed influence on cutting surface character in material laser cutting. MM Sci. J. 960–963 (2016). https://doi.org/10.17973/MMSJ.2016_09_201639
21. Coranic, T., Gaspar, S., Pasko, J.: Utilization of optimization of internal topology in manufacturing of injection Moulds by the DMLS technology. Appl. Sci. **11**(1), 262 (2021). https://doi.org/10.3390/app11010262
22. Kujawińska, A., Diering, M.: The impact of the organization of the visual inspection process on its effectiveness. Int. J. Adv. Manuf. Technol. **112**(5–6), 1295–1306 (2021). https://doi.org/10.1007/s00170-020-06543-9
23. Trojanowska, J., Kolinski, A., Galusik, D., Varela, M.L.R., Machado, J.: A Methodology of improvement of manufacturing productivity through increasing operational efficiency of the production process. In: Hamrol, A., Ciszak, O., Legutko, S., Jurczyk, M. (eds.) Advances in Manufacturing. LNME, pp. 23–32. Springer, Cham (2018). https://doi.org/10.1007/978-3-319-68619-6_3

# Augmented Reality as a Didactic Resource: A Proposal for Computer Equipment Maintenance Training

Fernando Banda(✉), Martín Luna Lázaro, Leticia Neira-Tovar, Ana Karen Luna Torres, and Said Gonzalez

Universidad Autónoma de Nuevo León, San Nicolas de Los Garza, NL, México
fernando.bandamn@uanl.edu.mx

**Abstract.** The didactic resources are mediators for the development and growth of the teaching-learning process, they support the interaction between the teacher, the student and the object of knowledge. Information technologies have come to transform education, through rapid access to information, and multiple sources of knowledge, facilitating communication, creativity, and innovation. One of the technologies that currently has the greatest momentum is extended reality, (XR), the augmented reality is one of its elements, based on the observation of the real world augmented with additional information generated by a computer, and with interaction that makes up the synthetic with images obtained from a real world. The objective of this work is to develop a didactic resource based on augmented reality techniques that makes it easier for technical high school students to learn the maintenance and configuration of computer equipment. This work was evaluated by means of an instrument that contemplates 3 dimensions: usability, design and satisfaction applying to students from the northeast of Mexico area, obtaining the following results, usability was evaluated by the students with 95.9%, design with 97.8% and satisfaction with 98.6%. It is concluded that augmented reality is an emerging technology that can be integrated into the planning as a didactic resource according to the context of the user and in case of social isolation supports us to give continuity to the educational process, its development should be with a multi-disciplinary approach which would allow obtaining a design where elements that recreate the teaching process - learning are integrated.

**Keywords:** Augmented reality · Didactic resource · Mobile learning

## 1 Introduction

AR is a technology that presents the user with a combined space of real and virtual objects using markers that the user can visualize in three-dimensional form through a mobile device [1–3]. The operation and configuration of computer equipment is the knowledge base for students of the technical career of Web Programming, these placements are carried out in laboratories following the teacher's instructions and with the support of printed materials to achieve the competences of the learning unit. Currently, they are


J. Trojanowska et al. (Eds.): MANUFACTURING 2022, LNME, pp. 253–266, 2022.
https://doi.org/10.1007/978-3-030-99310-8_20

made in a virtual way with the help of videos that the teacher uses as a reference to develop the practice.

Virtual education is transformed by promoting meaningful learning through the different strategies and didactic techniques where the teacher must promote digital, interactive and visual learning [4].

Among the six technologies and practices that will be carried out in 2021 according to the EDUCAUSE Horizon report, are the Open Educational Resources (OER) which contemplate materials for courses, learning objects, software, among others. They contribute, according to UNESCO, to quality education through knowledge sharing and capacity building [5].

The proposal is to make a prototype of an application with augmented reality (AR) in a technological guide of maintenance and configuration of computer equipment, the manual will have markers so that the student can visualize the main components of a computer equipment and the practice of assembly in Unity [6–8].

The purpose is to enrich knowledge and training with immersive experiences that produce greater interest in students, stimulating interaction and favoring the opportunity to learn by doing; through practical simulation, which reduces the risks of future mistakes in job performance [9–11].

This paper consists of six sections. Section 2 shows the objectives of this paper, followed by Sect. 3 that presents the methodology implemented for the development of the application. In Sect. 4, the results of the tests carried out on the use of the application are shown, followed by Sect. 5, where the discussion of the results is made comparing them to those of other researchers. The paper was completed in Sect. 6 followed by references.

### 1.1 Literature Review

The AR encompasses several technologies that allow user interaction with the world, integrating the real environment with the virtual through the different devices that contain a camera as a cell phone, laptop, generating a simulated or artificial learning environment as they allow viewing of images as if they were real also known as immersive learning [12–14].

Augmented Reality stimulates the human senses and favors virtual environments through user interaction with information. One of the advantages is not only to have information, but to create a certain representation of the world that is attractive to the user [15, 16]. Currently the AR is considered as an educational resource that has a greater impact on learning teaching processes. One tool that currently supports the educational process is mobile telephony [17–19].

## 2 Objective

Based on a preview work where a method to design tools for teaching with Virtual Reality (VR) are expose [20], in the present research the main objective was determined, which consists of developing an application of augmented reality as a didactic resource that facilitates students of technical baccalaureate to learn the maintenance and configuration of computer equipment.

The specific objectives that support the achievement of the primary objective are listed below:

- Design the test to evaluate the application of augmented reality.
- Evaluate the functionality of the application of augmented reality in students of technical baccalaureate.

## 3 Methodology

The teaching resources currently used in the training of students at upper secondary level are in printed form as manuals, learning guides and books containing information and images that support the thematic contents of the computer equipment maintenance and configuration learning unit. By using augmented reality as a learning resource that supports the development of competencies through meaningful learning [21].

To this end, literature information was searched in the different databases, using the different search criteria such as augmented reality, mobile learning, educational resource based on applications developed in the educational field, where it was demonstrated that the use of the same allows the acquisition of knowledge and improves the skills motivating the student and facilitating the learning teaching processes [22].

Based on the context of the technical baccalaureate and according to the needs of quality educational processes, the implementation of didactic resource in the use of technology on Augmented Reality is determined.

The Unity platform was chosen as a free-source and documentation platform for development, the database to be implemented on the platform is Vuforia SDK with the use of markers for the visualization of 3D models in the application performing recognition tests [23].

The tool that evaluates the augmented reality application for the maintenance and configuration of computer equipment, was developed in Google Forms, consists of 15 items which are divided into 3 dimensions: usability which evaluated the interaction in the management of the application by the user, consists of 4 items which were evaluated with Likert scale where 1 is represented as poor and 5 as excellent. In addition, 2 short-answer questions were asked. The design dimension evaluated the aesthetics of the different components of the application, consists of 5 items which were evaluated with Likert scale where 1 is represented as strongly disagree and 5 as strongly agree. Finally, the dimension of satisfaction measured the degree of fulfillment of the objective of the application, consists of 4 items which were evaluated with Likert scale, considering that the 1 implies the minimum degree of satisfaction and 5 the maximum [24–26].

It was requested the participation of 11 students of technical baccalaureate in an individual and personalized way, giving them to know the objective of the application AR for technological guide, as well as instructions to carry out the interaction at each of the stages with the app on their Android mobile device, giving each participant a time of 20 to 30 min, allowing them total freedom in the handling and interaction with the application and in case of doubts the pertinent clarifications were made. At the end of the test, the tool to assess implementation was applied, thanking them for their participation. The methodology that was implemented is shown in Fig. 1.

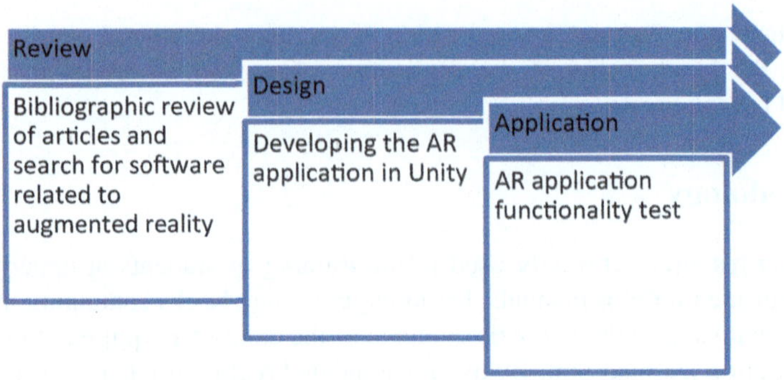


**Fig. 1.** Methodology used, Source: own elaboration.

### 3.1 Materials

The tools used for the design of platforms, applications and tests are listed below:

- Unity3D. Software used in the development of applications that incorporate augmented reality. This software allows the design of 2D and 3D scenes, as the application requires.
- Vuforia Engine is the most widely used platform for AR development, with support for major phones, tablets and glasses.
- Microsoft Visual Studio is a cross-platform programming environment, compatible with multiple programming languages such as C#.

### 3.2 Interface Development

The proposed interface was developed in Unity3D software in a Spanish version, using different UI elements (user interface) such as text and images, buttons that would be part of the scene. This interface consists of two buttons that redirect to the practice and exercise section as shown in Fig. 2.

**Fig. 2.** Home screen.

This section displays the buttons that identify the components of a computer, to redirect to the scene of the selected component. This interface consists of twelve buttons and an icon to return to the home screen as shown in Fig. 3.

**Fig. 3.** Components menu.

In Fig. 4, a scan area appears for the student to scan with an Android mobile device the selected component (found in the Maintenance and Computer Configuration manual) from the menu in Fig. 3, to view the associated 3D model as shown in Fig. 5.

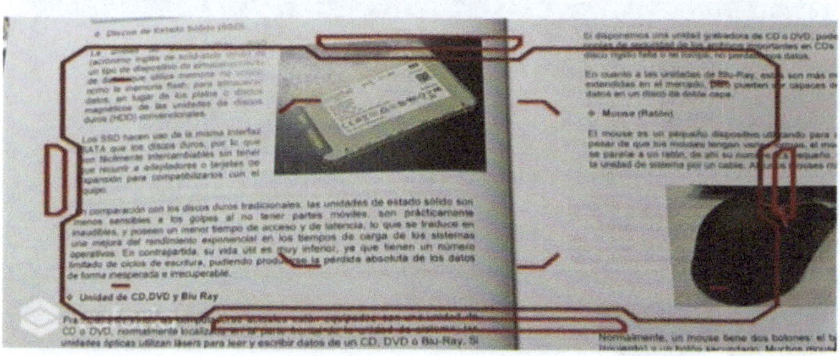

**Fig. 4.** Marker scanner display.

The screen has 7 buttons, on the left side there are 2 buttons to interact with the model (rotation of the model from bottom to top and from left to right), on the right side there are 2 buttons to visualize information (description and the parts that compose it), 1 button which displays a screen with the instructions of the buttons, as shown in Fig. 6.

**Fig. 5.** Selected component screen.

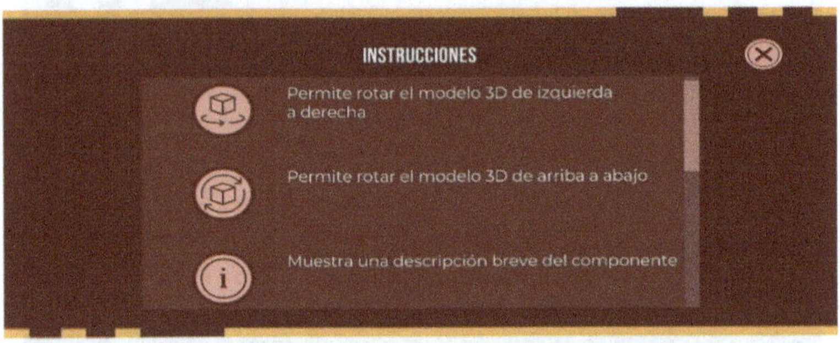


**Fig. 6.** Button instruction screen.

In Fig. 7 the two practices to be performed after the review of the components of the computer equipment are visualized. This interface consists of two buttons and an icon to return to the home screen.

**Fig. 7.** Practice menu.

In practice 1 (see Fig. 8), the student visualizes the essential components of a computer, in which he selects the name of the component and places it in the corresponding image. This interface consists of 7 labels, 7 models, 1 exit button and 1 button displaying instructions.

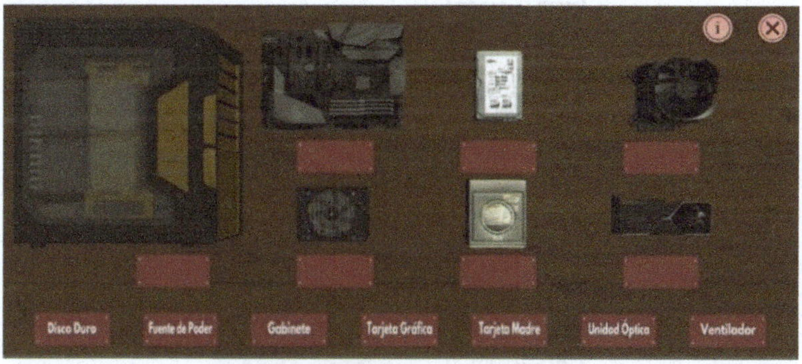


**Fig. 8.** Practice 1.

In practice 2 (see Fig. 9) the components of the computer equipment are displayed, where the student will select each one to integrate it into the computer tower in the position that corresponds to it. This interface consists of a list of 7 components, 1 exit button and 1 button displaying instructions.

**Fig. 9.** Practice 2.

The proposal to make this application of augmented reality in the manual of maintenance and configuration of equipment, is observed in Fig. 9 as the markers will be integrated so that the student interacts with the application (Fig. 10).

Escuela Industrial y Preparatoria Técnica "Álvaro Obregón"

❖ Microprocesador

El microprocesador (o simplemente procesador) es el circuito integrado central más complejo de un sistema informático; a modo de ilustración, se le suele llamar por analogía el "cerebro" de una computadora.

Es el encargado de ejecutar los programas, desde el sistema operativo hasta las aplicaciones de usuario; sólo ejecuta instrucciones programadas en lenguaje de bajo nivel, realizando operaciones aritméticas y lógicas simples, tales como sumar, restar, multiplicar, dividir, las lógicas binarias y accesos a memoria.

❖ Memoria RAM (Random Access Memory, Memoria de Acceso Aleatorio)

Componente que almacena temporalmente la información utilizada por la CPU mientras la computadora está siendo usada. La información almacenada en la memoria RAM es borrada cuando la computadora se apaga.

Prácticamente todos los otros componentes de una computadora están conectados por cables a la unidad del sistema. Los cables se encuentran conectados a las entradas (puertos) específicos, que se encuentran normalmente en la parte posterior de la unidad de sistema. El hardware que no forma parte de la unidad de sistema es, a veces, llamado **dispositivo externo o periférico.**

**Fig. 10.** Maintenance and equipment configuration manual incorporating markers.

## 4 Results

After applying the instrument to know the results for each dimension: usability, design and satisfaction. The following results were obtained.

Dimension results:

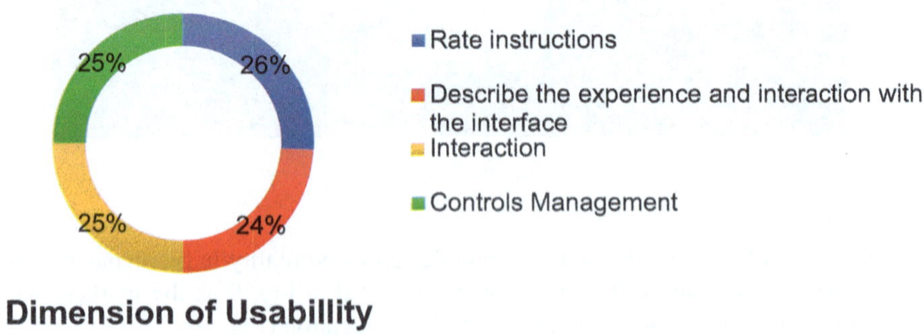


**Fig. 11.** Evaluation of the usability in the AR instrument for the maintenance and configuration of the computer equipment.

Figure 11 shows that the item that qualifies the instructions described in the application obtains 26% and the item where controls and interface interaction are evaluated was 25%, with 24% evaluating the interface experience.

When the instrument was applied to the students who participated, it can be seen in Table 1 that 45.45% did not present any situation, in the practice section they presented difficulties in 36.36% during the application you can see that they did not interact with the instruction button.

**Table 1.** Difficulties in mobile app interaction

| Responses | Frequency | % |
|---|---|---|
| No difficulty | 5 | 45.45% |
| Practices | 1 | 9.09% |
| Practice 1 | 1 | 9.09% |
| Practice 2 | 2 | 18.18% |
| Failure of marker detection | 1 | 9.09% |
| Connection failure | 1 | 9.09% |
| Total | 11 | 100% |

The participants identified as a unique experience at the time they observed in the application exercises and 3D models with 27.27%, the practice scene was evaluated with 27.27% as shown in Table 2, this evaluation is influenced by not having interacted with the instruction button.

**Table 2.** Moment you remember most in the interaction of the mobile app.

| Responses | Frequency | % |
|---|---|---|
| Display of the marker scanner | 1 | 9.09% |
| Exercises | 3 | 27.27% |
| 3D models | 3 | 27.27% |
| Practices | 1 | 9.09% |
| Practice 1 | 1 | 9.09% |
| Practice 2 | 1 | 9.09% |
| Rotation of 3D models | 1 | 9.09% |
| Total | 11 | 100% |

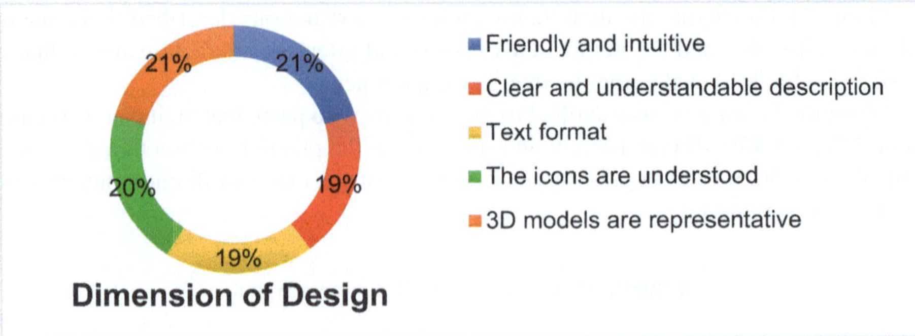


**Fig. 12.** Evaluation of the design in the AR instrument for the maintenance and configuration of the computer equipment.

Figure 12 shows that the items where the application describes how friendly and intuitive it was and the 3D models that were captured the result was 21%, with 20% were evaluated the icons that were used in the application and the text format and the clear and understandable description with 19%.

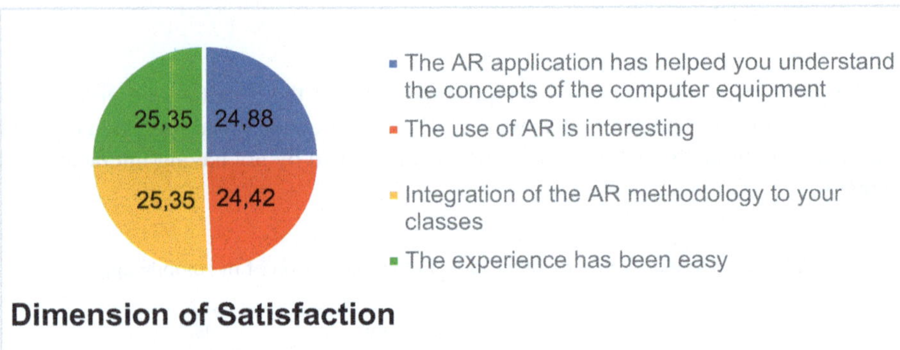


**Fig. 13.** Evaluation of the satisfaction in the AR instrument for the maintenance and configuration of the computer equipment.

The results of Fig. 13 show that the integration of the AR methodology in the classes and the experience that the student refers to is evaluated with 25.35%, The AR application has helped you understand the concepts of computer equipment was 24.88% and with 24.42% consider the use of AR interesting.

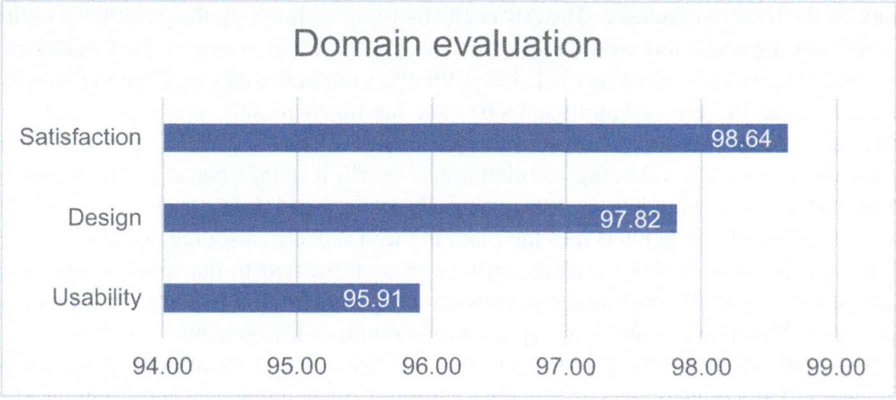


**Fig. 14.** Domain evaluation in the AR instrument for maintenance and configuration of the computing equipment.

As can be seen in Fig. 14, usability was evaluated by students with 95.91%, design with 97.82% and satisfaction with 98.64%.

## 5 Discussion

Consider AR as an emerging technology that can be used as a learning tool through the use of ICT and through innovation in education [27].

The satisfaction of the users in our study when interacting with the application in the tests that were carried out was obtained a result of 97.82% with the design, unlike the studies carried out by Veledo [28] where the failures were due to network connectivity, for Bezares Molina [25] the design was not intuitive which caused difficulties in the tasks that the students had to perform. In the usability domain, the results obtained in this research were 36.36% where there were difficulties in the practice section, these results are similar to those obtained by Encarnación De Jesús [24] where it was expressed difficulty to carry out the activities in Mixed Reality (MR) with 35%.

In order to carry out projects where the AR is applied, the context, the contents, as well as the design that best suits and can fulfill the purpose when using ICT must be considered [29]. Didactic resources alone do not guarantee learning, it is important that the teacher becomes familiar with the application, to share the experience with the students and make sure how it is used [30].

## 6 Conclusion

Technology plays an important role in our lives, it has permeated all areas so the educational environment is no exception, educational institutions must be transformed and prepared for the challenges of the future [9]. Situations of social isolation where mobility is not allowed and the educational process must be continued [31].

The tools or materials that make the development of the application possible are available, some are for free use and others have to be acquired which facilitates and shortens the time to conclude. The AR is an emerging technology that requires a multidisciplinary approach that would allow for a design that integrates elements that recreate the teaching process – learning [32]. The difficulties encountered in interacting with the application can be resolved as students become familiar with this type of resource. The practices that were carried out must be attractive to the students which motivate the individual or group work, allowing a collaborative work, it is indispensable the feedback which can be carried out between the group, the system and the teacher.

The AR can be integrated into the planning as a didactic resource according to the context of the user, in this case the application was directed to the level of Technical Baccalaureate in Web Programming for the development of competencies in the Unit of learning of Maintenance and Configuration of Computer Equipment.

The application will be presented to the directors of the Educational Institution to implement it as a didactic resource in the Learning Unit Maintenance and Configuration of computer equipment. The continuity of the project will be to evaluate the learning of the students with the use of the application.

**Acknowledgements.** In carrying out the present research work by undergraduate students at the School of Engineering, as well as Dr. Leticia Neira in charge of the virtual reality laboratory for her unconditional support, to MT Hector Francisco Berrones Cortez head of the Technical Bachelor in Web Programming for the opportunity granted to bring this research to a successful end.

## References

1. Martínez, N.M., Olivencia, J.L., Meneses, E.L.: Augmented reality as an emerging technology for educational innovation. Notandum (44), 125–140 (2017). http://www.hottopos.com/notand44/10MorenoLeivaLopez.pdf
2. Cano, M.M.B.: Diseño de herramienta software educativa con base en realidad aumentada y 3d, como estrategia pedagógica de apoyo al proceso de enseñanza-aprendizaje de las ciencias naturales en la ie valle de tenza [Tesis de pregrado, Universidad Cooperativa de Colombia]. Repositorio Institucional UCC (2021). https://repository.ucc.edu.co/handle/20.500.12494/32552
3. Garzón, J., Pavón, J., Baldiris, S.: Systematic review and meta-analysis of augmented reality in educational settings. Virtual Reality **23**(4), 447–459 (2019). https://doi.org/10.1007/s10055-019-00379-9
4. Cajo, B.G.H., Cajo, D.P.H., Chanalata, M.G.M., Cajo, I.M.H.: Augmented reality as a support resource in the teaching-learning process. Interuniversity Electron. J. Teacher Formation **24**(3) (2021). https://doi.org/10.6018/reifop.465451
5. Pelletier, K., Brown, M., Christopher, D., McCormack, M., Reeves, J., Arbino, N., Mondelli, V.: EDUCAUSE. Obtenido de EDUCAUSE: (26 de April de 2021). https://library.educause.edu/resources/2021/4/2021-educause-horizon-report-teaching-and-learning-edition
6. Neira-Piñeiro, M. del, R., Moral, M.E.D., Fombella-Coto, I.: Aprendizaje inmersivo y desarrollo de las inteligencias múltiples en Educación Infantil a partir de un entorno interactivo con realidad aumentada/ Immersive learning and multiple intelligences development in early childhood education by means of an interactive environment with augmented reality. Magister **31**(2), 1–8 (2020). https://doi.org/10.17811/msg.31.2.2019.1-8
7. Lou, D.: Two fast prototypes of web-based augmented reality enhancement for books. Library Hi Tech. News **36**(10), 19–24 (2019). https://doi.org/10.1108/LHTN-08-2019-0057

8. Tsuchida, S., Matsuura, S.: A role of augmented reality in educational contents: intermediating between reality and virtual reality. In: Lackey, S., Shumaker, R. (eds.) VAMR 2016. LNCS, vol. 9740, pp. 736–745. Springer, Cham (2016). https://doi.org/10.1007/978-3-319-39907-2_70
9. Martínez, O.M., Mejía, E., Ramírez, W.R., Rodríguez, T.D.: Incidence of augmented reality in the learning process of mathematical functions. SCIELO **32**(3) (2021)
10. Olalla, A., Toapanta, D.F., Vinicio, B.: Edu-Entertainment multimedia application for environmental education, using virtual reality and augmented reality techniques (6 de April de 2021)
11. Westerfield, G., Mitrovic, A., Billinghurst, M.: Intelligent augmented reality training for motherboard assembly. Int. J. Artif. Intell. Educ. **25**(1), 157–172 (2014). https://doi.org/10.1007/s40593-014-0032-x
12. Muñoz, G.P., Muñoz, J.C., Zambrano, Y.M., Cuadros, S.V.: Augmented reality as an interactive tool in the basic level teaching-learning process. Polo del Conocimiento **6**(1), 187–200 (2021)
13. Vila, R.R., Lledó, A.L., Andrés, S.M.: Perceived usefulness of augmented reality as a didactic resource in the infant education teacher degree. Campus Virtuales **8**(1), 19–35 (2019)
14. Díaz, V.M.: The augmented reality in the educational sphere of student of degree in cochildhood education: case study. Pixel-Bit. Revista de Medios y Educaci{\'o}n **51**, 7–19 (2017)
15. Bhorkar, G.: A survey of augmented reality navigation **8**(2), 73–272 (2017). https://doi.org/10.1561/1100000049
16. Córdoba, P.J.: augmented reality as a method of university learning. Int. J. Inf. Syst. Softw. Eng. Big Co. **5**(1), 95–100 (2018)
17. Ewais, A., Troyer, O.D.: A usability and acceptance evaluation of the use of augmented reality for learning atoms and molecules reaction by primary school female students in Palestine. J. Educ. Comput. Res. **57**(7), 1643–1670 (2019). https://doi.org/10.1177/0735633119855609
18. Tomara, M., Gouscos, D.: A case study: visualizing coulomb forces with the aid of augmented reality. J. Educ. Comput. Res. **57**(7), 1626–1642 (2019). https://doi.org/10.1177/0735633119854023
19. Cabero-Almenara, J., Vázquez-Cano, E., Villota-Oyarvide, W.R., López-Meneses, E.: La innovación en el aula universitaria a través de la realidad aumentada. Análisis desde la perspectiva del estudiantado español y latinoamericano. Revista Electrónica Educare **25**(3), 1–17 (2021)
20. Tovar, L.N., Castañeda, E., Leyva, V.R., Leal, D.: Work-in-progress—a proposal to design of virtual reality tool for learning mechatronics as a smart industry trainer education. In: 2020 6th International Conference of the Immersive Learning Research Network (iLRN), pp. 381–384 (2020). https://doi.org/10.23919/iLRN47897.2020.9155172
21. Peretó, N., Ghiotti, M., Pérez, B., Goría, C., Vargas, S.: Educación a distancia. Perspectiva de los alumnos acerca de los recursos educativos. Ciencia, Docencia y Tecnología **21**(40), 97–111 (2010)
22. García-Chitiva, M.D.P. y Suárez Guerrero, C.: Estado de la investigación sobre la colaboración en entornos virtuales de aprendizaje. Pixel-Bit: Revista de Medios y Educación **56**, 169–191 (2019)
23. Lobo Quintero, R.A., Santoyo-Díaz, J.S., Briceño-Pineda, W.: EducAR: using augmented reality for basic science learning in collaborative environments Revista Educación En Ingeniería **14**(27), 65–71 (2019). https://doi.org/10.26507/rei.v14n27.930
24. Encarnación De Jesús, L., Ayala Ramírez, S.: Didactic strategies through mixed reality, for theoretical-practical learning in high school students. RIDE Revista Iberoamericana Para La Investigación Y El Desarrollo Educativo **11**(22) (2021). https://doi.org/10.23913/ride.v11i22.922

25. Molina, B., Gabriel, F., Toledo, T.G., Acevedo, A.F., Mendoza, M.E.: Aplicación de realidad aumentada centrada en el niño como recurso en un ambiente virtual de aprendizaje. Apertura (Guadalajara, Jal.) **12**(1), 88–105 (2020). Epub 16 de septiembre de 2020. https://doi.org/10.32870/ap.vl2nl.1820
26. Kawamoto, A.L.S., Mezzomo, M.D.M.: Geomorphology classroom practices using augmented reality. In: Antona, M., Stephanidis, C. (eds.) UAHCI 2017. LNCS, vol. 10278, pp. 155–166. Springer, Cham (2017). https://doi.org/10.1007/978-3-319-58703-5_12
27. Osuna, J.B., Almenara, J.C.: Evaluation of the objects of the learning in augmented reality: pilot study in the degree of the medicine. Enseñanza Teach. **34**(2), 149–167 (2016). https://doi.org/10.14201/et2016342149167
28. San Pedro Veledo, M.B., Villalustre Martínez, L., Herrero Vázquez, M.: Diseño de un itinerario aumentado e interdisciplinar para la formación de maestros de educación primaria. Edutec. Revista Electrónica De Tecnología Educativa (68), 54–69 (2019). https://doi.org/10.21556/edutec.2019.68.1293
29. Marín-Díaz, V., Sampedro-Requena, B.E.: La Realidad aumentada en Educación Primaria desde la visión de los estudiantes. Alteridad **15**(1), 61–73 (2020). https://doi.org/10.17163/alt.v15n1.2020.05
30. Manrique-Losada, B., Zapata Cárdenas, M.I., Arango Vásquez, S.I.: Entorno virtual para cocrear recursos educativos digitales en la educación superior. Campus Virtuales **9**(1), 101–112 (2020). (www.revistacampusvirtuales.es)
31. Organización Mundial de la Salud. Consideraciones para la cuarentena de los contactos de casos de COVID-19: orientaciones provisionales, 19 de agosto de 2020 (2020). Organización Mundial de la Salud. https://apps.who.int/iris/handle/10665/333968. Licencia: CC BY-NC-SA 3.0 IGO
32. Jerez, M.C.M., Pérez, A.: Estudio de APPs de realidad aumentada para su uso en campos de aprendizaje en un entorno natural. Edutec. Revista Electrónica De Tecnología Educativa (62), 19–33 (a370) (2018). https://doi.org/10.21556/edutec.2017.62.1017

# Augmented Reality - A Tool to Support Learning in Engineering Schools. Automation Practice Case

Fernando Banda(✉), Francisco Javier Esparza, Jesús Melendéz, and Melany Ascacio

Universidad Autónoma de Nuevo León, Pedro de Alba S/N,
San Nicolas de los Garza, Nuevo León, México
fernando.bandamn@uanl.edu.mx

**Abstract.** Augmented reality is a technology that improves user experiences by allowing digital sensations to be integrated. While virtual reality is an immersive model based on simulated environments using glasses, these simulated environments are built through a server. Everything that the user can see or feel is artificially provoked through images and sounds. Both realities are technological tools that are adapted to the professional development of engineering learning, which have been of significant help during mobility challenges due to the global pandemic. Although virtual and augmented reality are expensive to implement in some countries, there are still opportunities to implement them by giving access to new learning tools and creating opportunities to develop learning even in times of challenges of learning due to separation and distance. This article collects information about virtual reality and augmented reality, both of which have been applied with remarkable success in America and Europe, as well as to propose these means as a learning tool to be used in the automation laboratory of engineering schools, using augmented reality techniques to support and improve the competences necessary for distance education in times of epidemic. The method applied was based on the observation of students from a high grade level as qualitative research, the results will support parallel research to include immersing reality.

**Keywords:** Virtual reality · Augmented reality · Distance learning · Automation

## 1 Introduction

There has currently been a significant delay in education globally, affected by the closure of schools at all levels and the economic recession due to measures to control the epidemic. In the last update at the end of April, the closure of schools in 180 countries was detected, causing 85% of students worldwide to be unable to attend schools [1]. Having school closings creates a delay and losses in learning in the case of engineering, since there is no access to schools for face-to-face classes, there is no access to laboratories and practices which are essential in the engineering area since they provide the possibility to understand and corroborate the hypotheses proposed in the scientific field.


J. Trojanowska et al. (Eds.): MANUFACTURING 2022, LNME, pp. 267–277, 2022.
https://doi.org/10.1007/978-3-030-99310-8_21

As a learning tool, it is currently sought to implement virtual reality and augmented reality in the automation area to be able to observe in a three-dimensional way and remotely access activities that cannot be felt in person [2].

Virtual reality seeks to recreate the reality in which we live through digital-based interactions. In the words of Memarsadeghi & Varshney [3] "virtual reality (VR) recreates the sensory world around us entirely through computer-generated signals of sight, sound, touch (and in some cases smell and taste)" in other words, it separates us from the reality to a high immersive experience on a digital created world in which we perceive the sensations as if it were real.

Therefore, augmented reality broadens the perception of what reality looks like through virtual technology. It is a technology that helps us and allows us to expand extra information to images when they are viewed on a device. Combining virtual and physical elements [4, 5].

On the other hand, automation is based on computer science, mechanics, and electromechanical processes that become functional with minimal human involvement. They help to optimize the operation of different processes and products [6]. Authors like Carrillo &Vásquez [7] define it as the reduction of labor and use of the necessary resources without wasting them. And the application of mechanical and electronic systems and computer bases to operate and control production. While it also can be described as a set of techniques associated with the application of mechanical/electronic systems and based on computers, whose objective is the operation and control of production.

Analyzing the different conceptual points of view on automation, we finally define automation as a system that, with the help of technology and command orders, reduces the use of labor resources. Avoiding a waste of costs and time, increasing the quality [8].

The face-to-face laboratories in this type of class are of the utmost importance since it is vital to be able to carry out physical tests to support your hypotheses. Many higher education institutions seek to implement vital strategies in times of pandemic, using virtual classroom models seeking greater participation and understanding of the exposed topics [9].

Currently there has been an increase in this type of virtual classroom, since they present greater accessibility and flexibility with the use of multiple devices, allowing similar experiences to be covered with virtual and augmented reality.

With these virtual experiences, it can be corroborated that by carrying them out, the need for high investments is reduced by substituting those infrastructures with easy-to-use tools, devices, and applications, giving greater access to the education sector, providing the same experiences virtually as in person. The virtual reality structures provide the student with learning tools and experiences that promote the exploration of information from real data in virtual media. [10, 11].

There have been cases in which the MR (Mix Reality) to learning processes in factories has been studied [12]. The conclusion reached is that the use of this technology extends the opportunities for learning significantly. It is well known that physicality of learning, and thus their connection to hardware, is the cornerstone for effective learning. While virtual environments on the other hand have a high degree of flexibility and are quickly adaptable, they are not bound by physical hardware. This allows learners to experience processes, methods and scenarios that are not available through conventional means. Combining the advantages of the real and virtual world unlocks enormous potential for enhancing learning successes. Other [13] authors have corroborated with the previously mentioned conclusions. They have conducted researched on the different areas where these technologies can be used, leading to the conclusion that the AR (Augmented Reality) technology shows a better use for the learning phase.

## 2 Methodology

This work was based on qualitative research that starts from documentary research, based on related articles, educational links books and didactic material used at the engineering school.

It is informative in nature because it shows valuable information and research on virtual and augmented reality. And on how these technologies are applied in the different learning styles in higher education in the engineering area.

Diverse sources were used for the research of useful teaching material on automatization topic, among which are:

- Printed documentation: books, research projects, case studies.
- Electronic documentation: materials found on the internet, such as books, specialized magazines and articles that are published in digital format.
- Audiovisual documentation: Videos and audios that contain information from interviews, presentations, and conferences.

After extensive research based on the hardware and software necessary to develop this application the ones mentioned below were selected:

- Unity: The version of unity used to develop the app was version 2018.4.15f1, for this to work it was necessary to use other add-ons like Vuforia and Visual Studio.
- Visual Studio: A programming IDE used to compile and to correct the functionality of Unity.
- Vuforia: An add-on used with Unity to create the markers and to implement the basic AR interactions for the app.
- Laptop: This device was used to run Unity with the minimum requirements.
- Smartphone: The smartphone used to run this app which had Android system API 19 (4.4 - KitKat) which is the minimum to run Vuforia and AR apps.
- Solidworks: The 3d models shown in the app were created using this software, they were chosen based on previous knowledge and experience used in the regular practices and how the students interacted with them.

Augmented reality was applied in laboratory practices related to automation through the developed application called ARLAB (Augmented Reality Laboratory) which a group of 30 students used to develop their experiments (See Fig. 1). Application of practical material was compared with theoretical material to support the theories and interpretations. A satisfaction survey was used based on Likert scale with two questions, the first one asking how much they liked the app and the second one asking about how easy it was to understand.

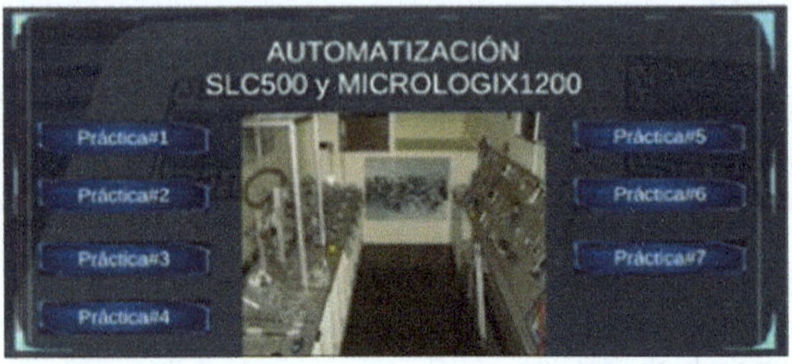


**Fig. 1.** Example of the app running during implementation in laboratory practices.

Practice one consisted of the student knowing the compact and modular PLC, which was presented by means of a QR code created using Vuforia packages (See Fig. 2), when pointing at the target, it showed the PLC 3D model allowing the students to disassemble its modules, just by dragging them to another location in the screen of their smartphones, successfully meeting the requirements of the educational program without having to attend their corresponding classes in person (See Fig. 3).

**Fig. 2.** QR code used.

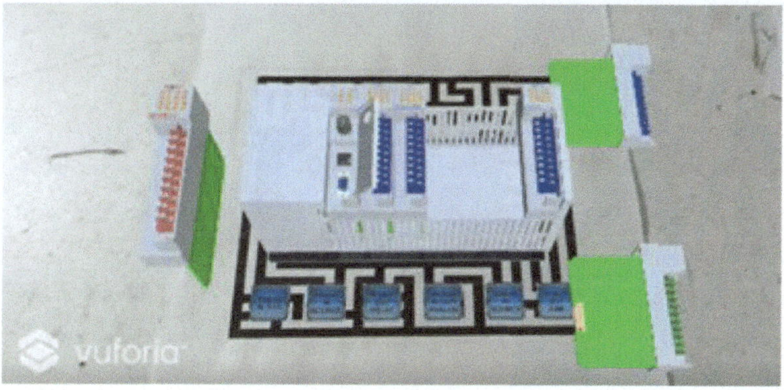


**Fig. 3.** Modular PLC 3D model showed.

In the second practice, in the same way as in the first, we worked with a modular and compact PLC, the applicable difference in this case was the information displayed (See Fig. 4), the student had to select the desired PLC.

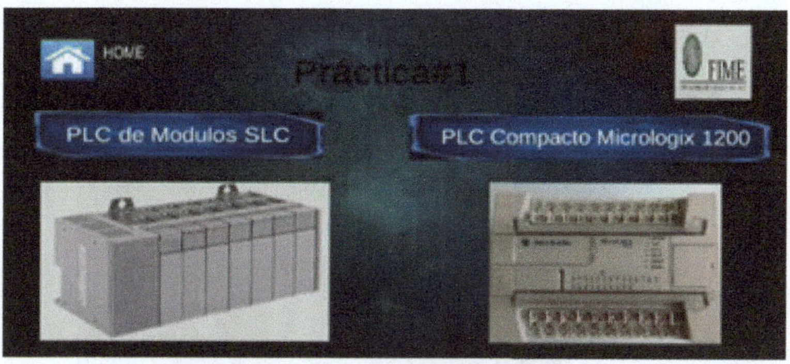


**Fig. 4.** PLC selection screen in the app.

After the selection of the desired PLC (See Fig. 5), the student, in addition to viewing the modular PLC, was able to look at theoretical information such as the I/O distribution, voltage necessary to make it work and some other technical information about the selected model.

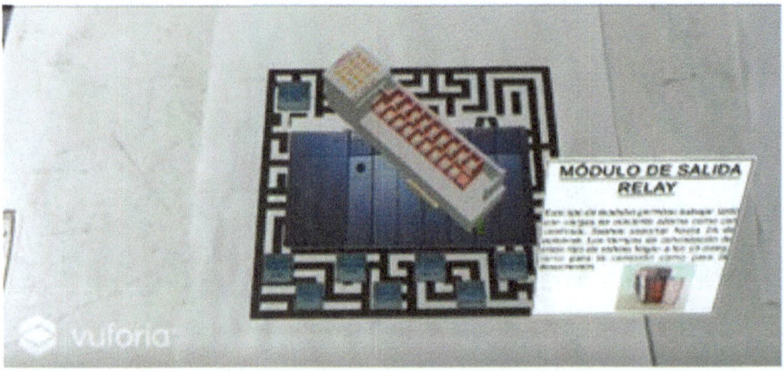


**Fig. 5.** Modular PLC information displayed.

For practice three, the students were asked to turn on different lights by means of a sensor connected to the PLC, which, thanks to augmented reality, was possible (See Fig. 6). The only thing that the student had to do was place his finger or hand under the sensor covering the QR code to turn on the lights.

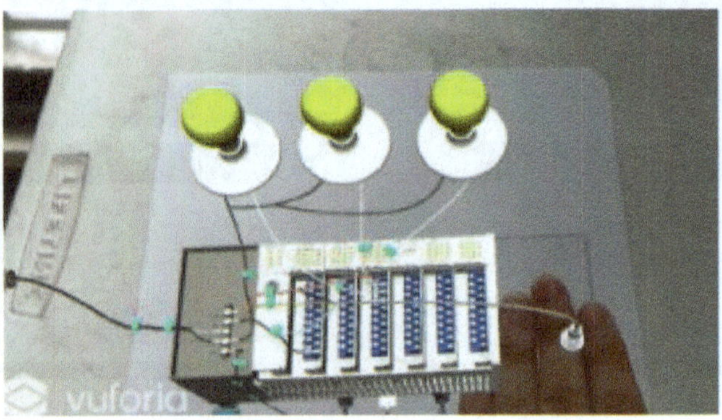

**Fig. 6.** Sensor and lamp connections.

Each time the students progressed with respect to their automation practices, the students were presented with a more detailed and advanced augmented reality case to be able to successfully develop what was requested. For these previous practices, the student just pointed at one marker at a time, once they became familiarized with this type of interaction, the students were asked to use more than one marker.

One case is practice four, which included a video tutorial so that the student could download different software which would make him capable of programming PLCs by augmented reality, the students just needed to touch the play icon in the GUI at the screen, to play the video, as a support the video was also provided to the students through a URL and shown during theoretical class (See Fig. 7).

The next step to perform consisted of placing the target image (QR code) in front of the camera, according to what section of the PLC they wanted to visualize. It should be noted that tutorial videos were reproduced by the augmented reality view, giving access to a more advanced type of visualization (See Fig. 8).

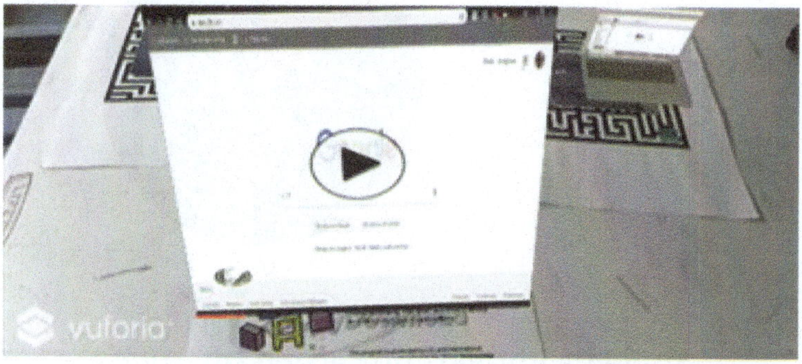

**Fig. 7.** AR view of the tutorial.

**Fig. 8.** Practical augmented reality view 4.

## 3 Results

After the use of this application in a group of 30 students, it was found that most of them believed this method to be more attractive and easier to understand than using the conventional method of showing the students the physical PLC. Also, the risk for the equipment to be damaged due to inadequate connections was avoided (See Fig. 9). This app was developed to be applied on future projects in order to analyze the usability of AR applications and the student's level of achievement during their study sessions.

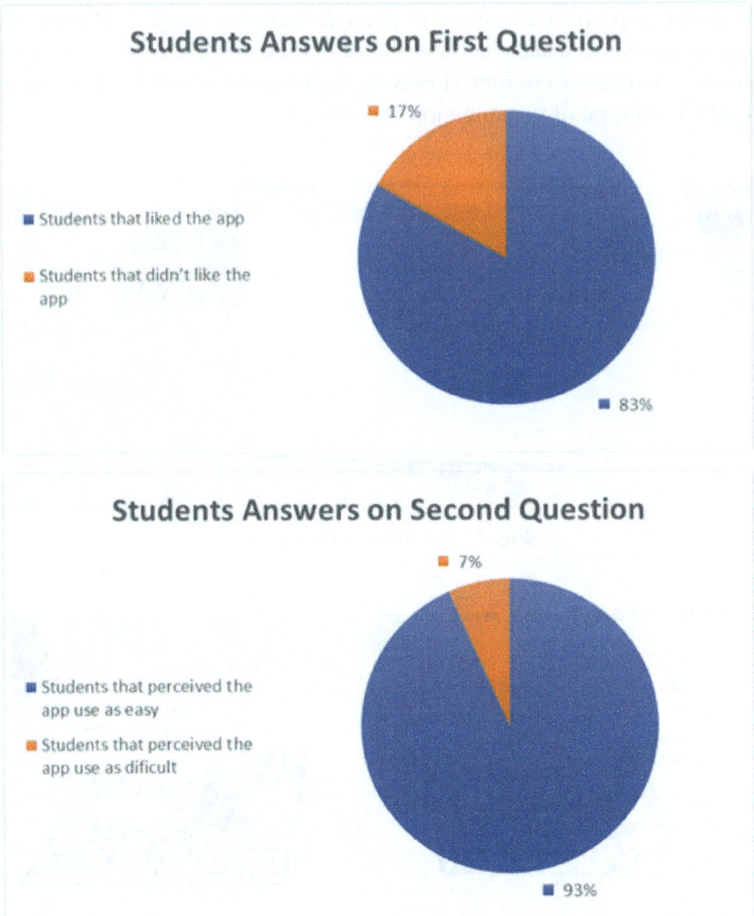


**Fig. 9.** Pie Chart showing the students answers to both questions asked in the survey.

Developing automation labs with augmented reality allows high school students and other educational levels access to additional tools by meeting their needs and implementing them in classrooms. The exception here is that during the pandemic, all the faculties and universities had been facing problems related to the restrictions of student access to the facilities. This tool was developed thinking of a hybrid educational plan on which the students will not be affected if they cannot attend their normal classes.

## 4 Discussion

Observing the obtained results, we noticed that the practices were carried out in a satisfactory way, where the student was able to see, perform, learn, and interact in a much more visual and effective way than just reading the theory and trying to understand the topics seen. In light of the uncertainty being experienced worldwide, this application is an extra tool that complements all the practices seen during the sessions of the online modality and we are convinced that it is an advance for aspiring engineers to have adequate training and not leave the issues seen unfinished, but all the practices to be done [14, 15]. It will ensure that a university level education be more independent of global inconsistencies.

It is also important to note that our tool, although it has only been applied in the automation area, is possible that in the future it will be used for many other subjects and practices of university colleagues [16].

## 5 Conclusion

To conclude with all the information that we have obtained so far, we clearly see that the implementation of augmented reality is necessary so that students can enjoy better-quality learning during times of unforeseen changes, Taking advantage of the online modality that is common today due to the pandemic would still remain in force at the UANL, as well as continuing with the visual aids and especially with the practices that are most useful if they are carried out interactively. Although the program continues to function in a stable way with the traditional class method, with the augmented reality, sessions would be complemented, inspiring more imaginative conceptual thinking, and knowledge would be better assimilated [17].

With all the results obtained, it should be noted that augmented reality was already in great development and was already being implemented in different universities around the world. With the pandemic came a new stage of education where it was no longer optional to acquire knowledge from a distance but became a necessity where students from their homes continued with their purpose of graduating as engineers while restricted. For that, augmented reality became a great ally that has had very positive results. It can continue to be part of the daily program post-pandemic in the Face-to-face modality. Some classrooms will have to be adapted for the implementation of an academic program that has augmented reality in it [18, 19].

All the information and results obtained have been very satisfactory for us who are engineering students, since it allows us to see a different panorama of how we perceive knowledge and we are utterly convinced that it is a particularly useful tool and a great complement in learning engineering. The greatest learning tool is experience, which is provided by use of this technology. We know that it will be of significant help to future generations who will be able to take advantage of this form of learning. It is also important to mention that its implementation will be a gradual step-by-step and can be adapted immediately to our needs today [20].

Finally, we would like to thank the Faculty of Mechanical and Electrical Engineering that allowed us to develop this project, which with this article and its application has a great positive impact and continues to innovate the way we acquire knowledge and that it always progresses to be of Best Quality [21].

## References


1. World Bank (2020). World Bank Education and COVID-19, interactive map, 24 April. https://www.worldbank.org/en/data/interactive/2020/03/24/world-bank-education-and-COVID-19
2. Cózar Gutiérrez, R., Hernández Bravo, J.A., De Moya Martínez, M., Hernández Bravo, J.R.: Digital education. Obtained from emerging technologies for the teaching of social sciences. An experience with the use of augmented reality in initial teacher training: https://dialnet.unirioja.es/descarga/articulo/5495910.pdf (June 2015)
3. Memarsadeghi, N., Varshney, A.: Virtual and augmented reality applications in science and engineering. Comput. Sci. Eng. **22**(3), 4–6 (2020). https://doi.org/10.1109/MCSE.2020.2987151
4. Dynatec. The uses of augmented reality in engineering: https://dynatec.es/2015/10/20/usos-realidad-aumentada-ingenieria/#:%7E:text=La%20realidad%20aumentada%20es%20utilizada,reality%20mix%20in%20time%20real. Accessed 21 Mar 2021
5. IAT. Obtained from augmented reality. What is it? Characteristics and types: https://iat.es/tecnologias/realidad-aumentada/#:~:text=Hardware%3A%20el%20procesador%20posibilita%20la,las%20im%C3%A1genes%20en%20realidad%20aumentada. 27 Nov 2019
6. Torres, C.E.T., Rodriguez, J.C.: Immersive learning environments and teaching cyber-generations. Obtained from research gate: https://www.researchgate.net/publication/333889294_Los_entornos_de_aprendizaje_inmersivo_y_la_ensenanza_a_ciber-generaciones. January 2019
7. Carrillo, R.D.A., Vásquez, M.J.L.: Automation of a Greenhouse with the S7–200 PLC. Autonomous University of Zacatecas, México (2008)
8. Vega, V.A.: Adaptive learning based on virtual reality simulators. Scielo. http://scielo.sld.cu/scielo.php?script=sci_arttext&pid=S2227-18992021000200138&lang=en. 1 June 2021
9. Delfino, C.Y.: Virtual Reality as a learning mediator: Development of a virtual reality mobile application aimed at children. Scielo. http://www.scielo.org.ar/scielo.php?script=sci_arttext&pid=S1850-99592020000300014&lang=en. December 2020
10. Gordon, AFDR. From classroom learning to virtual learning in times of pandemic. Scielo. https://www.scielo.cl/scielo.php?script=sci_arttext&pid=S0718-07052020000300213&lang=en. 26 October 2020
11. Work-in-progress—a proposal to design of virtual reality tool for learning mechatronics as a smart industry trainer education. In: IEEE Conference Publication, IEEE Xplore. https://ieeexplore.ieee.org/document/9155172. 1 Junio 2020
12. Juraschek, M., Büth, L., Posselt, G., Herrmann, C.: Mixed reality in learning factories. Procedia Manuf. **23**, 153–158 (2018)
13. Fast-Berglund, Å., Gong, L., Li, D.: Testing and validating extended reality (xR) technologies in manufacturing. Procedia Manuf. **25**, 31–38 (2018)
14. UPS, UP. Adaptability in education in times of COVID, 15 July 2020
15. Molina, B.F.G.: Child-centered augmented reality application as a resource in a virtual learning environment. Scielo. http://www.scielo.org.mx/scielo.php?script=sci_arttext&pid=S1665-61802020000100088&lang=en. 16 September 2020
16. Calderón, R., Arbesú, R.S., Rafael. ARLAB: laboratory with augmented reality (2013). https://repositorio.tec.mx/handle/11285/621353. 2 October 2021
17. Hugo, V., Luis, B., Rico, F., Llanos, A.C.: ICT tools as mediator elements in the self-learning process on automation fundamentals. International meeting of engineering education (2016). https://acofipapers.org/index.php/eiei/article/view/935. 2 October 2021
18. Electrotec Seminar: Introduction to Industrial Automation with PLC [Video File]. Youtube. https://www.youtube.com/watch?v=xQkfRamLso0. 28 March 2020

19. ACIEM Conference - Innovation in Automation [Video File]. YouTube. https://www.youtube.com/watch?v=V6c2NS0N2Zw. 27 October 2020
20. Unity Latam, Augmented reality with Unity and Vuforia - Unity Developer Day Chile 2017 [Video Archive]. Youtube.https://www.youtube.com/watch?v=2r5Oiyo9ORc (March 28, 2018)
21. CIDESI-CONACYT Differences between Augmented Reality, Virtual Reality and Mixed Reality [Video File]. YouTube. https://www.youtube.com/watch?v=bzFH0tizGHM

# A Design Proposal: Virtual Reality Environment for Safety Training in Electrical Substations

Edson E. Castañeda-Mancillas[1(✉)], Eduardo Sanchez-Rentería[1], Francisco Torres-Guerreo[1], and Paweł Buń[2]

[1] Universidad Autonoma de Nuevo Leon, Pedro de Alba S/N, Niños Heroes, San Nicolas de los Garza, Nuevo Leon, Mexico

[2] Poznań University of Technology, Piotrowo 3, 60-965 Poznan, Poland

**Abstract.** Virtual reality technology has developed at an exponential rate in the last decades, increasing in popularity due to its multiple benefits in learning and job training applications. One of these benefits is the ability to offer hands-on learning in virtual environments, which in real life are highly dangerous for inexperienced people, so it would be impossible to provide this type of learning through traditional teaching methods. In electrical engineering, it is common to encounter these situations, as is the case with electrical substations. There is a need for a trainer that allows trainees to acquire knowledge about these infrastructures with a closer approach. Being substation safety is one of the main factors in the training process in the electrical field. The trainer must have scenarios that first teach and promote personal safety, so topics such as personal protective equipment are considered. This paper introduces the process of a trainer for electrical substations simulator, specifically on following safety instructions, a development of a virtual practice using 2D and 3D is shown, which aims to instruct the user on personal protective equipment in an immersive environment. The process of the three-dimensional environment is presented, and the flow of the practice is described. The training tool teaches the users about the proper equipment to wear when going to an electrical substation and then shows what the user will find there. Tests were applied to the user to measure usability, satisfaction, and efficiency using the training tool.

**Keywords:** Virtual reality · Training · Electrical substation

## 1 Introduction

When it comes to the practical knowledge, traditional teaching methods are often insufficient in many areas, such as medicine [1, 2], construction [3], even chemistry [4] and engineering in general [5]. Although there are certain approaches to practical instruction, the lessons are often limited to observation by means of images or elements that are not in operation, or in their common environment. For subjects such as medicine, anatomical models that emulate the human body [6], or videos explaining procedures, are often used. Likewise, for construction [3], diagrams, pictures, and videos are used. It is very unusual that lessons are given in real situations, since the lack of experience of the


J. Trojanowska et al. (Eds.): MANUFACTURING 2022, LNME, pp. 278–291, 2022.
https://doi.org/10.1007/978-3-030-99310-8_22

trainee, and the environments in which these practices should be given would represent a considerable risk for the trainee, the infrastructure, and other personnel around him/her.

In engineering, specifically electrical engineering, the same situations apply. Trainees are taught about components and infrastructures that are not in operation or are isolated components. Such is the case of electrical substations, highly dangerous installations for inexperienced personnel, where each component is energized with high voltages and currents that represent lethal risks. In addition, substations are often located in remote places, which also implies transportation costs in case trainees wanted to be trained there.

Practical training at real substations is an obstacle for schools and students since there are some factors that impede his practical realization. There are situations such as safety of personnel, not being able to make any mistakes in matters of safety, transportation of personnel, the need to be guided by a host visit, technical requirements to train at real situations (schedule and times), and the cost of the equipment to train personnel [7].

Virtual reality allows facing the problem of the risks involved in the training of people with no or minimal training in different areas. In matters of safety, it is also known that there is a high rate of injuries related to the engineering and construction sciences that is the reason why the development of new technologies for implementation and experimentation has been a priority in recent years [8]. A relevant subject is also to provide the users the capability to assess their knowledge [9]. In the last decade alone, multiple proposals and projects have been developed for virtual reality systems for training in different areas. For example, in medicine, [1] describes the development of a non-immersive virtual reality system, using 3D vision, for the exploration of an anatomical model in breast cancer situations. In this work the evaluation of the users suggests an improvement in learning, being able to investigate the human body without the common restrictions, and being anatomically accurate models, in which it is possible to explore internally without the need to make cuts to a real person. In [2] is discussed the development of a virtual reality system for training in endovascular surgery, specifically to avoid collisions that could damage blood vessels when manipulating the catheter, due to the lack of the notion of the force applied in the teleoperated systems. The system warns beginners about possible collisions and assists the user in eliminating them. The evaluation of this system shows a reduction in the frequency of collisions and the distance that represents tissue damage. This means that it allows practical training of novice physicians in such procedures, without the need for trial and error with real patients, which would be highly irresponsible and dangerous. In [6] the procedures known as "T\&S" are described as procedures that require high levels of practice, which evidently cannot be obtained without real patients. Therefore, a virtual reality system is developed to facilitate the training of this procedure without the risk of involving patients who could be negatively affected by a poorly performed procedure. In terms of construction virtual reality is used to generate learning about safety in construction and engineering environments. The potential of this technology with HMD's (Head Mounted Displays) for visualization and planning in architecture and construction projects is also discussed [3, 10–12].

In electrical engineering, there are multiple projects focused on the development of virtual reality systems for training. In [13] a non-immersive virtual reality tool for maintenance of high voltage overhead lines is presented, where the non-immersive is justified because it enables the use of other more common peripherals for the user, reducing costs and training time in its use. More specifically about electrical substations, in [14] a virtual reality substation trainer is presented. In it we see great freedom of interaction and decision making for the user, giving him the ability to make decisions of free actions, even if incorrect, with visible consequences. The freedom of interaction translates into more realistic and meaningful experiences, in terms of learning. More recently in [15] the development of a so-called virtual reality training complex for power systems is shown. This is aimed at training and practice of personnel actions during common and emergency situations. It is proposed that the user will gain experience, eliminating the possibility of electric shock and equipment failure. It precisely states that if a practical approach to training is impossible due to lack of competence and potential danger, virtual reality environments can be used to acquire these competencies in a safe way. To create a trainer, procedures should be modeled in a way user decisions have to value for training. Instead of following a sequence of static steps, the model of any software needs to have options depending on the situation [16]. Many specific topics are addressed in the papers presented, such as maneuvering, exploration, etc.; however, few proposals focus on the personal safety of the user, who will eventually find himself personally in electrical substations, facing various lethal hazards.

The objective, given the above, is to develop a virtual reality trainer for electrical substations, specifically focused on electrical substation safety. The trainer will consist of multiple practices that will instruct the user in the safety measures to be followed when working in the facilities.

## 2 Materials and Methods

### 2.1 Methods

The method is focused on the development of an environment that the user could relate to a real workplace, based on different models already implemented [17, 18]. Prior to any development, the knowledge of experts, electrical engineers with extensive experience in the field, was sought to obtain a first approach to the environment to be developed as we can see in [19, 20] with related works. As a result of this approach, information was gathered on the environment and the development of common procedures in electrical substations, as well as safety issues corresponding to the environment. Subsequently, the main concept of the trainer, personal safety, was established.

The construction of different virtual environments began, making use of three-dimensional models for each object or room present in the trainer inspired also on models already implemented [21, 22]. In these virtual environments, the user can interact and perform a series of specific activities, which provide practical knowledge on safe practice.

After the development of the trainer, the proposal scenario has been completed, usability and heuristic tests were applied. A group of users did the virtual trainer exercises and then answered a survey used as a test. Tests were applied to 12 not experienced users and to 10 professional users. Both tests had questions designed according to specific aspects of the interaction between the user and the training tool. Each question used the

Likert scale, so the user selected one of five possible answers. Usability tests focus on how simple was for a user to understand what to do and do it. Heuristic tests evaluated how the user felt while using the task, in terms of satisfaction, motivation and how effective they could complete the objectives of the task. After development, the training tool was tested.

### 2.2 Materials

The scenarios are built in Unity 3D software. The models used for each scenario are three-dimensional models with textures and materials that provide a more realistic visualization. The 3DS Max software was used. Events and interactions are generated through C# programming, and through Unity 3D tools, codes are assigned to models to allow interactions with them. Two platforms were designed, one using a VR Oculus Quest SDK, to provide greater immersion of the user in the environment. The second platform was a desktop version that will be maintained for a wider scope. These tools were used in different projects using Virtual Reality and Virtual Environments [23, 24].

### 2.3 Definition of Practices

One of the most important parts when performing maneuvers in an electrical substation is the proper use of personal protective equipment, and it complies with certain specific characteristics to avoid any accident. The first practice consists of the trainee becoming familiar with and correctly identifying the personal protective equipment to be used, selecting the correct equipment from among various models of personal protective equipment.

For a second practice, in a virtual electrical substation influenced [25], the user will learn about safe distances to maintain from certain elements of a substation to minimize the risk of electrical shock [5]. The practice consists of a guided walk through the electrical substation, in which the user is instructed to go to certain areas but alerted when an area is considered unsafe [26]. The consequences of not respecting these distances, such as an electric shock, are also included.

## 3 Results

### 3.1 Equipment Selection

The exercise for the first practice consists of indicating to the engineers which equipment they need to go to an electrical substation. The flow of interactions for the first practice has also been designed, from the start and user prompts to the final evaluation of the user's performance. Each part considered can be found listed below. In the beginning, the trainee will appear in front of a screen, on which he/she will be able to read the objective of the practice. This screen will change, giving context and indications on how to proceed.

The user will be instructed, through the screen, to explore the warehouse in search of personal protective equipment for an electrical substation, which should consist of a helmet, goggles, gloves, and boots. The user will find on the shelf distinct types of equipment for each item, i.e., different helmets, glasses, gloves, and boots. The user must select from each item the correct one to be used in a substation. Once the user has finished selecting the equipment, he/she considers appropriate, he/she must trigger the evaluation through a control (Model that will serve as a trigger for the evaluation).

If all selections have been correct, positive feedback will be given to the trainee, otherwise negative feedback will be shown. The latter will consist of a message explaining that, due to a bad choice of equipment, a serious accident happened while in the electrical substation, so that a strong impression is generated in the user about what not using the right safety equipment could imply.

The scene simulates a warehouse. The user can navigate the room and choose the equipment he needs to go to a substation. The first exercise teaches the user what to do before he is in an electrical substation.

In this warehouse, you can find furniture, such as tables, shelves, racks, doors, lockers. Figure 1 shows a section of the virtual warehouse with tables and lockers. There is also a screen, on which objectives and instructions are indicated.

**Fig. 1.** Screen with objective of the first practice.

In Fig. 2 a view of the room is shown from another point. You can see the tables, some decorative elements, and in the background, you can see shelves containing some of the models of personal protective equipment. These elements are added to simulate a real room.

**Fig. 2.** View of the warehouse of the first practice.

The warehouse has different models of personal protective equipment in 4 categories: helmets, gloves, goggles, and boots. These are located on shelves and racks. As shown in Fig. 3 there are multiple models of gloves of different materials and textures, as well as multiple models of glasses. For the user to have information about the models, when he focuses on a model, a sign with information and a description of the object appears (Fig. 4).

**Fig. 3.** Different models of gloves and lenses.

**Fig. 4.** Information signs of the model displayed in front of the user. Spanish version.

As well as gloves and glasses, there are multiple models of safety helmets and boots. The different helmets represent different class helmets, in accordance with NOM-115-STPS-2009. Similarly, each model for the boots represents different degrees of protection or use. Information regarding these elements is given to the user through signage, as shown in Fig. 5 and Fig. 6.

**Fig. 5.** Information signage on class E helmet. Spanish version.

**Fig. 6.** Information signage of industrial cap boots. Spanish version.

### 3.2 Electrical Substation Tour

After finishing the first practice, the user could continue at the second practice, which is a tour in a small transformer substation (Fig. 7). In this exercise, the user walls across the substation get to know the components of an electrical substation.

**Fig. 7.** Multiple models for substations.

Although each substation is different, there are a series of common components. In a substation, the user can find elements as electrical towers, voltage, and current transformers, and switches, all modeled and included in the virtual environments. These elements are shown in Fig. 8. Entries and exits are also included in the scene (Fig. 9).

**Fig. 8.** Switchgear models, current and voltage transformers, and electrical towers.

**Fig. 9.** Portal to leave simulator.

### 3.3 Evaluation

This step represents the development of the trainer proposal, it has been determined to apply usability tests to electrical engineers with previous experience in interacting with electrical substations. In these tests, the users will be trying the interface (Fig. 10), the different environments, the interactions, and the practices in general. Once the test is finished, they will be asked to fill out a survey, based on which statistical analysis will be made. Comments will also be received regarding approaches to be considered in the simulator. The final objective of this analysis will be to identify areas of opportunity to improve and to have feedback to subsequently provide the trainer with an industrial approach.

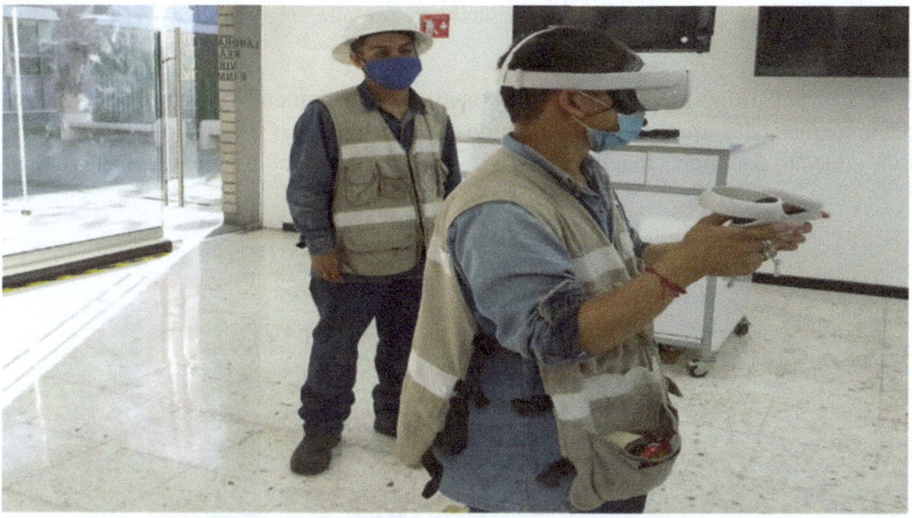

**Fig. 10.** Users testing the immersive environment.

To know the acceptance and experience that the simulator could give to the trainee, three tests using the Likert scale were conducted: usability test, composed of guidance quality and interaction experience, and a heuristic test that evaluated satisfaction, motivation, efficiency, and effectiveness.

For most users, interactions and guidance were acceptable or better. As shown in Fig. 11, no users evaluated interactions as poor or worse, and more than 60% of users answered the interactions were good. In comments the indicated the interface was easy to use. However, guidance can be improved. Around 30% users evaluated the guidance was not complete and require additional indications in how to do the exercises.

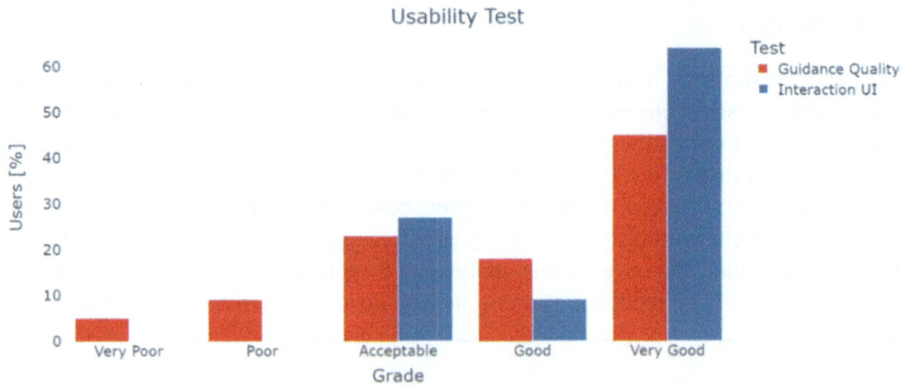


**Fig. 11.** Usability test results.

The virtual training environment generated user engagement. The heuristic test shows that more than 65% users graded motivation and satisfaction as very good (Fig. 12). Some users added to comments that the use of the virtual training environment was entertained.

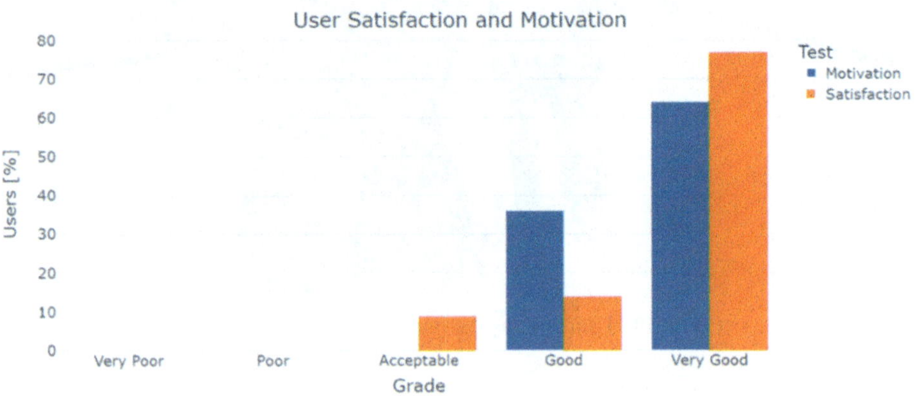


**Fig. 12.** Satisfaction and motivation results.

In the case efficiency questions, all users indicated that the game was good or better (Fig. 13). This indicated the users have no problem doing what the activity required. No bugs were reported in the comments.

**Fig. 13.** Efficiency and effectiveness results.

## 4 Discussion

The guidance could be improved. After efficiency and usability were evaluated, the virtual reality tool was easy to use and attractive for the users. Most users enjoyed the way they interact with virtual elements. However, a group of users felt lost in the second exercise. In the first one, there is a text with instructions while moving in the warehouse. On the other hand, the second exercise only indicated what to do at the start of the actions. For training, indications should be modified in the second exercise.

For a virtual reality simulation to keep improving, developers need to keep track of how technology changes. New technology generates new methodologies. In fields like manufacturing, there are constant proposals in how to improve the design of elements [27]. Electrical substation maintenance can also be affected by new proposals. Although the results of the current version of the virtual reality environment were considered effective by the users and are based on current methods, once there are new elements in the methods, the current version is no longer useful for learning. Changes in how to do a specific task should be also modified in a virtual environment to keep being useful for learners.

## 5 Conclusions

Virtual environments currently represent a solution to support the acquisition of practical experience, without the need to worry about risks to the health or life of the trainee, or to produce failures in the installations, caused by poor operation due to lack of experience. It has been decided to follow a safety approach, because, despite the previous existence of other works on electrical substations, few consider the user's safety in the first place. Ensuring the safety of personnel is fundamental, and many of the accidents that occur in these environments are due precisely to lack of training or experience. The exposed VR simulator has allowed the user to learn about safety procedures at an electrical substation, from the validation of personal protective equipment to interaction with the equipment in operation in electrical substations. As a result of a good level of acceptance, represented by the usability results, it is expected to continue the next practice that will be of significant help in generating practical learning. Improving the guidance support and the efficiency, incorporating the next scene will permit to know the level of learning using an evaluation process that will be store the trainees' actions into a database.

## References


1. Charissis, V., Ward, B.M., Naef, M., Rowley, D., Brady, L., Anderson, P.: An enquiry into VR interface design for medical training: VR augmented anatomy tutorials for breast cancer. In: McDowall, I.E., Dolinsky, M. (eds.) The Engineering Reality of Virtual Reality 2008, SPIE (2008)
2. Wang, Y., Guo, S., Li, Y., Tamiya, T., Song, Y.: Design and evaluation of safety operation VR training system for robotic catheter surgery. Med. Biol. Eng. Compu. **56**(1), 25–35 (2017). https://doi.org/10.1007/s11517-017-1666-2
3. Sacks, R., Perlman, A., Barak, R.: Construction safety training using immersive virtual reality. Constr. Manag. Econ. **31**(9), 1005–1017 (2013)
4. Greenberg, H.: Risk Assessment and Risk Management for the Chemical Process Industry. Wiley, New York (1991)
5. Xue, Z., Gao, W., Li, X., Wu, W.: Research on operation, repair and simulation training method of electrical equipment in substation based on virtual reality technology. J. Phys. Conf. Ser. **1345**(6), 062035 (2019)
6. Tang, Y.M., Ng, G.W.Y., Chia, N.H., So, E.H.K., Wu, C.H., Ip, W.H.: Application of virtual reality (VR) technology for medical practitioners in type and screen (t&s) training. J. Comput. Assist. Learn. **37**(2), 359–369 (2020)
7. Silva, A.C., Rosentino, A.J.P., Cardoso, A., Lamounier, E.A.: Use of virtual reality as learning object of electrical substations. In: Rocha, Á., Adeli, H., Dzemyda, G., Moreira, F., Correia, A.M.R. (eds.) WorldCIST 2021. AISC, vol. 1368, pp. 141–150. Springer, Cham (2021). https://doi.org/10.1007/978-3-030-72654-6_14
8. Hernández, Y., Pérez-Ramírez, M.: Virtual reality systems for training improvement in electrical distribution substations. In: 2016 IEEE 16th International Conference on Advanced Learning Technologies (ICALT), pp. 75–76 (2016)
9. Wang, W., Li, G.: Virtual reality in the substation training simulator. In: The 2010 14th International Conference on Computer Supported Cooperative Work in Design, pp. 438–443 (2010)
10. Hilfert, T., König, M.: Low-cost virtual reality environment for engineering and construction. Vis. Eng. **4**(1), 1–18 (2016). https://doi.org/10.1186/s40327-015-0031-5
11. Li, X., Yi, W., Chi, H.L., Wang, X., Chan, A.P.: A critical review of virtual and augmented reality (vr/ar) applications in construction safety. Autom. Constr. **86**, 150–162 (2018)
12. Kassem, M., Benomran, L., Teizer, J.: Virtual environments for safety learning in construction and engineering: seeking evidence and identifying gaps for future research. Vis. Eng. **5**(1), 1–15 (2017). https://doi.org/10.1186/s40327-017-0054-1
13. Ayala García, A., Galván Bobadilla, I., Arroyo Figueroa, G., Pérez Ramírez, M., Muñoz Román, J.: Virtual reality training system for maintenance and operation of high-voltage overhead power lines. Virtual Reality **20**(1), 27–40 (2016). https://doi.org/10.1007/s10055-015-0280-6
14. Nasyrov, R.R., Excell, P.S.: New approaches to training of power substation operators based on interactive virtual reality. In: Pammer-Schindler, V., Pérez-Sanagustín, M., Drachsler, H., Elferink, R., Scheffel, M. (eds.) EC-TEL 2018. LNCS, vol. 11082, pp. 551–555. Springer, Cham (2018). https://doi.org/10.1007/978-3-319-98572-5_43

15. Strizhichenko, A.V., Zenina, E.G., Kochergin, D.S., Efimov, V.V.: Development of a training complex of elements of power systems and metering of electric power with application of VR technologies. IOP Conf. Ser. Mater. Sci. Eng. **1035**(1), 012039 (2021)
16. Lazár, I., Husár, J.: Validation of the serviceability of the manufacturing system using simulation. J. Effic. Responsib. Educ. Sci. **5**(4), 252–261 (2012)
17. Carvalho, A., et al.: A methodology for reducing the time necessary to generate virtual electric substations. In: 2016 IEEE Virtual Reality (VR), IEEE (2016)
18. Dai, W., et al.: A model and application of collaborative simulation training system for substation based on virtual reality. In: Park, J.J., Loia, V., Yi, G., Sung, Y. (eds.) CUTE/CSA -2017. LNEE, vol. 474, pp. 610–616. Springer, Singapore (2018). https://doi.org/10.1007/978-981-10-7605-3_100
19. Barata, P.N.A., Filho, M.R., Nunes, M.V.A.: Consolidating learning in power systems: virtual reality applied to the study of the operation of electric power transformers. IEEE Trans. Educ. **58**(4), 255–261 (2015)
20. Cardoso, A., et al.: Vrcemig: A virtual reality system for real time control of electric substations. In: 2013 IEEE Virtual Reality (VR), pp. 165–166 (2013)
21. Fanqi, M., Yunqi, K.: An improved virtual reality engine for substation simulation. In: 2010 2nd International Conference on Future Computer and Communication, vol. 1, pp. V1–846–V1–849 (2010)
22. Liao, X., Niu, J., Wang, H., Du, B.: Research on virtual reality simulation training system of substation. In: 2017 International Conference on Virtual Reality and Visualization (ICVRV), pp. 413–414 (2017)
23. Mattioli, L., Cardoso, A., Lamounier, E.A., do Prado, P.: Semi-automatic generation of virtual reality environments for electric power substations. In: Rocha, A., Correia, A.M., Costanzo, S., Reis, L.P. (eds.) New Contributions in Information Systems and Technologies. AISC, vol. 353, pp. 833–842. Springer, Cham (2015). https://doi.org/10.1007/978-3-319-16486-1_83
24. Ribeiro, T.R., et al.: AGITO: Virtual reality environment for power systems substations operators training. In: De Paolis, L.T., Mongelli, A. (eds.) AVR 2014. LNCS, vol. 8853, pp. 113–123. Springer, Cham (2014). https://doi.org/10.1007/978-3-319-13969-2_9
25. Liao, X., Wang, H., Niu, J., Xiao, J., Liu, C.: Research on simulation training system of immersive substation based on virtual reality. IOP Conf. Ser. Mater. Sci. Eng. **486**(1), 012112 (2019)
26. Chang, Z., Fang, Y., Zhang, Y., Hu, C.: A training simulation system for substation equipments maintenance. In: 2010 International Conference on Machine Vision and Human-machine Interface, pp. 572–575 (2010)
27. Kaščak, J., et al.: Design of an atypical construction of equipment for additive manufacturing with a conceptual solution of a printhead intended for the use of recycled plastic materials. Appl. Sci. **11**(7), 2928 (2021)

# Implementation of Virtual Reality for Training in Education and Industry: A Forklift Case

Leticia Neira-Tovar(✉), Edson E. Castañeda-Mancillas, Guillermo Rossano, Jorge Cupich, and Hector Cruz

Universidad Autonoma de Nuevo Leon, Pedro de Alba S/N, Niños Heroes, San Nicolas de los Garza, Nuevo Leon, Mexico
leticia.neiratv@uanl.edu.mx

**Abstract.** Virtual reality industry training environments proposals are more common in recent years. Due to the versatility of a virtual environment, the interaction with specialized and expensive equipment could be simulated. This type of solution can be applied to many fields and specific processes. Lots of proposals of virtual reality usage have similarities in how they aim for learning to be meaningful for a user. In any teaching process, either in industry or school, there are learning concepts and abilities related to some specific tasks. By a review of proposals in several fields, this work identifies concepts and characteristics that should be taken into consideration when creating a training virtual reality simulator, and then those concepts are applied to a real case. The case implemented is virtual reality for forklift operator training. The review showed the advantages of applications in different fields and highlighted two key concepts to apply in a proposal: fidelity and flexibility. Based on fidelity and flexibility, there are modifications of interactions and instructions. Those concepts were applied to the virtual environment for forklift operation, changing visual elements in the 3D environment and modifying interactions by adding a driving wheel controller. After the development stage, usability and functionality test were applied to a group of students to evaluate user experience and guidance. Tests also allowed to identify what to improve in the virtual environment. These types of tools should stick to reality as much as possible, but extra objects are wanted when those elements improve the learning process.

**Keywords:** Virtual reality environments · Industry training · Forklift training

## 1 Introduction

In the last decades, there have been multiple proposals of virtual reality environments for education. Specifically, since 2013 there has been a greater amount of virtual reality projects due to the higher availability of hardware (HMD devices). These proposals show an environment focused on a specific field, such as medicine, school education, industrial plant operation, etc. Many of those investigations compare traditional teaching (books, class explanations, and real practice) with these types of solutions. Virtual reality can be applied in different stages of a process [1]. One of the main scopes of those applications


J. Trojanowska et al. (Eds.): MANUFACTURING 2022, LNME, pp. 292–305, 2022.
https://doi.org/10.1007/978-3-030-99310-8_23

is related to the teaching process. In industry, this process occurs in training, when a user learns needed skills. The person in charge of a task must have the knowledge and skills required to perform it. A virtual reality application can guide users or help them have a better understanding of a specific task.

Usually, a virtual reality environment proposal includes a first evaluation of implementation. The general aim of these solutions is to be proven useful for helping in the learning process of the task they are designed for. This evaluation can focus on how the user feels after using the training environments, and a skill evaluation related to the content of the environment [1–3]. Results show if there is a similar o greater level of comprehension from the users related to the task he learned about. When the learning process is good enough, it can be considered that the virtual reality environment is effective for teaching those tasks.

Although each proposal of virtual reality training environment has its own goal and approach, there are some general good practices or key concepts that would be useful for new projects. This paper analyses a set of virtual environments and related reviews aiming to identify advantages of this type of solution, as well as characteristics an effective design should consider. Those characteristics are later applied to a real case of a virtual reality training environment: forklift operation. Good practices from previous work aim to improve interaction in terms of guidance and ease of use of the interface. Once the environment is enhanced, a group of users tests the tool to evaluate the user experience.

## 2 Method

The project has three main stages: related work analysis, development, and testing. A review of related work help identifying characteristics of different proposal for virtual reality training environments in industry. Once there is a definition of good practices, those practices are applied to the forklift operation training virtual environment. To evaluate new features of the virtual reality environments, 20 users interacted with the tool and answered a survey in which they indicated how easy or hard was to use the tool.

### 2.1 Related Work Analysis

The review process is shown in Fig. 1. In the first step, four different database research engines were considered: Web of Science, Springer Link, IEEE Xplore, ACM Digital Library. Most of them are multidisciplinary, so the research would get information from multiple fields. In the case of IEEE Xplore, most papers are related to engineering, which is an area highly related to industrial training both in general and to specific used considered for this paper. In all engines, the search included the next list of keywords: virtual reality (AND) training, virtual reality (AND) education, and virtual reality (AND) forklift.

**Fig. 1.** Research method applied in this overview.

Part of the aim of this study is to focus on recent applications. Because of it, the literature research was limited to papers from the previous decade (2010–2020). Another limitation in the research was to consider only full-text papers, so this review can further evaluate how each proposal was developed.

For the analysis of information, a multi-step review process was applied. From several collected papers, there was an initial review of abstracts and conclusions. This initial review aimed to define if the paper is related to training and is related to this research, or the virtual reality environment from that paper considered other types of application, such as monitoring. From papers defined as useful in this first review, a second review of the entire paper. In total, 25 different papers were fully analyzed with the second review of this method.

### 2.2 Development of Interface

To apply the identified characteristics, a functional project was selected. First, interactions are modified to be closer to reality. In the selected virtual reality environment, the objective was to drive a forklift across a warehouse, move in reverse and stop the forklift. Hardware was updated to improve the simulation of a real forklift. After changing interactions, instructions were modified. Additional instructions were added in the form of audios, to simulate someone was explaining in the room. Text instructions are also included at the beginning of exercises. Most elements that do not exist in real facilities are removed, such as 3D messages.

### 2.3 Testing

The modified application was tested by a group of 20 students. Each user of the group used the virtual environment, only guided by instructions inside the interface. At the end of the exercise, users filled a survey that contained usability and functionality tests. Those tests include questions related to provided instructions and how easy was to use the interface to accomplish the goal of the training environment. To have a numeric evaluation, the answers options of the survey were either yes/no or use the Likert scale. There was also an additional question to add a custom opinion on how each user felt and compile recommendations of how to keep improving the tool.

# 3 Characteristics of Virtual Reality Training Applications

## 3.1 Virtual Reality Applications for Training

The range of virtual reality applications as a training tool is wide. Table 1 shows some examples of applications that have already been proposed in the literature. Applications can be identified both in medicine and in the manufacturing industry. Similarities between those areas are they both involve complex tasks and complex equipment. Not only industry, but universities already look at how this kind of technology would improve the performance of learners. A tool related to experiment with electrical equipment for engineering students would allow them to assemble and disassemble complex equipment, which is an interactive learning experience they are not usually allowed to do [3]. Many fields are experimenting on how virtual reality can improve the performance of a user in a certain task.

**Table 1.** Virtual reality training applications examples.

| Field | Application examples |
|---|---|
| Industrial plant | Power grid operation [4, 5], Forklift driving [6, 7], Manufacturing [8, 9], Smart factory [1] |
| Construction | Driving simulator [10], Safety [11] |
| Medicine | Anatomy [12], Surgery [13] |
| Other approaches | Railways [14], Disaster response [15], Familiarization with equipment [3] |

One of the reasons for this type of implementation is accident prevention. In a virtual environment, you can create hazardous situations for the user, and he will not be in real danger. A constant aim of fields as constructions is the reduce as much as possible the number of accidents at work [11]. There cannot be driving accidents in an industrial plant if the user only drives a virtual vehicle in a safe space. The user can learn from the implications of different accidents and learn how to interact without the need of a previous real experience [6, 7]. As for a forklift operator, hazard management is based on a constant awareness of other forklift operators, any user walking, and unsafely placed items in the warehouse. In a virtual environment the user could learn these specific skills [16]. A virtual reality environment is a safe environment to learn skills.

Virtual reality is also used to teach more deeply the task the user needs to learn. A clear example can be identified in medicine, where the traditional approach a traditional way to teach medical producers was through videos. By using a virtual reality simulation, a student can experiment with complex procedures such a surgery and learn by doing [13, 17]. At the same time, in industry, it is possible to identify that the initial lack of interaction with control systems and complex machines is related to the cost of that equipment. Virtual reality provides an environment in which interaction is close to the real one, without risking equipment. Those types of interactions are meaningful in a learning process [8]. In a virtual reality environment, the user can see things he would not see in the real environment.

### 3.2 Virtual Reality Training Advantages

There are several advantages a business considers when implementing a virtual reality proposal. Although trainee learning would be an important factor in applying this technology to a specific process, research shows other factors which potential value can be of great interest for the industry. The two other most significant advantages are increasing safety and minimizing the costs of training. All these factors are considered when defining the utility of a virtual reality application proposal.

The most relevant advantage of a virtual reality simulator is safety. It is said that that learners are safe when using a virtual reality application. When the trainee has just started to learn how to interact with equipment, it would be risky to be on the operational side of an industrial plant. One of the main causes of lack of safety in industrial procedures is lack of training. When training is not enough, it is hard to identify risky situations, and more accidents a generated [18]. The maintenance of a power station involves tons of safety procedures related to switching on and off equipment, and by not being careful both equipment and the user are at risk [5]. There are cases in which safety is so important that the study proposes virtual remote operation environments to avoid any danger from the workplace [14]. When the users have not enough experience related to the process, is a risk for them to be near the real process, and a virtual reality environment in a safe place becomes an interesting solution for training.

In addition, safety is improved also for real life produces. If the content design of a virtual reality training environment includes safety protocols of the task, the user will already have experiences about what to do to keep him and his colleagues safe. This is an advantage of the learning by interaction approach of virtual environments. However, the safety procedures indicated in a real environment will only be as good that the real procedures they are based on. Therefore, validation of those procedures in real life is needed before teaching them to users [19]. In a proposal focused on disaster response [15], hazards are simulated so the user can train and learn how to respond to those events in real life. If some of the considered disasters occur, the user would have already practiced an effective way of how to deal with the problem. Due to the learning by interaction approach, safety is improved not only by keeping the users in virtual reality environments where there is no harm possibility, but by facilitating a safe interaction with hazards.

At the same time, most literature from this review indicated immersion as an important advantage in the learning process. The increased immersion in recent years seems to be compatible with different learning approaches, especially for activities that involve the user to understand their surroundings and specific interaction with equipment [12, 20]. Virtual reality projects enhance the learning outcome in tasks that require visualization or design of objects [2]. By including simulations of surgeries, a learner has a better view of what steps he should follow in the process and a visual reference of how to do it [13]. If the facilities in a virtual environment are similar to the real ones, the user will already be familiar with the location of the different elements of the plant. The trainee gets a prior knowledge of the real process he could apply once training ends.

In the case of industry, finances benefit from solutions with virtual reality. Although a virtual reality environment will require a higher initial investment than alternatives, the benefits shown later as no needing new equipment in the future are an advantage in expense. At the same time, the risk is reduced for an improvement of experience in users, the costs of dealing with accidents also decrease. There is no need to use equipment from production and production can go on normally [1, 21]. Specialized equipment used for practical exercises is expensive, so a trainee is not usually allowed to use it [3]. Although the user will need certain guidance from an expert as an instructor this guidance will also be reduced compared to a classic teaching method, so the instructor can carry his operative functions as normal. The expert will invest a higher time in production than in training.

### 3.3 Characteristics of Efficient Virtual Reality Environments

Virtual reality projects are often described in terms of two concepts: immersion and presence. Those concepts relate to the sense of presence and interaction of the user when using the training environment. From an immersion perspective, this concept indicates how deep a user is inside the interface: Is the user wearing an HMD or is he looking at the virtual environment through a computer screen? On the other hand, presence relates to how similar the user behavior is between the real and virtual environment [9, 20]. How interaction is produced relates to the complexity of the model. Some studies analyze that virtual environments should incorporate real data [21]. That way the user will have a chance to interact safely with complex procedures directly related to a case that happens in production. Many of the projects mentioned before in the paper require a range of motor skills, therefore for the user to feel like he is performing the task is necessary for effective learning. Immersion and presence can help the learning process in virtual reality training environments.

Although immersion and presence are important characteristics of a virtual reality design, more characteristics should be considered when creating a virtual environment: fidelity and flexibility. These concepts are related to the similarities between the real environment and a virtual one. After training, a user must know how to perform a task in the real facilities of an industrial plant. These two concepts will be further explored next.

A virtual reality training environment design needs fidelity to the real environment. This means the virtual reality environment is based on real facilities. Fidelity can be analyzed from different perspectives, such as hardware, 3D designs, and content. Hardware selection must be based on how the user interacts with the real components. If the environment is related to driving, a steering wheel controller is more like the real interaction that a keyboard interface [6]. Also, when designing 3D environment fidelity must be considered. Operators need to know how much distance there is between the different places of a factory involve in the process [5]. The design of 3D elements should be based as well on real objects. A virtual forklift should have the same type of movement in the real fabric as in the training simulation: forks move up and down but not sideways. By that consideration, model design should base on the real elements that the task involves [1, 22]. By the consideration of this concept, there is a guarantee the trainee will already be familiarized with the real facilities and processes.

Virtual reality design must also be flexible. Industry changes over time, which means the virtual reality training environments must be capable to adapts to those changes and the different situations can happen in the real process. Virtual reality allows scalable designs: if modifications are needed, there is no need to replace equipment [1]. As seen in the Table 1, there are multiple proposals focused on the driving training industry. In this field of application, the user needs to be constantly aware of their surroundings, which involves training for a long time [10]. In traditional methods, machinery would be required, but with a virtual reality application, there is no need to use real machines in training. If a step or a tool of the process changes while the user trains, the environment can be adapted to the new elements without getting new expensive equipment. If the elements of an industrial plant are reordered, these changes can be adapted in the environment so is always up to date. On the other hand, virtual reality must be flexible to hazardous situations. Hazards can occur at any time. These are abnormal situations, to which a virtual environment should prepare the users for [19]. A linear script of a virtual environment helps to user to know how to proceed in an ideal scenario but, accidents are always a possibility in real life. This type of solution should adapt to get the user in virtual hazard situations so he can learn how to react to the risk. Virtual reality allows flexibility in terms of adaptation to changes and the interaction with hazard situations.

In content design, these two concepts are integrated as one. On one hand, the content must be based on the real task. Not only the way things are shown but the way things are told matter in a virtual reality application. The design of the content of a training environment involves a definition of what the user can do around the environment, what can happen after his actions, and what he is supposed to do. An effective content design allows the user to identify situations that in real life would be harder to understand [23]. Applying real data, such as variables measured by sensors, or experiences from previous operators helps to improve a virtual reality design. By incorporating this information in the design of the objects and the configuration of the environment would be useful for learning when it is time to interact in a real environment. The users get a feeling they already did the task [23]. However, content must be flexible. To apply the fidelity concept does not forbid adding nonreal elements to a virtual element. In the analyzed proposals, some researchers add audio and/or text information if they consider that would improve learning from the user [9]. Fidelity and flexibility should no be limitations in a virtual reality environment design, virtual reality is supposed to expand the capabilities of real training.

## 4 Results

### 4.1 Improvements of the Forklift Operator Training Environment

A virtual reality forklift driving training system would need some of the features to enhance the learning process. Some of those features are indicated in Fig. 2. On one side, there are elements from the real task that should be recreated in the virtual environment for the user to be meaningful. The virtual environment would require extracting features from the ideal operation case. This involves digitizing machines and the rest of the facilities into 3D models to include in a virtual environment. On one hand, conceptualization should be based on the real procedures the industry follows. On the other hand, the virtual environment should also create risky situations the user may find in real life. Fidelity and flexibility can be applied to the virtual forklift training system.

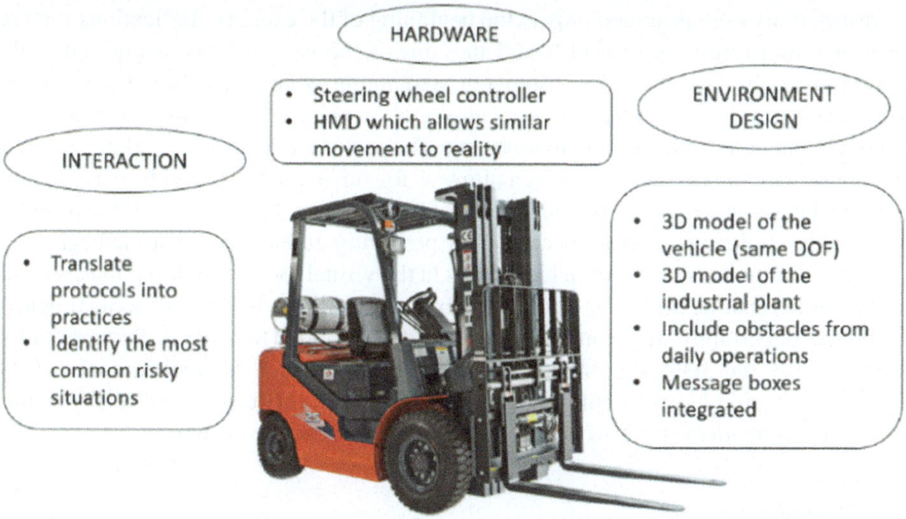


**Fig. 2.** Fidelity and flexibility in a virtual training environment for forklift operation.

Fidelity was improved in hardware. As indicated before, for the user to relate the knowledge from virtual environments to real life, controllers must be similar to what he is going to use in the real industrial plant. In a first approach, the virtual training interface could be controlled either by a gamepad controller staring at a screen or by an HMD headset with the controllers the headset included. The first option made the training tool portable but was not immersive enough for the learning process. By the use of HMD, immersion is improved because the user will be able to turn his head as he would normally do. The user is deep in the virtual reality environment. However, in terms of presence, it is not meaningful to replace driving movements with a joystick controller. To increase fidelity, those joystick controllers were replaced was replaced with a steering wheel controller for the user to feel like driving (Fig. 3). Controllers should be similar to what the user is going to use in the real task.

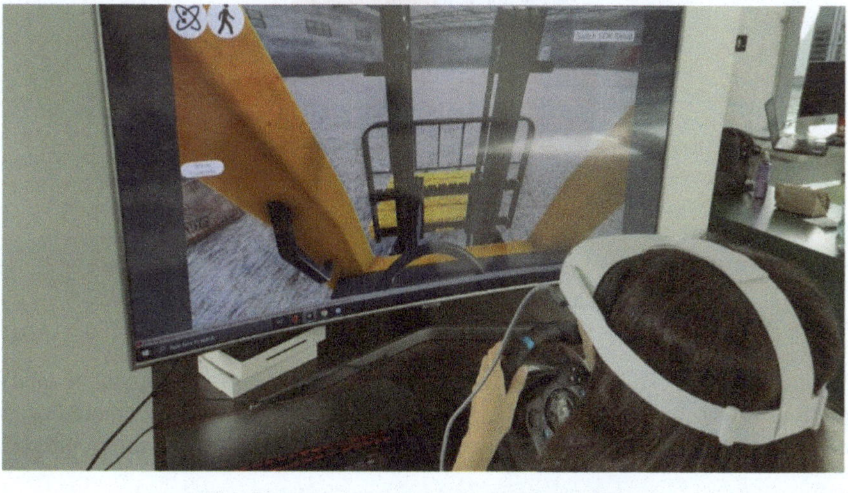

**Fig. 3.** HMD in virtual training environment.

Instructions were provided only at the beginning of the exercise. Indications showed the user how to move the forklift and the aim of the exercise. By using audio, the environment simulated there was someone explaining what to do. For desktop view, there was also a menu where the user can see all available actions as a new window, without being part of the virtual environment. The use of elements not available in a real warehouse was low because the design aim was for surroundings to be close as possible to the real case. However, at the beginning of the exercise, there is also 3D text on the virtual room indicating available actions. By providing guidance only at the beginning, it is aimed for the user to focus on his actions in the virtual tool when doing the exercise.

Flexibility can be added by adding uncertainty to all the features described below. To add this uncertainty, accidents and risk should be included at nonspecific times of the training. Also, the 3D design should consider obstacles, such as blocked zones of the facilities, he would have to deal with when using a real forklift. A virtual reality trainer can combine the ideal scenario and its flaws to get an effective design.

### 4.2 Evaluation

Questions in surveys were organized by topic. The two main topics were instructions and user experience. All results of Likert answers for a topic were sum by category. This sum results in an overall summary of how good the topic was. Percentages were obtained from the categories of the answers to express how well or poorly did the improvements enhance user interactions and learning. User experience and provided guidance were evaluated tests.

By including all instructions at the beginning of the exercise, the user understating of what he must do in the virtual environment decreases. Figure 4 shows that only 20% of users in test group considered provided instructions were complete.

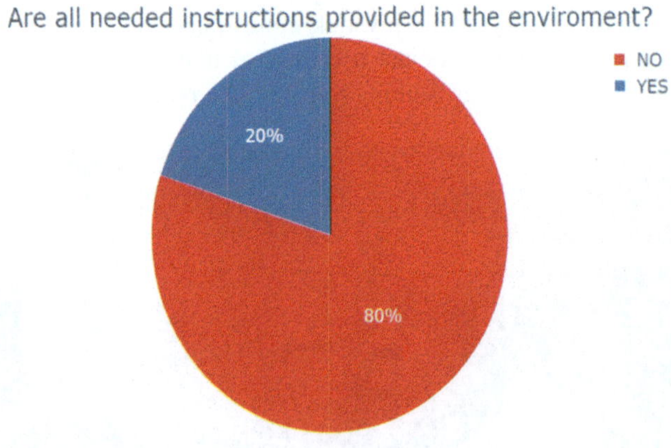


**Fig. 4.** Results for instruction availability.

When grading those instructions, as shown in Fig. 5 most user graded the interface instructions lower than acceptable. In a comment section of the surveys more than 50% of users indicated that they understood how to move in the virtual environment and available actions, but specific objectives of each exercise were not clear enough. In this application, instructions needs improvements.

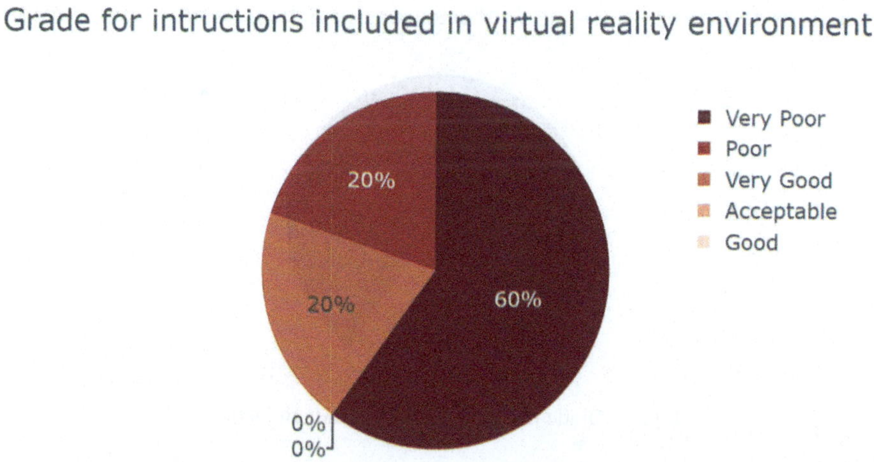


**Fig. 5.** Grade for included instructions.

Interaction results shows user felt interactions in the VR interface were better than instructions. More than half the users in test considered all actions were easy to perform (Fig. 6). Including the driving wheel help the users feel like they were driving a vehicle.

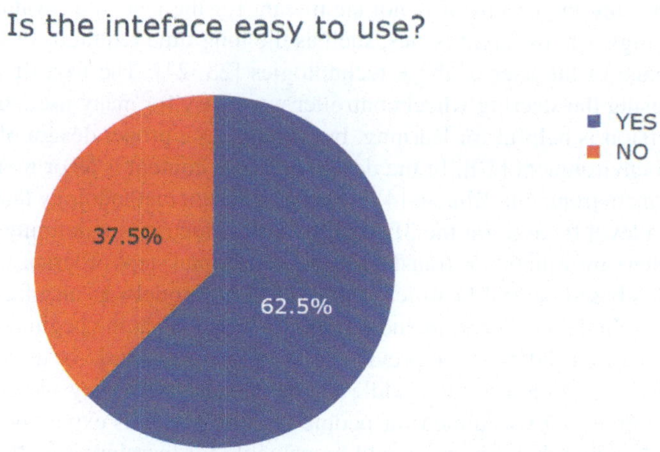


**Fig. 6.** Results for interactions in VR environment.

Grade for interactions and user experience, shown in Fig. 7, highlight some missing elements. The instructions issue caused in some cases a user did not know what to do next. This caused discomfort to user. Being that the case, user experience can be entirely good. The addition of fidelity elements make training feel real, but problems in guidance can generate bad user experience.

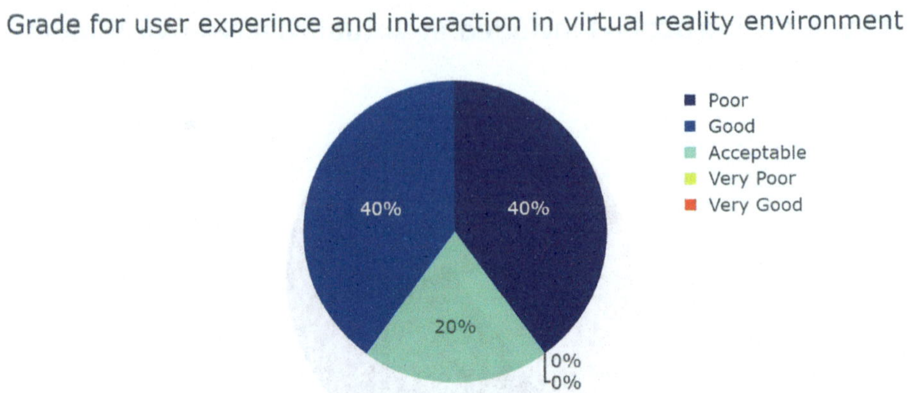


**Fig. 7.** Grade for user experience and interactions.

## 5 Discussion

Although immersion and the presence of virtual reality applications can help in the learning process of motor skills, there are still challenges in terms of implementation. In some of the studies, there is not a significant improvement in learning outcomes. Some studies identify the improvement is not significant for the task after evaluation. Some other studies highlight different issues, such as the long time required for training, and the inexperience of the user of those technologies [23, 24]. The forklift training tool interactions using the steering wheel controller were easy for many users of the testing group. Immersion is helpful for learning, but it requires a proper design of the content of the virtual environment [20]. In the design process, content is as or more important than the visual environment. Blümel [4] proposes a design methodology that considered three levels: a level focused on the 3D design, a level focused on defining the content, and a level where information is translated into educational steps or lessons. The forklift warehouse was based on real facilities, and the forklift model was also based on a real sample of the vehicle. However, including only instructions at the beginning generated confusion as the user did not have previous experience in the task. In the tests, 60% of users indicated they forgot at the middle of the exercise what they were supposed to do. As the training tool was aimed for people with no previous experience, additional guidance while doing the exercise would be valuable for learning. An effective virtual reality platform design is not only related to the visual design or the interaction, but the content design is also an important factor in improving the learning process.

Fidelity would be also helpful in learning improving, but here a time definition for development is needed. In one case, the 3D design aimed to imitate as best as possible a real case of a power station. However, this proposal identifies they require more than 1000 h for getting the proposal ready. This type of imitation can take a long time [5].

An imitation as faithful as possible would always require a higher amount of time, from the start of the project there must be a defined end time. If someone proposes a faithful design, there should be a balance between the development needs and the work capacity a group has. 3D is an important feature but is neither the only one nor the most important. All other stages of design need time for an effective practice to be designed. Although the objective of the forklift training tool was to be close to reality, the development stage focused on the content of the exercise and interactions. A project should focus on being as close as a real environment depending on how much time the developers have for implementing a proposal.

Several concepts and advantages of this type of simulation have been defined in this work. However, the efficiency of each virtual environment relies on the evaluation the study applies. Even if evaluation is a quiz after using the interface, designing the evaluation should consider what is important to evaluate in that specific application, how can that characteristic be evaluated, and a proper number of users to test the information [25]. This type of test allows research to verify how useful a virtual reality proposal can be, and how the proposal con improve. For these results to be significant, an evaluation must be designed based on the most important features of the task.

## 6 Conclusion

Fidelity and flexibility would help a user have a better understanding of the simulated task. On one hand, fidelity would help the user relate the elements in a virtual environment with the real ones and know how the elements are set up in the real facilities of an industrial plant, and how and to where he should go to get the task done. In the additional comments of users of the forklift training tool, the users indicated they liked how real the environment feel and that they got the feeling they were driving a vehicle. The user experience was good for forklift driving due to interaction. However, to be faithful to a real environment does not mean that the real environment is a limit. Instructions failed to provide proper guidance to most of the users of the testing group. When instructions were only provided at the beginning, users with no previous experience failed to remember what the goal of the task was. At the beginning of learning a new skill, supervision is needed when doing the task. The virtual environment should always add information and scenarios if they provide important information in the learning process. On the other hand, flexibility allows the adaptation to change at minimal costs and to prepare the user for risky situations. A flexible virtual reality environment should introduce possible hazards of processes so the user will know how to proceed when this type of situation happens in real life. For future works, flexibility and fidelity are concepts that will be considered for a model to design a virtual reality application.

Another important factor to keep improving in a virtual simulation is user concentration. Concentration is very important in any type of work, and vital when the activity impacts directly to the physical integrity of anyone, directly or indirectly. The use of a forklift is no exception in terms of how important it is to be focused when operating it, so it is though it was important to consider some technique to help us in this regard. Approaches that improve the concentration of the user in the specific activity would facilitate meaningful knowledge.

Virtual reality has a wide range of possible applications related to training environments. Many processes require complex knowledge of procedures, such as a power station operation, driving trainers for factories, manufacturing lines. Procedures that involve complex equipment or involve risky situations could benefit from a virtual safe training environment. When driving a real forklift, there is a risk of hitting something or someone and the forklift operator must be aware of his surroundings. In virtual reality, there is no risk for trainees or the equipment. This is useful for people with no previous experience. Training systems could benefit from virtual reality environments.

## References

1. Żywicki, K., Zawadzki, P., Górski, F.: Virtual reality production training system in the scope of intelligent factory. In: Burduk, A., Mazurkiewicz, D. (eds.) ISPEM 2017. AISC, vol. 637, pp. 450–458. Springer, Cham (2018). https://doi.org/10.1007/978-3-319-64465-3_43
2. Halabi, O.: Immersive virtual reality to enforce teaching in engineering education. Multimedia Tools Appl. **79**(3–4), 2987–3004 (2019)
3. Kamińska, D.: Virtual reality-based training: case study in mechatronics. Technol. Knowl. Learn. 1–17 (2020)
4. Blümel, E., Haase, T.: Virtual reality platforms for education and training in industry. In: Grundspenkis, J., Kirikova, M., Manolopoulos, Y., Novickis, L. (eds.) ADBIS 2009. LNCS, vol. 5968, pp. 1–7. Springer, Heidelberg (2010). https://doi.org/10.1007/978-3-642-12082-4_1
5. Nasyrov, R.R., Excell, P.S.: Creation of interactive virtual reality scenarios as a training and education tool. In: Earnshaw, R., Liggett, S., Excell, P., Thalmann, D. (eds.) Technology, Design and the Arts---Opportunities and Challenges. SSCC, pp. 353–369. Springer, Cham (2020). https://doi.org/10.1007/978-3-030-42097-0_19
6. Yuen, K.K., Choi, S.H., Yang, X.B.: A full-immersive CAVE-based VR simulation system of forklift truck operations for safety training. Comput-Aided Des. Appl. **7**(2), 235–245 (2010)
7. Choi, M., Ahn, S., Seo, J.O.: VR-Based investigation of forklift operator situation awareness for preventing collision accidents. Accid. Anal Prev. **136**, 105404 (2020)
8. Novak-Marcincin, J., Janak, M., Barna, J., Novakova-Marcincinova, L.: Application of virtual and augmented reality technology in education of manufacturing engineers. In: Rocha, Á., Correia, A.M., Tan, F.B., Stroetmann, K.A. (eds.) New Perspectives in Information Systems and Technologies, Volume 2. AISC, vol. 276, pp. 439–446. Springer, Cham (2014). https://doi.org/10.1007/978-3-319-05948-8_42
9. Matsas, E., Batras, D., Vosniakos, G.-C.: Beware of the robot: a highly interactive and immersive virtual reality training application in robotic manufacturing systems. In: Emmanouilidis, C., Taisch, M., Kiritsis, D. (eds.) APMS 2012. IAICT, vol. 397, pp. 606–613. Springer, Heidelberg (2013). https://doi.org/10.1007/978-3-642-40352-1_76
10. Wu, Y.-F., Zhang, Y., Shen, J., Peng, T.: The virtual reality applied in construction machinery industry. In: Shumaker, R. (ed.) VAMR 2013. LNCS, vol. 8022, pp. 340–349. Springer, Heidelberg (2013). https://doi.org/10.1007/978-3-642-39420-1_36
11. Li, R.Y.M., Li, R.Y.M.: Virtual reality and construction safety. In: An Economic Analysis on Automated Construction Safety, pp. 117–136. Springer Singapore (2018). https://doi.org/10.1007/978-981-10-5771-7_6
12. Chen, S., et al.: Can virtual reality improve traditional anatomy education programmes? A mixed-methods study on the use of a 3D skull model. BMC Med. Educ. **20**(1), 1–10 (2020)

13. Izard, S.G., Juanes, J.A., García Peñalvo, F.J., Estella, J.M.G., Ledesma, M.J.S., Ruisoto, P.: Virtual reality as an educational and training tool for medicine. J. Med. Syst. **42**(3), 1–5 (2018). https://doi.org/10.1007/s10916-018-0900-2
14. Qi-Zhong, L.: Application and research of virtual reality technique on the safety of transportation of dangerous goods. In: Zhou, Q. (ed.) ICTMF 2011. CCIS, vol. 164, pp. 84–91. Springer, Heidelberg (2011). https://doi.org/10.1007/978-3-642-24999-0_12
15. Nguyen, V.T., Jung, K., Dang, T.: Vrescuer: a virtual reality application for disaster response training. In: Proceedings - 2019 IEEE International Conference on Artificial Intelligence and Virtual Reality, AIVR 2019, pp. 199–202. Institute of Electrical and Electronics Engineers Inc. (2019)
16. Lehtonen, E., et al.: Learning game for improving forklift drivers' safety awareness. Cogn. Technol. Work **23**(4), 743–753 (2020)
17. Shi, Y.-L.: Application of virtual reality technology in medical education. In: Li, S., Jin, Q., Jiang, X., Park, J.J.H. (eds.) Frontier and Future Development of Information Technology in Medicine and Education. LNEE, vol. 269, pp. 467–476. Springer, Dordrecht (2014). https://doi.org/10.1007/978-94-007-7618-0_46
18. Li, X., Yi, W., Chi, H.L., Wang, X., Chan, A.P.: A critical review of virtual and augmented reality (VR/AR) applications in construction safety. Autom. Constr. **86**, 150–162 (2018)
19. Patle, D.S., Manca, D., Nazir, S., Sharma, S.: Operator training simulators in virtual reality environment for process operators: a review. Virtual Reality **23**(3), 293–311 (2018). https://doi.org/10.1007/s10055-018-0354-3
20. Jensen, L., Konradsen, F.: A review of the use of virtual reality head-mounted displays in education and training. Educ. Inf. Technol. **23**(4), 1515–1529 (2017). https://doi.org/10.1007/s10639-017-9676-0
21. Kózlak, M., Kurzeja, A., Nawrat, A.: Virtual reality technology for military and industry training programs. Studies in Computational Intelligence **481**, 327–334 (2013)
22. Juang, J.R., Hung, W.H., Kang, S.C.: Using game engines for physics-based simulations - a forklift. Electron. J. Inf. Technol. Constr. **16**, 3–22 (2011)
23. Berg, L.P., Vance, J.M.: Industry use of virtual reality in product design and manufacturing: a survey. Virtual Reality **21**(1), 1–17 (2016). https://doi.org/10.1007/s10055-016-0293-9
24. Gavish, N., et al.: Evaluating virtual reality and augmented reality training for industrial maintenance and assembly tasks. Interact. Learn. Environ. **23**(6), 778–798 (2015)
25. Checa, D., Bustillo, A.: A review of immersive virtual reality serious games to enhance learning and training. Multimedia Tools Appl. **79**(9–10), 5501–5527 (2019). https://doi.org/10.1007/s11042-019-08348-9

# Mixed Reality Training in Electrical Equipment Operating Procedures

Paweł Buń[1(✉)], Jakub Gapsa[1], Jozef Husár[2], and Jakub Kaščak[2]

[1] Faculty of Mechanical Engineering, Poznan University of Technology, Piotrowo 3, 60-965 Poznań, Poland
pawel.bun@put.poznan.pl

[2] Faculty of Manufacturing Technologies with a Seat in Prešov, Technical University of Košice, Bayerova 1, 080 01 Prešov, Slovakia

**Abstract.** The implementation of Industry 4.0 in the enterprise forces increased digitization, production flexibility, improvement of employees' competences and the integration of employees and IT systems. For this purpose, cutting-edge IT systems and solutions such as Virtual, Augmented and Mixed Reality (VR/AR/MR) are increasingly used. These technologies can be used, in training processes of ma-chines and equipment operating procedures. The paper is a case study on the possibility of using MR applications in training in the use of electrical devices (ARS 2 Pro switch disconnector). The aim of the work is to compare MR and standard training and evaluate their effectiveness in this type of training. The paper briefly describes developed applications and preliminary test-results.

A group of 20 people in the age group 20 to 40. The participants were divided in-to 2 subgroups in order to properly analyze the effectiveness of individual training tools.

**Keywords:** Mixed reality · Industrial training · Operating procedures

## 1 Introduction

The use of Virtual, Augmented and Mixed Reality (VR/AR/MR) in education, industrial training and as remote assistance tool is becoming common. Integrating those systems with company Information Technology (IT) systems such as decision support systems, logistic [1], production planning [2] and design systems [3] in order to support both factory workers and engineers at the workplace [4] in Smart Factories [5] is one of key features of Industry 4.0.

The above-mentioned technologies are described as Extended Reality (XR) [6] as distinction between them is often unclear. The term VR describes a real-time computer generated, immersive virtual environment with which user can interact using various interaction devices such as controllers [7], hand and head tracking systems. In most of VR systems user is completely isolated form real environment. In AR, the real world is extended (augmented) by the virtual contents displayed by devices such as tablets, smartphones [8] or smart glasses. The user has the possibility to interact with the real


J. Trojanowska et al. (Eds.): MANUFACTURING 2022, LNME, pp. 306–316, 2022.
https://doi.org/10.1007/978-3-030-99310-8_24

world as well as with the virtual world and to manipulate the virtual con-tents [9]. MR can be defined as an immersive fusion of the virtual and real world in which both user and real content can interact with virtual environment [10]. Some of the publications introduce the concept of reality-virtuality-continuum [11] that goes from real environment to virtual environment (virtual reality) which with an increasing share of virtual elements.

This paper discusses possibility of using XR technologies to increase effectiveness of training in electrical equipment operating procedures. The work undertaken was part of a project to develop a flexible training system for employees dealing with maintenance and service of electrical infrastructure [12]. The authors developed VR and MR training applications to check how chosen technologies affect process of learning transfer and how they can be used to improve training process of operating procedures.

## 2 Materials and Methods

### 2.1 Learning Transfer in XR

Learning transfer is the amount of knowledge gained during training that can be applied to a new task when there is task similarity and a given person is able to perceive the similarity [13]. Properly prepared XR applications can reach a sufficiently high degree of similarity for learning transfer to occur. At the same time they offer many training benefits, such as:

- simulating training in circumstances, that exclude traditional methods such as dangerous situations or situation when there is a risk of damaging equipment [14];
- it allows making mistakes while learning procedural tasks, without any real consequences and at the same time giving feedback to user [15];
- it enables avoiding downtime of machines [16].

### 2.2 Interactions in XR

Interaction with the virtual environment or virtual elements used to augment the real environment takes place through software and sensors embedded in the target device. Depending on the used device, a specific interaction pattern should be adopted that allows for an impact on the virtual environment. In most of the cases XR interactions with virtual objects can be divided into: pointing, grasping, moving and releasing of objects [17]. Many publications [18–20] focus on interactions in VR as controllers used by VR HMDs and possibility of adding additional input devices such as haptic devices [21]. In the case of HMDs and controllers compatible with the OpenVR [22] standard, despite the differences in shape, all of them must meet certain standard which also helps to develop interactions.

In most of MR application dedicated for Microsoft HoloLens [23] user can inter-act with virtual environment using voice commands, pointing at object with ray going from middle of user's eyes and Air Tap Gesture which is presented at Fig. 1.

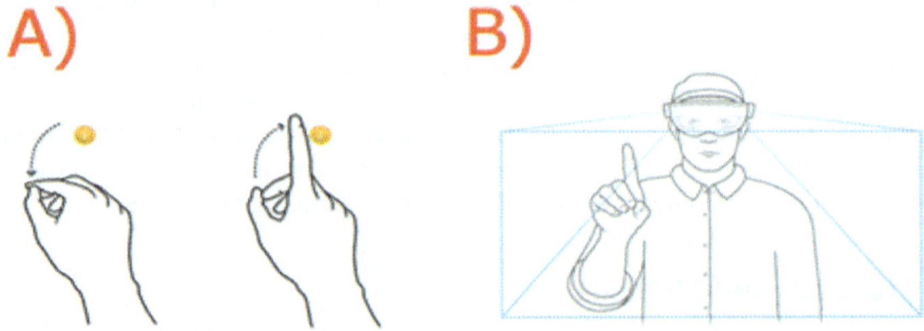


**Fig. 1.** Interaction using gestures in HoloLens a) AirTap gesture, b) the area in which gestures should be made [24].

### 2.3 Hardware and Software

Both applications were prepared in the Unity 3D [25] environment, which is currently one of most popular game engines, capable of developing XR applications. Interactions in the VR application were based on SteamVR [26] with additional scripts extending the range of possible interactions. The use of this solution allows to run the application on any HMD compatible with OpenXR. For test purposes HTC Vive Pro [27] was used. The mixed reality application was dedicated to the Microsoft HoloLens [23] and uses the Mixed Reality Toolkit for Unity [28]. Autodesk Inventor [29] and Autodesk 3ds Max [30] were used for modification of 3D models, such as dividing meshes into separate objects or simplifying mesh, which was necessary for MR Application.

### 2.4 MR Application Preparation

The Virtual Scene was generated in the Unity 3D environment. Figure 2 presents a virtual low voltage switchgear that already has all the necessary elements inside (including ARS 2 Pro switch disconnectors) and is a representation of the actual station. The models were previously prepared as part of the project described in [12].

During the preparation of the MR application, it was necessary to simplify the models of 3D elements. The need to introduce a simplification was dictated by the fact that the application is intended for HoloLens with incomparably lower computing power than a VR workstation. Optimization was performed with the mentioned modeling software. In MR, only the ARS 2 Pro switch disconnector is virtual, as well as the information board and the button terminating the application. Due to the fact that the virtual world overlaps the real world, it is not necessary to design or use already prepared environment.

**Fig. 2.** Base scene prepared for project described in [12].

As the MR applications was developed for Microsoft HoloLens 1 number of possible interactions were limited, that is why simplified interactions have been implemented in the training application. The interactions were developed using Mixed Reality Toolkit for Unity. In order to grasp an object, indicate it using the pointer coming from the point between the eyes, then perform the AirTap gesture [24]. The object is now attached to the pointer. Releasing the object takes place after performing the AirTap again. In case of objects that should be placed in a specific place, an additional script is attached to the objects, which allows to identify the type of the object (e.g. a fuse). The zone in which object should be placed also contains script that takes identifies all objects which enter zone. As soon as the object is in its zone, which recognizes the script attached to the object, it will be automatically released by the user and it will take the position and orientation defined by the zone. The snapping process is presented on Fig. 3. On objects such as doors or levers, there are scripts that allows to run a predefined animation, which is triggered by pointing it with head gaze and performing AirTap gesture.

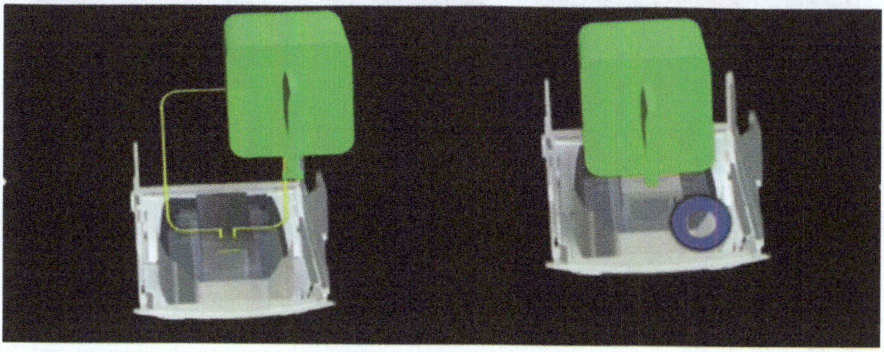

**Fig. 3.** Object snapping into zone [31].

During the development of MR application, several elements were added to this scene to help throughout the training process, one of the first them was board that presents user instruction about current task and interaction methods (Fig. 4).

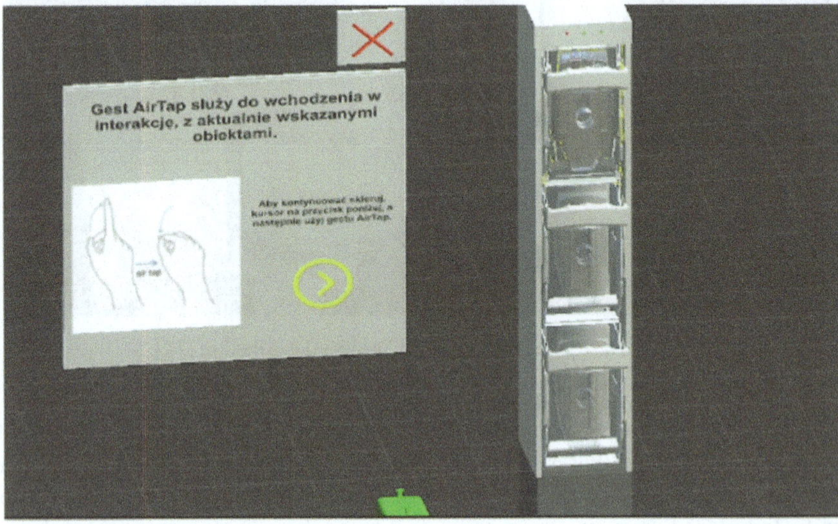


**Fig. 4.** Information board and z scene models in MR application [31].

The other element added to scene was voltage meter that was used in one of scenario steps. Model of a voltage meter was very simplified in order to make application run smoothly.

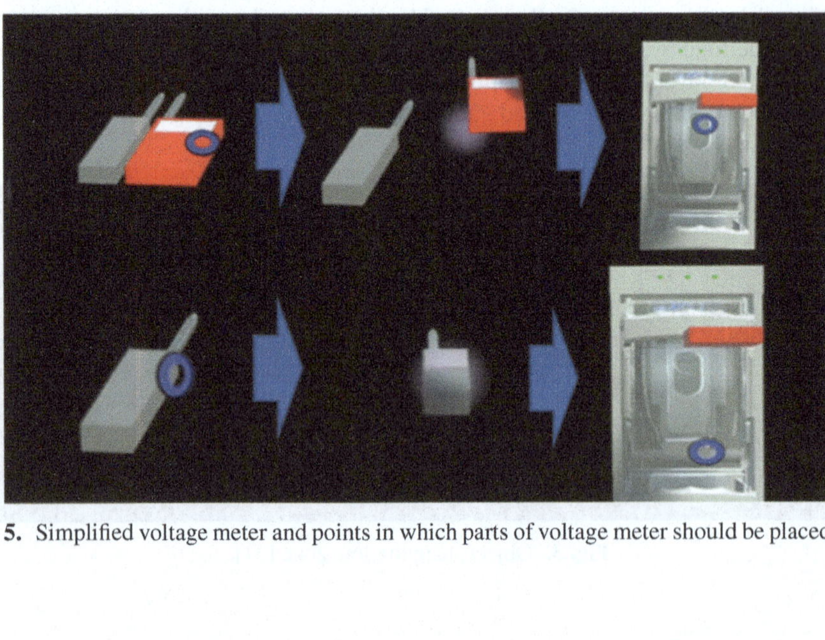

**Fig. 5.** Simplified voltage meter and points in which parts of voltage meter should be placed [31].

### 2.5 Application Scenario

In developed training application user had to replace defective fuse in electrical de-vice. The training scenario was divided into steps. Only after correctly preforming current step the user can go to next step. The instruction were changing automatically. The steps the user had to take to complete the scenario were as follows:

1. Introduction showing the basics of interaction
2. Assessment of fuse efficiency by looking at the diode
3. Disconnecting the voltage in a given segment of the switch by pulling the lever
4. Removal of the fuse cartridge to which the blown fuse was attached
5. Removing the blown fuse element from the insert and replacing it with a functional one
6. Putting fuse cartridge back and switching on the voltage by pulling the lever back to a start position
7. Measure the voltage by placing the electric meter at the two ends of the fuse (as presented at Fig. 5)

## 3 Research Plan

The research was conducted on 20 people, aged between 20 and 40 years (teachers and students from Poznan University of Technology). The participants were di-vided into 2 groups of 10 people. The first group trained using MR application. Group 2 had opportunity to try traditional training – watching movie before replacing real fuse and try MR application.

For each user time of using MR application and time of performing fuse replacement was registered. The training questionnaire, basing on System Usability Scale [32] that the users were asked to fill contained following questions with Likert scale from strongly disagree to strongly agree (1–5):

Q1. The guidance I received during the training was sufficient to complete it without any problems.
Q2. Interacting with objects was not a problem for me.
Q3. The quality of graphics and mapping of elements did not have a negative impact on the training process.
Q4. During the training I felt safe/at ease moving around the stage.
Q5. The training process prepared me for the fuse replacement in real object.
Q6. I believe mixed reality applications work well as a teaching tool.
Q7. I believe that virtual reality applications should be widely used because they work well as a training tool (for manual tasks).
Q8. The use of mixed reality technology equipment did not cause me any problems during the training.

## 4 Results and Discussion

Table 1 contains average results from user questionnaires that were completed after training with the use of MR technology. Based on the averaged results, it can be concluded that the reception of the application was positive, and most participants would be happy to use the XR application for training in the future. It is worth noting that the interacting with objects (Q2) and ease of use of all the questions (Q8) were evaluated worst. 30% of study participants said they had problems using Microsoft HoloLens. This could be due to the small field of view of the device and a relatively small field in which a AirTap gesture should be made for it to be correctly interpreted by the system. Answers to questions Q1 and Q3 indicate that both the proposed instructions and the level of graphics were sufficient to complete the task. Based on the answers to the remaining questions, it can be concluded that the respondents positively assessed the proposed solution in the context of training and preparation for per-forming the task in reality.

**Table 1.** Average questions ratings.

| | Q1 | Q2 | Q3 | Q4 | Q5 | Q6 | Q7 | Q8 |
|---|---|---|---|---|---|---|---|---|
| MR | 4.60 | 3.5 | 4.6 | 4.9 | 4.3 | 4.7 | 4.5 | 3.9 |

**Fig. 6.** Average training completion time of MR application [31].

The average training completion time is presented in Fig. 6. It can be seen here that group that did traditional training first and replaced the real fuse and then tried the MR application did training faster, probably because they already knew what to do. Maximal completion time for MR group was 648 s, while minimum was 248 s, median 369 s. For control group the maximal time was 552 s, while minimum was 159 s, median 307 s.

Average fuse replacement time, is presented in Fig. 7. The group that first watched the movie replaced fuse faster that group that trained using MR application. It may be due the fact, that simplifications in MR application did not represent a complex manual operation accurately enough.

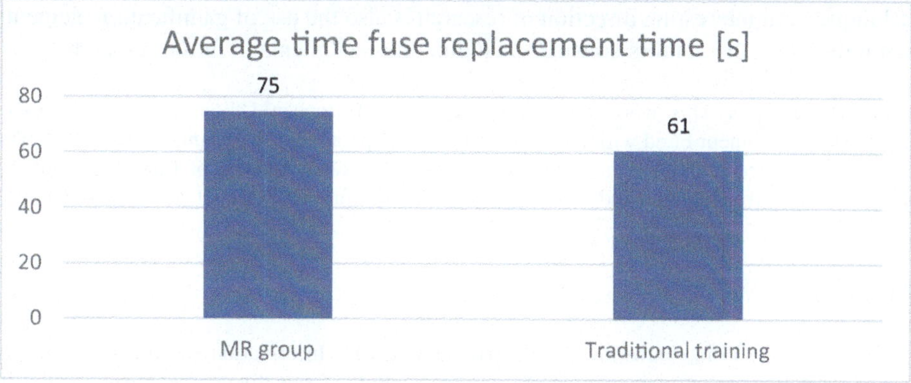


**Fig. 7.** Average training completion time of MR application [31].

Most of the publications on the use of XR technology in training on performing procedural tasks and manual work focus on the use of VR [33, 34] and AR [35]. The examples of using MR for training described in the literature focus on medical training [36, 37]. In case of many VR applications [34] virtual training is not as effective tool as traditional training, but on the other hand it can be repeated as many times as it is necessary without additional costs and hazards. The same principle seems to apply for MR applications.

## 5 Conclusions

The main problem of using XR technology is the need to simplify some actions by using gestures to interact with objects (in the case of MR) and the use of controllers (in the case of VR). For trainings based mainly on very accurate manual activities, given solution may not be positively received. However, the continuous development of software and hardware allow for the wider use of VR and MR as well as possibility of developing more complex interactions. The dynamic development of these technologies makes them a promising training solution. The new version of HoloLens [38], offers possibility of doing more complex interactions. Another solution is to connect the MR device to an external interaction device such as Leap-Motion [39].

The research showed that the forms of training, based on XR techniques, appealed to the participants and most of them want to try it again in the future. This is probably due to the fact that it is an interesting and so far not very common solution. The interaction requirement means that the user actively goes through the training process [40], instead of being passive, which means that he or she can acquire knowledge better and more willingly. Contact in the virtual world with the object and possible presentation of dangerous events, familiarize the course participant with workplace, which leads to greater self-confidence and a sense of security in a real environment.

The research was conducted on a relatively small group, and it is planned to repeat the research for a larger group in the future and compare traditional training to other XR techniques. An interesting direction of research is also the use of gamification elements in training [41], the authors consider including these elements in future research.

**Acknowledgments.** This work was supported by the European Union from the European Regional Development Fund within the Smart Growth Operational Programme 2020–2024. The project is executed within the priority axis "Support for R&D Activity of Enterprises" of the National Centre for Research and Development under the contract POIR.01.01.01-00-0463/18.

## References

1. Semenov, I., Filina-Dawidowicz, L., Trojanowski, P.: Information resources structuring to support decision-making process for organization of sensitive cargo transportation in the COVID-19 crisis. In: Sustainable Economic Development and Advancing Education Excellence in the era of Global Pandemic Proceedings of the 36th International Business Information Management Association Conference (IBIMA), pp.9881–9890 (2020)
2. Kotliar, A., et al.: Ensuring the economic efficiency of enterprises by multi-criteria selection of the optimal manufacturing process. Manag. Prod. Eng. Rev. **11**(1), 52–61 (2020)
3. Karpus, V.E., Ivanov, V.A.: Choice of optimal configuration of modular reusable fixtures. Russ. Eng. Res. **32**(3), 213–219 (2012)
4. Marino, E., Barbieri, L., Colacino, B., Fleri, A.K., Bruno, F.: an augmented reality inspection tool to support workers in industry 4.0 environments. Comput. Ind. **127**, 103412 (2021)
5. Żywicki, K., Buń, P.: Process of materials picking using augmented reality. IEEE Access **9**, 102966–102974 (2021)
6. Fast-Berglund, Å., Gong, L., Li, D.: Testing and validating extended reality (xR) technologies in manufacturing. Procedia Manuf. **25**, 31–38 (2018)
7. Bun, P., Trojanowska, J., Ivanov, V., Pavlenko, I.: The use of virtual reality training application to increase the effectiveness of workshops in the field of lean manufacturing. In: 4th International Conference of the Virtual and Augmented Reality in Education, VARE 2018, pp. 65–71 (2018)
8. Ivanov, V., Pavlenko, I., Trojanowska, J., Zuban, Y., Samokhvalov, D., Bun, P. Using the augmented reality for training engineering students. In: 4th International Conference of the Virtual and Augmented Reality in Education, VARE 2018, pp. 57–64 (2018)
9. Flavián, C., Ibáñez-Sánchez, S., Orús, C.: The impact of virtual, augmented and mixed reality technologies on the customer experience. J. Bus. Res. **100**, 547–560 (2019)
10. Kunnen, S., Adamenko, D., Pluhnau, R., Loibl, A., Nagarajah, A.: System-based concept for a mixed reality supported maintenance phase of an industrial plant. Procedia CIRP **91**, 15–20 (2020)
11. Juraschek, M., Büth, L., Posselt, G., Herrmann, C.: Mixed reality in learning factories. Procedia Manuf. **23**, 153–158 (2018)
12. Gorski, F., Grajewski, D., Bun, P., Zawadzki, P.: Study of interaction methods in virtual electrician training. IEEE Access **9**, 118242–118252 (2021)
13. Carlson, P., Peters, A., Gilbert, S.B., Vance, J.M., Luse, A.: Virtual training: learning transfer of assembly tasks. IEEE Trans. Visual Comput. Graphics **21**(6), 770–782 (2015)
14. Kaplan, A.D., Cruit, J., Endsley, M., Beers, S.M., Sawyer, B.D., Hancock, P.A.: The effects of virtual reality, augmented reality, and mixed reality as training enhancement methods: a meta-analysis. Hum. Factors **63**(4), 706–726 (2021)
15. Rehman, U., Cao, S.: Comparative evaluation of augmented reality-based assistance for procedural tasks: a simulated control room study. Behav. Inf. Technol. **39**(11), 1225–1245 (2020)

16. Radhakrishnan, U., Chinello, F., Koumaditis, K.: Immersive virtual reality training: three cases from the danish industry. In: 2021 IEEE Conference on Virtual Reality and 3D User Interfaces Abstracts and Workshops (VRW), pp. 1–5 (2021)
17. Balani, M.S., Tümler, J.: Usability and user experience of interactions on VR-PC, HoloLens 2, VR cardboard and AR smartphone in a biomedical application. In: Chen, J.Y.C., Fragomeni, G. (eds.) HCII 2021. LNCS, vol. 12770, pp. 275–287. Springer, Cham (2021). https://doi.org/10.1007/978-3-030-77599-5_20
18. Shao, X., Feng, X., Yu, Y., Wu, Z., Mei, P.: A natural interaction method of multi-sensory channels for virtual assembly system of power transformer control cabinet. IEEE Access **8**, 54699–54709 (2020)
19. Rogers, K., Funke, J., Frommel, J., Stamm, S., Weber, M.: Exploring interaction fidelity in virtual reality: object manipulation and whole-body movements. In: Proceedings of the 2019 CHI Conference on Human Factors in Computing Systems, pp. 1–14 (2019)
20. Jung, J., Park, H., Hwang, D., Son, M., Beck, D.: A review on interaction techniques in virtual environments. In: Proceedings of the 2014 International Conference on Industrial Engineering and Operations Management Bali, Indonesia, 7–9 January 2014
21. Grajewski, D., Buń, P., Górski, F.: Examination of effectiveness of a performed procedural task using low-cost peripheral devices in VR. In: Chen, J.Y.C., Fragomeni, G. (eds.) VAMR 2018. LNCS, vol. 10909, pp. 403–415. Springer, Cham (2018). https://doi.org/10.1007/978-3-319-91581-4_30
22. OpenVR documentation. https://partner.steamgames.com/doc/features/steamvr/openvr. Accessed 10 Jan 2022
23. Microsoft HoloLens specification. https://docs.microsoft.com/en-us/hololens/hololens1-hardware. Accessed 10 Jan 2022
24. HoloLens 1 usage. https://docs.microsoft.com/en-us/hololens/hololens1-basic-usage. Accessed 26 Oct 2021
25. Unity3D documentation. https://docs.unity3d.com/2019.3/Documentation/Manual/index.html. Accessed 10 Jan 2022
26. SteamVR Plugin for Unity3D. https://assetstore.unity.com/packages/tools/integration/steamvr-plugin-32647. Accessed 10 Jan 2022
27. HTC Vive Pro main page. https://www.vive.com/eu/product/vive-pro/. Accessed 10 Jan 2022
28. Mixed Reality Toolkit Documentaton. https://docs.microsoft.com/en-us/windows/mixed-reality/design/gaze-and-commit. Access 26 Oct 2021
29. Autodesk Inventor documentation. https://knowledge.autodesk.com/support/inventor/learn. Accessed 10 Jan 2022
30. Autodesk 3ds Max documentation. https://help.autodesk.com/view/3DSMAX/2020/ENU/. Access 10 Jan 2022
31. Gapsa, J.: Training of operation procedures of electrical equipment in virtual and mixed reality (in Polish), Master Thesis, Poznan University of Technology, Poznan (2021)
32. Lewis, J.R.: The system usability scale: past, present, and future. Int. J. Hum. Comput. Interact. **34**(7), 577–590 (2018)
33. Grassini, S., Laumann, K., Rasmussen Skogstad, M.: The use of virtual reality alone does not promote training performance (but sense of presence does). Front. Psychol. **11**, 1743 (2020)
34. Abich, J., Parker, J., Murphy, J. S., Eudy, M. A: review of the evidence for training effectiveness with virtual reality technology. Virtual Reality 1–15 (2021)
35. del Amo, I.F., Erkoyuncu, J.A., Roy, R., Palmarini, R., Onoufriou, D.: A systematic review of augmented reality content-related techniques for knowledge transfer in maintenance applications. Comput. Ind. **103**, 47–71 (2018)
36. Azimi, E., et al.: Can mixed-reality improve the training of medical procedures?. In: 2018 40th Annual International Conference of the IEEE Engineering in Medicine and Biology Society (EMBC), pp. 4065–4068. IEEE (2018)

37. Hu, H.Z., et al.: Application and prospect of mixed reality technology in medical field. Curr. Med. Sci. **39**(1), 1–6 (2019)
38. HoloLens 2 basics. https://docs.microsoft.com/en-us/hololens/hololens2-basic-usage. Accessed 26 Oct 2021
39. Ababsa, F., He, J., Chardonnet, J.-R.: Combining hololens and leap-motion for free hand-based 3D interaction in MR environments. In: De Paolis, L.T., Bourdot, P. (eds.) AVR 2020. LNCS, vol. 12242, pp. 315–327. Springer, Cham (2020). https://doi.org/10.1007/978-3-030-58465-8_24
40. Nguyen, D., Meixner, G.: Gamified augmented reality training for an assembly task: a study about user engagement. In: 2019 Federated Conference on Computer Science and Information Systems (FedCSIS), pp. 901–904. IEEE (2019)
41. Cazzolla, A., Lanzilotti, R., Roselli, T., Rossano, V.: Augmented reality to support education in Industry 4.0. In: 2019 18th International Conference on Information Technology Based Higher Education and Training (ITHET), pp. 1–5. IEEE (2019)

# Hand Tracking in Extended Reality Educational Applications

Paweł Buń[1(✉)], Jozef Husár[2], and Jakub Kaščak[2]

[1] Faculty of Mechanical Engineering, Poznan University of Technology, Piotrowo 3, 60-965 Poznan, Poland
pawel.bun@put.poznan.pl

[2] Faculty of Manufacturing Technologies with a Seat in Prešov, Technical University of Košice, Bayerova 1, 08001 Prešov, Slovakia

**Abstract.** Virtual, Augmented, and Mixed Reality (VR/AR/MR) or Extended Reality (XR) applications may be used as an innovative educational tool. Devices such as Oculus Quest 2 offer the possibility of using passthrough view and hand tracking which enables user to interact with virtual environment and see his surroundings. These functions enable the preparation of training applications, in which the user is able to simultaneously see his workplace and the number of generated job instructions or practice work-related activities without losing contact with the environment. This opens up a number of opportunities related to broadly understood education. The article briefly discusses the developed concept that allows for training in XR, in which user interactions with virtual elements are as close to the real ones as possible. The development of interactions using hand tracking and passthrough are described.

**Keywords:** Mixed reality · Education process · Object manipulation

## 1 Introduction

The development of the virtual, augmented and mixed reality (VR/AR/MR) industry in recent years has focused mainly on games, marketing and sales support [1, 2]. However, as the technology matured, companies changed the field of application and began to use it to optimize critical internal processes, a good example of which is Smart Factories [3]. Many industrial applications focus on production support [4] and training. The development of both the hardware and software may lead to wide use in the educational process.

Virtual, augmented and mixed reality are often referred to as Extended Reality (XR) [5] because they interpenetrate and with the emergence of new solutions, the boundary between them may blur. Because of that the concept of reality-virtuality-continuum [6] had been introduced. This concept places applications on the real-to-virtual axis with an increasing proportion of virtual components. VR is a term that describes a digitally generated, virtual world that the user can interact with using a variety of devices such as user tracking systems, controllers [7] and/or hand tracking systems [8]. Most of today's


J. Trojanowska et al. (Eds.): MANUFACTURING 2022, LNME, pp. 317–325, 2022.
https://doi.org/10.1007/978-3-030-99310-8_25

VR solutions are based on HMD, which, completely obscure the real world for the user. AR solutions can be based on smartphones, tablets [9], smart-glasses, thanks to which the user can see the real world, which is enriched (augmented) with virtual content. Due to this, the user can perform activities (interact) with the real world, as well as manipulate virtual content [1]. In Mixed Reality virtual and real world are combined and both the user and the real content can interact with the virtual environment [10].

This paper discusses the possibility of using XR technologies to increase the effectiveness of educational processes. This is the early stage of work aimed at using new possibilities of devices such as Oculus Quest 2 to enable natural interaction with virtual content in order to increase effectiveness of classes and industrial training. The use of new technologies may have a positive impact on the engagement [11] of users in the education/training process, which in turn may potentially translate into higher efficiency of the process.

# 2 Extended Reality Learning Application

## 2.1 Training in XR

If XR training is to be effective, the user should be engaged, and at the same time knowledge transfer must take place. It can occur only when the tasks or knowledge in the field covered by the training are similar to the actual situation and the person is able to perceive an analogy between them [12]. In the case of XR applications, in order to ensure the occurrence of this phenomenon, it is necessary to map the actual conditions as accurately as possible. This can be done both by reconstructing the real environment in the virtual world and by providing the most natural possible interactions with virtual elements. A significant contribution to this can be, among others, the use of natural interaction with the user's hand thanks to the use of hand tracking.

If the conditions for knowledge transfer are met, the benefits for both education and industrial training are significant. Training with the use of XR has many advantages, including the possibility of simulating life-threatening conditions or those with a high risk of damage to the equipment [13]. Users can make mistakes while performing a procedural task without consequences [14], such as machine downtime [15].

## 2.2 Interactions and Object Manipulation in XR

All interactive applications assume the need to influence digitally generated elements. The method of interaction is primarily influenced by the input devices and interface prepared in the application, which can be in the form of both Graphic User Interface and natural interaction. In XR applications, the impact on virtual objects can be divided into pointing, grabbing, moving, and releasing objects [16]. The development of effective and natural methods of interaction is extremely important [17]. Manipulation of virtual objects is difficult for many users, as most XR applications do not offer force feedback or any other "sense" of a virtual object [18]. The solution to this problem may be the use of haptic devices [19], however, these devices increase the costs of the station and may negatively affect the mobility of the station. Most VR applications focus on the use of controllers as the default interaction device [20, 21]. Currently, most of these controllers are compatible with the OpenXR standard. Although hand recognition systems integrated with HMD may change this.

The issue of manipulating virtual objects and the broadly understood interaction with the virtual environment, which may translate into effective knowledge transfer [22], has been raised many times by the authors. So far, the focus has been on interactions using controllers [23], tangible GUIs [24], gloves [25], and haptics [19]. This time, the authors decided to check how the use of hand tracking and passthrough affects the effectiveness of the educational application. For this purpose, work was undertaken to adapt the previously prepared educational application [26] to the use of the solutions mentioned above.

### 2.3 Hardware and Software

The application was prepared for the Oculus Quest 2 device. They are currently one of the most popular VR googles equipped with inside-out tracking. Unlike the Oculus Rift and other devices with outside-in tracking, there is no need to set up additional sensors or beacons that enable tracking, the whole thing is done by HMD. The device can work in both standalone mode and when connected to a computer for more demanding VR applications. These features and the relatively low price contributed to the popularization of the device. The full specification can be found in the table below (Table 1).

**Table 1.** Oculus quest 2 parameters [27].

| Parametr | Value |
|---|---|
| VR category | Standalone and PC based |
| Tracking | Oculus Insight inside-out tracking. 6 DOF |
| Input | Two oculus touch controllers |
| Resolution | 1832 × 1920 per eye |
| Supported refresh rate | 72 Hz (default) |
| Display panel | Fast-switch LCD |
| Lens distance | Adjustable - 3 preset IPD adjustments |
| Audio | Integrated positional audio |
| Memory (RAM) | 6 GB |
| Processor | Qualcomm® Snapdragon™ XR2 Platform |

The device allows the user to view the surroundings without removing the helmet, after turning on passthrough mode. It makes it easier, among others, to raise the controllers even when they are not turned on and tracked. The view of the surroundings is generated on the basis of the images from the tracking system cameras placed on the helmet. The image of objects that come too close to the helmet becomes distorted. A natural step in the development of the software was the use of cameras used in the inside-out tracking system for hand tracking and recognition. Due to this, it is possible to interact more naturally with the virtual environment. Unfortunately, it comes with some limitations. These restrictions include the fact that the user's hands must be within the field of view of the cameras. In addition, it is necessary to adapt the application to recognize specific gestures that are to replace the standard interactions performed with buttons on the controllers. One of those gestures can be seen on Fig. 1.

**Fig. 1.** Pointer pose gesture that can be used to point and select element [28].

Oculus provides an SDK and an integration package for Unity 3D. The package includes ready-made scripts that allow, among other things, to track the position of the head and controllers. The package allows to build both an application intended for use with a computer and a standalone application (Android). It is worth mentioning that in order to use the passthrough function (which allows you to see the surroundings), application must be used in standalone mode.

## 3 Preparation Mixed Reality Application

The application was prepared in the Unity 3D environment on the basis of the application described earlier in [24]. It is worth mentioning that the previous version of the application was prepared with the use of Virtual Reality Toolkit 3.3 in mind, which is not fully compatible with the new solutions. The assumption was that the newly prepared application would use all the possibilities offered by Oculus Quest 2, therefore the Oculus Integration v34 package was used. Among other things, this package allows to get an image of the surroundings. Thanks to this, the user using the application has the ability to move freely in the real environment and can pick up real objects without having to remove the helmet each time. In addition, to ensure the most natural interaction with the virtual environment and to minimize the need to learn how to use the application, it was decided to use hand-tracking. During the development of the application, it turned out that the combination of these two features causes some implantation problems.

Using the application in passthrough mode requires switching it to the OpenXR backend (as presented in Fig. 2). The software producer is gradually switching to this solution in his SDK, but in the v34 version not all functions are available there.

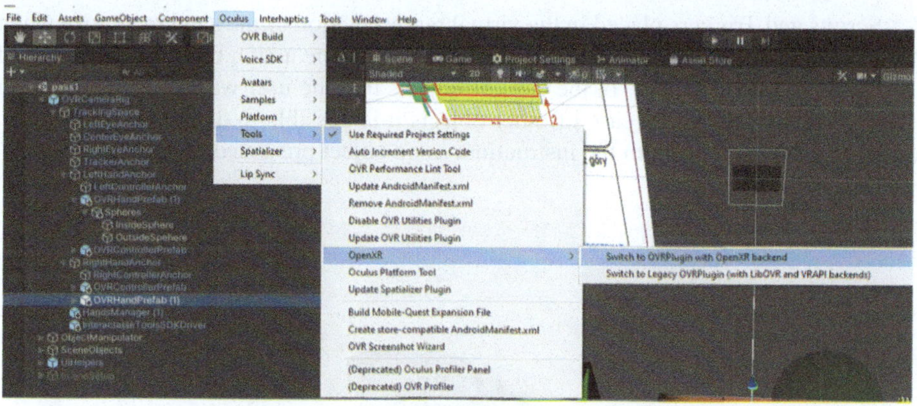


**Fig. 2.** Changing of backend in OVRPlugin to enable passthrough option.

It is possible to enable hand tracking, but the API for recognizing the bending state of individual fingers is not available. Therefore, as part of the work on the application, it was necessary to develop own methods of interaction with the virtual environment based on hand movements. Simple interaction with objects is possible thanks to collider objects placed on a virtual representation of the phalanges and the palm of the hand. Thanks to them, it is possible to move/push objects, but gripping the object causes some problems. It is related to the aforementioned lack of force feedback. Therefore, in order to facilitate the manipulation of virtual objects, it was necessary to write scripts and add Trigger objects to determine if the user is pinching their fingers as can be seen on Fig. 3.

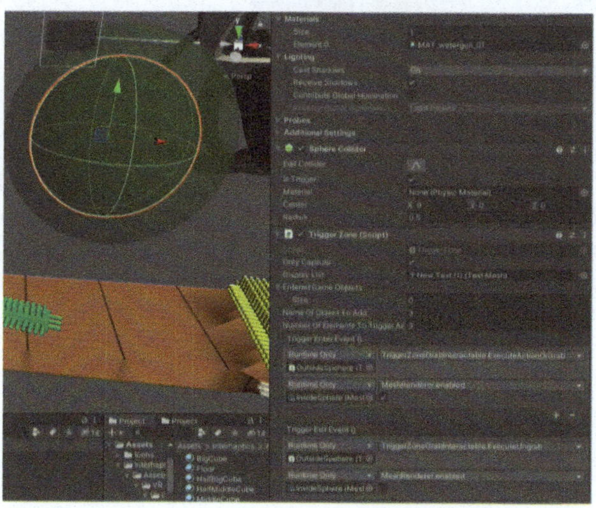

**Fig. 3.** Trigger area and scripts added to hand in order to determine if user is pinching fingers.

Scripts and Triggers placed in the virtual representation of the hand recognize when the hand is clenched sufficiently to trigger an action. The action that will be triggered is defined by added scripts. In the case of objects that the user wants to grab, such an object is picked up, in the case of other objects it is possible to define another action, e.g. switching the slide with the instruction, as on object presented in Fig. 4.

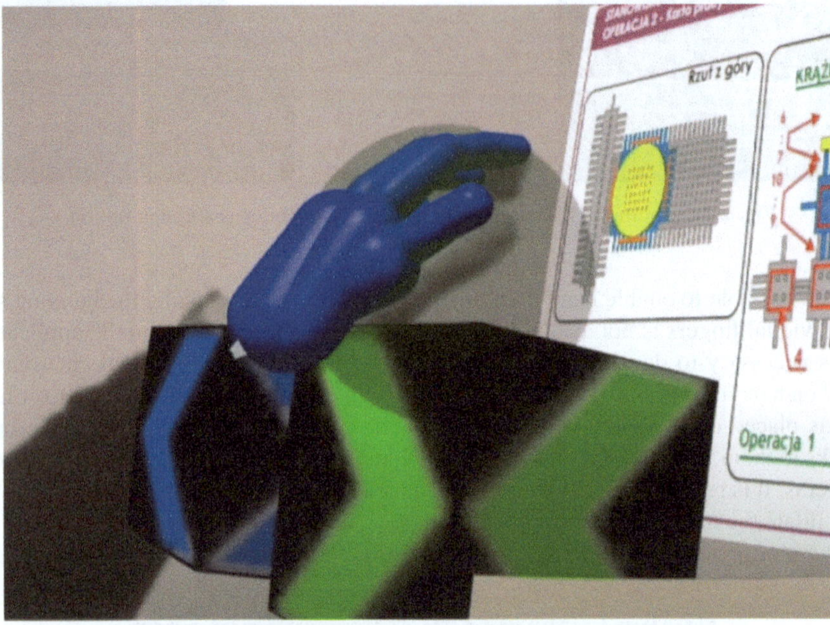


**Fig. 4.** Hand interaction in mixed reality. In order to interact with cube to change instruction image user must touch cube and make a fist gesture.

As in the base application, the elements that the user has to grasp are enlarged in relation to their real counterpart as presented on Fig. 5. In the course of the development of the application, the authors will try to give them a real size, but you should check whether such small collision zones will not make it impossible to grasp the element. However, to determine this, it is necessary to conduct research with a test group and with real elements.

**Fig. 5.** Virtual element grabbed by virtual hand representation and real object grabbed by hand. Virtual object is enlarged in order to simplify manipulation.

## 4 Conclusions and Future Work

Due to the pandemic situation and the need to maintain the sanitary regime, it has not been possible to test the described solution on a sufficiently large group of students. However, preliminary results of internal tests suggest that the proposed solution may interesting to users. Thanks to the significant reduction in the cost of XR devices, it is possible that in the future most students will be able to participate in the classes in a virtual form, just as they are participating in them via teleconference now. However, conducting classes in this form requires properly prepared applications, for this it is necessary to conduct research and improve the methods of interaction in XR applications. The application must be prepared in a way that allows it to be used without additional instructions from the teacher, and the interactions are natural enough for the students to have no problems with completing the tasks assigned to them. For this purpose, it is necessary to introduce some simplifications related to interactions. In order to grasp a virtual object, bring the virtual representation of a hand close to the object, and then press and clench your hand. Releasing the subject takes place after releasing user fingers. Virtual objects that should be placed in a specific place have an additional script attached, which is responsible for marking the position in which the object should be. When the object is in the correct position and orientation (with the set tolerance), it is automatically released and takes the desired position. It should be tested how these restrictions will affect the knowledge transfer process.

Currently the virtual object is 4 times bigger than in reality to make collision detection easier. However authors plan to develop other detection method in order to make elements real size and check does it affect learning transfer. The future update of Oculus software may also fix the problem with finger pinch/gestures detection, which may help to improve application performance.

The applications that uses both passthrough and hand tracking can be used to teach both manual skills, procedural task or product inspection tasks [29].

Popularizing XR solutions and making the application available to students for home use may contribute to a better use of time during classroom classes for substantive analysis of issues and open discussion. Therefore, research on the possibility of using XR technology in industrial education and training is planned. It is also planned to develop a multiplayer version of the application, which will require students to collaborate or have elements of gamification [30].

## References

1. Flavián, C., Ibáñez-Sánchez, S., Orús, C.: The impact of virtual, augmented and mixed reality technologies on the customer experience. J. Bus. Res. **100**, 547–560 (2019)
2. Hariharan, A., Pfaff, N., Manz, F., Raab, F., Felic, A., Kozsir, T.: Enhancing product configuration and sales processes with extended reality. In: Jung, T., tom Dieck, M.C., Rauschnabel, P.A. (eds.) Augmented Reality and Virtual Reality. PI, pp. 37–50. Springer, Cham (2020). https://doi.org/10.1007/978-3-030-37869-1_4
3. Żywicki, K., Buń, P.: Process of materials picking using augmented reality. IEEE Access **9**, 102966–102974 (2021)
4. Marino, E., Barbieri, L., Colacino, B., Fleri, A.K., Bruno, F.: An augmented reality inspection tool to support workers in industry 4.0 environments. Comput. Ind. **127**, 103412 (2021)
5. Fast-Berglund, Å., Gong, L., Li, D.: Testing and validating extended reality (xR) technologies in manufacturing. Procedia Manuf. **25**, 31–38 (2018)
6. Juraschek, M., Büth, L., Posselt, G., Herrmann, C.: Mixed reality in learning factories. Procedia Manuf. **23**, 153–158 (2018)
7. Bun, P., Trojanowska, J., Ivanov, V., Pavlenko, I.: The use of virtual reality training application to increase the effectiveness of workshops in the field of lean manufacturing. In: 4th International Conference of the Virtual and Augmented Reality in Education, VARE 2018, pp. 65–71 (2018)
8. Craig, A., Krishnan, S.: Fusion of leap motion and Kinect sensors for improved field of view and accuracy for VR applications. Virtual Reality Course Report (2016)
9. Ivanov, V., Pavlenko, I., Trojanowska, J., Zuban, Y., Samokhvalov, D., Bun, P.: Using the augmented reality for training engineering students. In: 4th International Conference of the Virtual and Augmented Reality in Education, VARE 2018, pp. 57–64 (2018)
10. Kunnen, S., Adamenko, D., Pluhnau, R., Loibl, A., Nagarajah, A.: System-based concept for a mixed reality supported maintenance phase of an industrial plant. Procedia CIRP **91**, 15–20 (2020)
11. Allcoat, D., von Mühlenen, A.: Learning in virtual reality: effects on performance, emotion and engagement. Res. Learn. Technol. **26**, 2140 (2018)
12. Carlson, P., Peters, A., Gilbert, S.B., Vance, J.M., Luse, A.: Virtual training: learning transfer of assembly tasks. IEEE Trans. Visual Comput. Graphics **21**(6), 770–782 (2015)
13. Kaplan, A.D., Cruit, J., Endsley, M., Beers, S.M., Sawyer, B.D., Hancock, P.A.: The effects of virtual reality, augmented reality, and mixed reality as training enhancement methods: a meta-analysis. Hum. Factors **63**(4), 706–726 (2021)
14. Rehman, U., Cao, S.: Comparative evaluation of augmented reality-based assistance for procedural tasks: a simulated control room study. Behav. Inf. Technol. **39**(11), 1225–1245 (2020)

15. Radhakrishnan, U., Chinello, F., Koumaditis, K.: Immersive virtual reality training: three cases from the Danish industry. In: 2021 IEEE Conference on Virtual Reality and 3D User Interfaces Abstracts and Workshops (VRW), pp. 1–5 (2021)
16. Krichenbauer, M., et al.: Augmented reality versus virtual reality for 3D object manipulation. IEEE Trans. Visual Comput. Graphics **24**(2), 1038–1048 (2018)
17. Caputo F.M., Giachetti A.: Evaluation of basic object manipulation modes for low-cost immersive virtual reality. In: Proceedings of the 11th Biannual Conference on Italian SIGCHI Chapter, pp. 74–77. ACM (2015)
18. Leonardis, D., Solazzi, M., Bortone, I., Frisoli, A.: A wearable fingertip haptic device with 3 DoF asymmetric 3-RSR kinematics. In: 2015 IEEE World Haptics Conference (WHC), pp. 388–393. IEEE (2015)
19. Grajewski, D., Buń, P., Górski, F.: Examination of effectiveness of a performed procedural task using low-cost peripheral devices in VR. In: Chen, J.Y.C., Fragomeni, G. (eds.) VAMR 2018. LNCS, vol. 10909, pp. 403–415. Springer, Cham (2018). https://doi.org/10.1007/978-3-319-91581-4_30
20. Shao, X., Feng, X., Yu, Y., Wu, Z., Mei, P.: A natural interaction method of multi-sensory channels for virtual assembly system of power transformer control cabinet. IEEE Access **8**, 54699–54709 (2020)
21. Rogers, K., Funke, J., Frommel, J., Stamm, S., Weber, M.: Exploring interaction fidelity in virtual reality: Object manipulation and whole-body movements. In: Proceedings of the 2019 CHI Conference on Human Factors in Computing Systems, pp. 1–14 (2019)
22. Levac, D.E., Huber, M.E., Sternad, D.: Learning and transfer of complex motor skills in virtual reality: a perspective review. J. Neuroeng. Rehabil. **16**(1), 1–15 (2019)
23. Gorski, F., Grajewski, D., Bun, P., Zawadzki, P.: Study of interaction methods in virtual electrician training. IEEE Access **9**, 118242–118252 (2021)
24. Husar, J., Knapčíková, L.: Exploitation of augmented reality in the industry 4.0 concept for the student educational process. INTED **2019**, 4797–4805 (2019)
25. Buń, P., Górski, F., Lisek, N.: Comparison of manual assembly training possibilities in various virtual reality systems. In: Machado, J., Soares, F., Veiga, G. (eds.) HELIX 2018. LNEE, vol. 505, pp. 398–404. Springer, Cham (2019). https://doi.org/10.1007/978-3-319-91334-6_54
26. Buń, P., Trojanowska, J., Rewers, P.: VR and AR in lean manufacturing classes. In: Trojanowska, J., Ciszak, O., Machado, J.M., Pavlenko, I. (eds.) MANUFACTURING 2019. LNME, pp. 342–351. Springer, Cham (2019). https://doi.org/10.1007/978-3-030-18715-6_29
27. Oculus Quest 2 Specification. https://developer.oculus.com/resources/oculus-device-specs/. Access 06 Dec 2021
28. Oculus Quest 2 hand tracking documentation. https://developer.oculus.com/documentation/unity/unity-handtracking/. Accessed 10 Jan 2022
29. Kuric, I., Kandera, M., Klarák, J., Ivanov, V., Więcek D.: Visual product inspection based on deep learning methods. In: Tonkonogyi, V. et al. (eds.) Advanced Manufacturing Processes. InterPartner-2019. Lecture Notes in Mechanical Engineering, pp. 148–156. Springer, Cham (2020). https://doi.org/10.1007/978-3-030-40724-7_15
30. Nguyen, D., Meixner, G.: Gamified augmented reality training for an assembly task: a study about user engagement. In: 2019 Federated Conference on Computer Science and Information Systems (FedCSIS), pp. 901–904. IEEE (2019)

# Author Index





J. Trojanowska et al. (Eds.): MANUFACTURING 2022, LNME, pp. 327–328, 2022.
https://doi.org/10.1007/978-3-030-99310-8

The manufacturer's authorised representative in the EU is Springer Nature Customer Service Centre GmbH, Europaplatz 3, 69115 Heidelberg, Germany. If you have any concerns regarding our products, please contact ProductSafety@springernature.com

Printed and bound by CPI Group (UK) Ltd, Croydon, CR0 4YY
30/09/2026
02225320-0013